MASTERMINDING NATURE

The Breeding of Animals, 1750–2010

In *Masterminding Nature*, Margaret E. Derry examines the evolution of modern animal breeding from the invention of improved breeding methodologies in eighteenth-century England to the application of molecular genetics in the 1980s and 1990s. A clear and concise introduction to the science and practice of artificial selection, Derry's book puts the history of breeding in its scientific, commercial, and social context.

Masterminding Nature explains why animal breeders continued to use eighteenth-century techniques well into the twentieth century, why the chicken industry was the first to use genetics in its breeding programs, and why it was the dairy cattle industry that embraced quantitative genetics and artificial insemination in the 1970s, as well as answering many other questions. Following the story right up to the present, the book concludes with an insightful analysis of today's complex relationships between biology, industry, and ethics.

MARGARET E. DERRY is an adjunct professor in the Department of History at the University of Guelph.

Masterminding Nature

The Breeding of Animals, 1750–2010

MARGARET E. DERRY

UNIVERSITY OF TORONTO PRESS
Toronto Buffalo London

© University of Toronto Press 2015
Toronto Buffalo London
www.utppublishing.com
Printed in the U.S.A.

ISBN 978-1-4426-4904-0 (cloth)
ISBN 978-1-4426-2652-2 (paper)

Printed on acid-free, 100% post-consumer recycled paper with vegetable-based inks

Library and Archives Canada Cataloguing in Publication

Derry, Margaret E. (Margaret Elsinor), 1945–, author
Masterminding nature : the breeding of animals, 1750–2010 /
Margaret E. Derry.

Includes bibliographical references and index.
ISBN 978-1-4426-4904-0 (bound)   ISBN 978-1-4426-2652-2 (pbk.)

1. Livestock – Breeding – History.   I. Title.

SF105.D47 2015   636.08'209   C2014-907177-9

University of Toronto Press acknowledges the financial assistance to its publishing program of the Canada Council for the Arts and the Ontario Arts Council, an agency of the Government of Ontario.

Canada Council   Conseil des Arts
for the Arts   du Canada

ONTARIO ARTS COUNCIL
CONSEIL DES ARTS DE L'ONTARIO
an Ontario government agency
un organisme du gouvernement de l'Ontario

University of Toronto Press acknowledges the financial support of the Government of Canada through the Canada Book Fund for its publishing activities.

Contents

Acknowledgments   vii

Introduction   3

1 Artificial Selection Theory and Livestock Breeding, 1750-1900   13

2 Early Developments in Genetics   40

3 Practical Breeding via Theoretical Population Genetics   71

4 New Directions: Artificial Insemination Technology and Quantitative Genetics   94

5 Molecular Genetics, the Rise of Genomics, and Livestock Breeding   129

6 Biology, Industry Needs, and Morality in Livestock Breeding   160

Conclusions   179

Notes   193

Glossary   235

Bibliography   241

Index   291

Acknowledgments

The conundrum of what is science and what is merely practice and/ or culture in animal breeding fascinates me, perhaps partially because I have bred purebred beef cattle for over twenty-five years. I have also written about the topic of animal breeding from many angles in a number of books and a few articles. In this book, I again pursue the subject but develop it within a broader framework, particularly as to variety of species addressed and as to modernity of time frame.

I was very lucky in having a number of scholars read over this work before I submitted it to the University of Toronto Press for peer review. The readers provided many wonderful suggestions for improvement and I spent considerable time revising the book as a result. I have the pleasure of thanking these readers now. Bert Theunissen, a historian of science (and also a sheep breeder), at the Department for History of Science, Descartes Centre, Utrecht University in the Netherlands, provided invaluable help. He is very knowledgeable about animal breeding, from the practical, scientific, and historical points of view. Ian Hacking, Professor Emeritus from the University of Toronto in the philosophy of science brought a different perspective to the work, a thoughtful and nuanced approach to the overall topic. He also introduced the book to several other people who gave wonderfully helpful comments. Rasmus Winther, professor of philosophy with the Department of Philosophy at the University of California at Santa Cruz, and interested in the history of genetics, made many good suggestions; as did Dr. Sophie Petersen of Portland, Oregon, a scholar with a PhD in cell biology from University of California at Berkeley and trained as a veterinary surgeon at University of California at Davis. These two readers provided knowledge of the history of genetics and the science of genetics itself. Marian

Horzinek, Professor Emeritus of the Veterinary Research Council at Utrecht University in the Netherlands, also brought a scientific point of view to his review of my book. An expert in the genetics of animal diseases, Marian added thoughtful ideas concerning the role of "art" versus "science" in the development of knowledge.

After revision as a result of the input of the above readers, I submitted the book to the University of Toronto Press. I wish to thank the two unknown readers for their strong support of the book. They made me rethink how I had organized certain concepts and led me to undertake further refining revision. I also wish to thank Len Husband for his continuing interest in my work, as well as the people at the University of Toronto Press who brought the book to fruition.

MASTERMINDING NATURE

The Breeding of Animals, 1750–2010

Introduction

"All my notion about *how* species change are derived from long-continued study of the works of (& converse with) agriculturalists. & horticulturists," wrote Charles Darwin to Asa Gray, a Harvard University botanist, in 1857.[1] Darwin suggested that the process of breeding domestic animals and plants had been critical to developing his theory regarding the effects of natural selection on evolution. This linkage of practical artificial selection to scientific questions was cemented by Darwin, and would be ongoing from that time. Because artificial selection illustrated that changes in species were possible, it was a natural tool in the study of theories of heredity and evolution. Some scientists argued that mutation explained why artificial selection successfully altered type in domestic animals, while others believed Darwinism (the gradual change of one species into another) did so. In turn, theory could drive practice. Mutation suggested different approaches to breeding than Darwinism did. Animal and plant breeding, both crucial aspects of agriculture, became two sides of one coin – as vehicles to prove theory, and as practices to be modified by theory. Ultimately artificial selection methodology itself would be looked upon as a science in its own right. This book is about the evolution of practice and theory in artificial selection methodology as applied to animal breeding from about 1750 until 2010 in Europe and North America, and is designed to give the reader a solid background on the subject by providing an overview of major trends. At the heart of the matter is the fusion of aspects of science with traditional practice.

A history of artificial selection theory and practice is enriched by information on various farm animal industries, as well as the organization of breeding and breeding associations; and correspondingly, by information on plant hybridist traditions, quantitative genetic theory, and strategies made possible by genomics. By emphasizing the agricultural/farm perspective in attempting to explain how scientific innovations shifted breeding practices, I take a somewhat unusual scholarly approach to the question of scientific agriculture. I believe it is impossible to understand how science affected practice without an appreciation of the fundamental organization and dynamics of differing farm industries. Also central to this story of agriculture and science is the rise of technologies such as artificial insemination and the BeadChip, which profiles SNP arrays in DNA. My approach, then, might be described as a discussion concerning the convergence of three paths: namely, the structure of agriculture, developments in genetics, and technological innovations. This strategy is the only way to see how or if evolving scientific theory in conjunction with technology could be made applicable to practical breeding, therefore altering breeding strategies followed by breeders, or whether modern approaches could work in consort with traditional methods.

Contributions to artificial selection might have emanated from modern science, evolving technology, and traditional breeding methods (or what is often defined as craft methodology), but these components did not combine in such a way as to form a seamless union when it came to the practice of animal breeding. Animal breeding has always been (and still is) often hybrid in nature when it comes to the science/technology/tradition question. At any county fall fair where livestock are exhibited, one can see animals that reflect myriad past and present breeding strategies: some that rely on the science of livestock genetics, often in conjunction with technology, some that emanate out of selection practices that are hundreds of years old and have been impervious to inputs from science, and others that demonstrate that a sort of combined scientific/traditional approach to breeding has been followed. This hybrid characteristic is fundamental for any understanding of animal breeding; and also makes it an interesting subject to study on a number of levels. The evolution of animal breeding practices provides an impressive example of how complex the development of human thought can be, as well as showing that newer ideas do not necessarily supersede and replace older ones. Instead, new and old often coexist and sometimes even work in concert with one another. The history of animal breeding demonstrates that a strange – or perhaps incomplete – scientific/technological revolution has evolved over the past centuries. Even so, the idea that a serious cleavage existed (or exists) between what is often described as uneducated practice on the one hand and scientific views on the other is often based on imaginary rather than real issues. I hope this book will go some way in revealing why this was/is the case. Because a central part of the story is the role that technology played, first in shaping genetics and second in influencing how animals are bred, it is often less a question of the affect of genetics on animal breeding, than of technology/genetics combined; namely the effects of biotechnology on animal breeding.

Aspects of the dairy, beef, chicken, and pig industries emerge in this book in order to make clear how relevant (or non-relevant) various principles concerning artificial selection emanating out of developing livestock genetics and technology were to them at any particular point in time. Conditions within the worlds of genetics/technology and chicken breeding in the 1940s and 1950s, for example, promoted a breeding trend that favoured innovation. The same could not be said about the dairy industry within that time frame, a significant detail in view of the fact that by the late 1950s shifts in genetic thinking and the advent of new forms of technology would change the picture entirely. By 1960 genetic and technological innovation fit more naturally with traditional breeding of dairy cattle, and could, therefore, work collaboratively with older ways.

In the first chapter I look at how eighteenth and early nineteenth century practical breeders approached the problem of artificial selection and what theories they put forward to improve its effectiveness. I review the culture and thinking of the early Thoroughbred horse breeders, the work of Robert Bakewell and Sir John Sebright, and then the rise of purebred breeding and standardbred breeding over this period. An outline of philosopher/naturalist concerns with artificial selection is also provided in the chapter. The naturalists developed quite different breeding strategies with different ends in mind. The breeders wanted to increase the presence of desired characteristics, and the naturalists hoped to understand the dynamics of evolution and speciation. Darwinism complicated the situation, but also initiated an academic field that became critical to future artificial selection theory: biometry. Under biometric theory, breeding practices could alter the inheritance structure of populations.

Chapter 2 addresses patterns in genetics as applied to animal breeding up to 1940. The chapter begins with the rediscovery of Mendel's laws in 1900, which rocked an increasingly divisive academic biological world, but also brought the breeding of livestock and agricultural plants to the fore of scientists' minds. I address the work of R.A. Fisher and his view that the inheritance of any characteristic resulted from the interaction of many genes, W. Castle and his studies of inbreeding and artificial selection, and Sewall Wright's theories concerning inbreeding and populations shifts. J.L. Lush used the ideas of the aforementioned to formalize artificial selection strategies aimed at improving farm animals. The development of artificial selection theory during these years, however, had virtually no impact on the way practical breeding of livestock proceeded. Purebred breeding theory, methodology, and culture continued to dominate prevailing views in the farming world on how to improve livestock until at least the late 1930s.

In chapter 3 I review how early livestock genetic theory concerning artificial selection worked with practical livestock breeding between 1940 and 1960, using chickens, beef cattle, and pigs as examples. I focus on characteristics of the chicken industry in order to show how its breeding culture dovetailed agreeably with theoretical approaches espoused by livestock geneticists in the late 1930s and 1940s. Innovative views towards selection theory were not at the heart of the matter. Market forces could revolutionize artificial selection strategies – as chicken breeding made evident. It is also interesting to question how much the authority of science, over that of practical breeders, shaped a breeding revolution when it came to chickens. The selection strategies of scientists did not play as major a role in beef cattle breeding or in pig breeding. In the latter two cases, traditional approaches, which emanated out of purebred breeding, stayed entrenched. The biology of the animals and the structure of their industries made the breeding ideas of the geneticists impractical, thereby supporting the continuing dominance of older ways.

The implementation of progeny testing, promoted by geneticists (and espoused by early livestock breeders as far back as the eighteenth century), would become feasible on a more comprehensive scale after the advent of artificial insemination (AI), which allowed for the regulation and testing of breeding males. Chapter 4 begins by describing the background to regulations originally put in place to control public breeding along the male lineage side. Artificial insemination organizations would evolve from these older structures and regulations. Data collection on cows and the contingent registration of pedigrees meant that the cows could be linked to their sires, a situation that made it possible for AI organizations to select bulls that produced superior daughters. Information arising from AI organizations – in conjunction with data generated by the dairy cattle industry – provided geneticists with valuable tools. This did not happen in any other livestock industry, partially because the structure of the dairy breeding industry was unique, and partially because bovine semen worked better than that of other species under AI conditions. (Why beef cattle breeding was not as affected by AI/genetics as dairy cattle breeding is addressed in this chapter.) The 1950s through to the 1970s were the heyday of quantitative genetics in livestock production, largely thanks to its interplay with the dairy industry. The story of dairy cattle breeding illustrates how principles emanating from genetics could work with and enhance purebred breeding. Dairy cattle breeding came to reflect a combination of modern science and traditional breeding.

The approaches of livestock genetics to artificial selection up until the 1970s were dominated by the theoretical underpinnings of classical genetics (viz. population and quantitative genetics based on Mendelian theory), which argued that breeding results could be predicted without any understanding of the gene structure involved. Molecular genetics, aimed primarily at understanding heredity at the DNA level, would change that outlook, and came to play a particularly significant role in artificial selection methodology as applied in livestock breeding near the end of the twentieth century, when breeding by DNA became possible in certain animals. In the fifth chapter, I briefly outline the background of aspects of molecular genetics that would ultimately become important to geneticists interested in livestock breeding. It is, primarily, a story of the rise of genomics. The development of the dairy breeding industry that had occurred between 1950 and the late 1970s – as a result of the combined input of quantitative geneticists and purebred breeders – put the industry in a position to capitalize on specific advances in genomics in a way that would be impossible in other livestock breeding industries. Genomic templates of superior bulls could not have been generated without the data that had accumulated from years of collecting information under the principles of quantitative genetics and that, in turn, would not have been possible without the public records issued out of traditional purebred breeding practices.

Chapter 6 looks at the entangled issues of biology, industry-needs, contemporary technology, and ethics. Chronicling how male young in the dairy and table egg chicken industries have been handled since the late nineteenth century illustrates the way an interplay between biology/technology and the needs of industry triggers ethical concerns. Breeding

plays a role in this complex interaction because its outcome often intro-
duces a problem where one had not previously existed. What should
one do with excessive male calves or excessive young chicken cocks,
when females do all the producing in both industries? Select killing, the
rise of new separate industries such as the veal industry, and the advent
of technology that could divide semen by sex were all methods used to
solve this problem and none is without its ethical criticisms. Sometimes,
when genetics provided the knowledge on how to select biologically
in order to solve an ethical/industry-need issue, it was not used. The
story of dehorning cattle from the nineteenth century until today illustrates that pattern. This chapter concentrates on farmer reactions and
response to these problems and to pressures from animal rights groups
and humane societies, but provides only an overview commentary on
the historical background of the groups and societies. Comprehensively
reviewing the huge amount of literature dealing with modern animal
rights or the rise of humane societies is beyond the scope of this book.

Many complicated subjects, deserving more in-depth study, emerge
in the book. What, for example, does the fact that breeding today presents science/craft unevenly tell us about the nature of biotechnologi-
cal revolutions? Will the main contributions to animal breeding in the
future come from technology and not genetic theory? How important
will the study of genetic architecture – today studied by bioinformatics
or functional genomics – be to future animal breeding, either in theory
or in practice? In other words, will geneticists conclude that it is it just
as effective to see the genetic make-up of animals as being encased in
a "black box" as to understand how the transmission of traits actually
works? Is the story of livestock genetics' interaction with mathematics
useful to academics trying to understand the role of mathematics in
biology, or in biotechnology?[2] Historians have also shown an increasing
interest in trying to unite the history of mathematics with the history
of science.[3] And finally, will modern genetics and biotechnology undermine the prevalence of purebred breeding in livestock species (and one
might add in pet animals like dogs)?

My view that the history of artificial selection in animal breeding should be approached with a dual emphasis on practice and sci-
ence/technology is not common in academic historical literature. The
hybridity of animal breeding methods from a science/tradition standpoint encourages writers to address the activity from either one or the
other point of view, but not from both. There is a considerable amount
of material devoted solely to the culture and craft aspects of breed-

ing. Animal breeding has attracted the attention of many writers who were/are not concerned with how its roots evolved from science. What these people have to say is valuable, and, ultimately, does much to contribute to a richer way of seeing the whole. Works of this nature often indicate, for example, even if they are not intended to do so, how parallel much of practical breeding was/is with genetic ways to breed. The importance of cultural input to breeding methodology also emerges from this material. There are studies involving horse, cattle, sheep, and dog breeding – too many to list in totality, but some important ones are mentioned in the notes.[4] (It should be noted that the subject of animals and science has stimulated much research, and not only on breeding. Animal disease is but one example of topics addressed under this vast subject.)[5]

Historians have traditionally addressed the subject of science in agricultural breeding by looking at plants – specific plants at that, and within a specific timeframe – and not at animals at all.[6] Hybrid corn breeding, for example, has commanded considerable attention. The phenomenal success of hybrid corn in the market place has made the subject of its breeding particularly interesting. Historians have explained how genetics led to hybrid-corn growing in the American Midwest states of Iowa and Illinois by the 1930s.[7] Hybrid corn and its use have interested scholars on other levels. The way it spread among Midwest farmers in the late 1930s, for example, triggered seminal sociological studies in the early 1940s, out of Iowa State University, on how innovation diffusion evolved in this particular case.[8] Studies that address agricultural colleges, their plant research, and sometimes also its applicability to farming in Germany or France, but in conjunction with farming in the United States, add to the story.[9] There are numerous other articles that deal with early efforts at scientific plant breeding. Excellent material on early Mendelism and plants in the United States, France, and Germany, for example, appeared in the *Journal of the History of Biology*'s special 2006 issue.[10] A book on agriculture and genetics, *New Perspectives on the History of the Life Sciences and Agriculture*, due out in 2015 and edited by Sharon Kingsland and Denise Phillips, will contain articles that focus on plants rather than animals, within a framework of practice/science.[11] Compared with the history of plant genetics there is a dearth of information on both livestock genetics itself, and its relationship to practical animal breeding.

The subject of animal breeding and the science/practical connection is, however, beginning to attract the serious notice of scholars. It has become a growing field, and an international one at that, over the past ten years. The area received some earlier attention: a review of Mendelist research at an American experiment station before 1920, which dealt with experimental patterns in chicken breeding, for example.[12] Chicken-breeding practices also attracted scholars in order to learn more about the enforcement of trade secrets or "patenting," meaning the intellectual property protection of biology.[13] A good article dealing with livestock genetics in farm breeding, and from a somewhat historical perspective, was written by scientists and appeared in the *Journal of Animal Science*.[14] In keeping with many studies on plant breeding, some historians focused on the farm/science divide from the animal perspective by looking at it through a certain lens: namely the rise of agricultural colleges and the attitudes of the scientists at them regarding research and the applicability of their work for agriculture.[15] An excellent book looking at research on sheep breeding in Moravia in the pre-Mendel period appeared in 2001, initiating what has become a more robust interest in the general topic of science and practical animal breeding.[16]

Newer studies have further expanded knowledge in the field. A good article, for example, assessed early nineteenth century livestock breeding in central Europe before the advent of Mendel.[17] Another looked at dairy cattle breeding in the Netherlands and general genetic theory over the twentieth century.[18] There are academic studies dealing with the historical effects of reproductive technology on livestock in various countries. A particularly important set of papers appeared in a 2007 volume of the journal *Studies in History and Philosophy of Biological and Biomedical Sciences* on this general theme.[19] The papers addressed cattle breeding, pig breeding, and AI technology. The topic of AI and the dairy industry has been looked at elsewhere as well.[20] A PhD thesis from Yale University, completed in late 2011, is another example of modern research that looks at (what the author describes as) the "borderlands" between craft and science in animal breeding. The thesis is unusual in that it deals with various large farm animal species and compares the situation across two countries: the United States and Germany between 1860 and 1914. The focus, like many other studies of craft and science in breeding, is on agricultural colleges, as well as education and research in relation to farm activities.[21] My *Art and Science in Breeding: Creating Better Chickens*, published in

2012, assessed the way genetics interacted with traditional chicken breeding internationally speaking, and provided a brief outline of major trends in livestock genetics. When it came to the larger story of how inputs to artificial selection theory evolved over centuries in relation to various livestock and in different places, the book only revealed the tip of the iceberg.

The complicated subject of the agricultural breeding of plants and animals, in relation to both science and culture, has increasingly attracted the attention of scholars interested in the export of ideas from the metropole to the periphery, or from an imperial centre to outlaying colonies. For plants, see S. McCook's book, *States of Nature*, for example, which addresses the implications of coffee production and other crops in the Spanish Caribbean between 1760 and 1940.[22] B. Charnley, in "Experiments in Empire-building," addresses the complex topics of plant breeding, Mendelism, and Imperialistic thinking.[23] Excellent work has been done on horse breeding and equine importations into Southeast Asia and southern Africa.[24] Here the reader sees that importation of breeding animals and breeding theory often does not result in the expected improvement of stock. Environmental conditions ultimately dictate what type of animal will prevail – which, in turn, provokes interesting questions around Darwinism. In fact, over the past two hundred years it has been difficult to separate concepts concerning artificial selection and natural selection. To some degree it remains so today.

Masterminding Nature: The Breeding of Animals, 1750–2010 looks at the development of artificial selection theory and how it was used over the years in the breeding of various species (generally speaking beef and dairy cattle, pigs, chickens, and horses to some degree) in various European countries and North America over two and a half centuries. In effect, therefore, it reviews the historical rise of animal breeding practices across nations, species, and extended time. I know of no other book that attempts to address the various practical and scientific roots of breeding and breeding theory in more or less equal depth over such a lengthy period. The coverage of late twentieth and early twenty-first century affairs in animal breeding is also unusual.

This book deals with a number of highly specialized subjects, and I attempted to handle them in a user-friendly way without the use of jargon. But it was impossible to completely remove language that might be foreign to some readers. Those with a lot of knowledge of genetics, for example, but not traditional animal breeding, might find language used to describe breeder methodology strange. And those who come from a background of animals and culture might profit from definitions of language used to describe genetic principles. While most terms are described and defined within the text, it seemed advisable to add a glossary at the end. Readers, therefore, can easily refresh their minds on terminology that is new to them at their convenience.

Chapter One

Artificial Selection Theory and Livestock Breeding, 1750–1900

Animal breeding, via artificial selection, is an ancient human occupation, dating back to the time of domestication at least fourteen thousand years ago. The continued artificial selection of animals for breeding led not only to marked physical changes from their wild progenitors, but also to distinctive types. Varieties of dogs, cattle, and horses evidently existed as early as four thousand years ago.[1] How much attention was given to the effectiveness of one breeding method over another is not known, but it is likely that most practices were quite haphazard in nature. In many parts of Europe it was difficult to control the mating habits of livestock in the Middle Ages because farmland was not enclosed or fenced in any way. Mating was random and variation in type tended to relate to geographic districts. Enclosure played an important role in the rise of more formalized and widespread attempts to modify animals via artificial selection in Britain and continental Europe (and to articulate how that should be done). While evidence of more organized breeding practices could be found by the late seventeenth century, it was the eighteenth century that saw real development in artificial selection theory and extension of its use. Further expansion in attitudes to theory and on how theory should be practiced emerged in the nineteenth century. Organizational structures were put in place to orchestrate breeding, and in the process professionalized the occupation. In the same period, naturalists became interested in artificial selection because the effects it had on domestic animals began to reveal some information on hereditary laws. These people undertook breeding experiments as well, and developed certain selection strategies. Charles Darwin believed parallels existed between artificial and natural selection, thereby even more closely linking animal breeding with questions of science. The connection has been ongoing ever since.

In this chapter, I review attitudes towards artificial selection that arose from the practical breeders and naturalists between roughly 1750 and 1900. This period saw the birth of quite distinct approaches to animal breeding within the two groups. I start by discussing the rise of structured breeding methodology developed by practical breeders, beginning with the breeding of the English Thoroughbred because the way it was created would profoundly influence not just animal breeding patterns, but also attitudes to heredity for the next two hundred years. The rise of purebred breeding was central to the evolution of artificial selection strategies designed to develop breeds of farm animals over the nineteenth century, and therefore its background is important to this story. Views regarding how to breed for experimental reasons were developed by naturalists within this period, thereby establishing the foundations of what might be described as "normal science," namely a framework within which to approach the problem.[2] A breeding method, which came to be seen as the scientific way to breed, was born in the late eighteenth century and would become increasingly entrenched over the nineteenth century. The work of Charles Darwin made each group increasingly aware of the other, but did not increase their understanding of each other's point of view. Ultimately the widespread use of purebred breeding and the hardening of attitudes within the ranks of purebred breeders widened the distance between them and increasingly fractured scientific attitude towards breeding generally.

Artificial Selection and Thoroughbred Horse Breeding

Attention to horse breeding in the late sixteenth century, particularly in Britain, initiated a complex theoretical approach to artificial selection and the nature of the hereditary process itself – important factors in shaping future attitudes of farm breeders to animal breeding and heredity. Until the end of the sixteenth century, horse breeders in Britain – generally speaking the nobility and royalty – selected animals on the basis of size and strength. That breeding strategy changed dramatically with the importation of Arabians, the first in 1616 by King James I, with more horses over the next century.[3] Distinctly different in type from the local animals (horses in hot environments like Arabia would deviate from horses indigenous to the colder climate of northern Europe[4]), Arabians fascinated the British breeders. Selective breeding practiced by Arab people on the horses intensified the distinctiveness already embedded in them by the environment, and perhaps more importantly, encouraged the stock to reproduce consistently to type in a way not seen in British horses. When Arabians were bred to British stock, the progeny over generations changed in phenotype (physical looks) profoundly and with greater consistency than had occurred previously in Britain. The results of this simple breeding strategy – crossing pure Arabians (generally speaking these were stallions) on local horses (most often mares) – led to considerable theorizing; specifically, that the purity of the Arabian enabled it to stamp type and, in doing so, change horses drastically. The progeny could almost be seen as a different species. Because the idea that species were immutable dominated the thinking of the breeders, they believed they were in effect hybridizing. It was the purity of the Arabian, almost as a separate race, that allowed this process to take place. A horse was "thoroughly bred" – a Thoroughbred – if it resulted from the crossing of a pure Arabian on any other horse. These Thoroughbred animals were tested for stamina by racing them. Racing, however, soon required some form of regulation, a detail that resulted in the establishment of a pedigree-keeping system.

In 1791 James Weatherby compiled identification information for racing horses, and published them in what was called the General Stud Book (GSB). If Weatherby had any genealogical records, he included them. The GSB was designed to stop falsified data (such as the age of the horse) being used for entry in races.[5] Three features in the GSB proved to be crucial in shaping much of future artificial selection theory and practice in farm animal breeding. First, having a pedigree in the GSB made a horse more valuable because being able to pin down the animal's identity provided a form of guarantee to the buyer. Second, the separation of horses by type from all others, through this identification or a guarantee process by the GSB, initiated a system that eventually protected intellectual property in living things. Third, the GSB maintained records in a public registry, something quite different from private record keeping by breeders. Huge ramifications would arise from these simple facts. Importantly, the significance of these features did not lie in the articulation of breeding theory. No strategy for artificial selection was embedded in this public registry.

While the original value of the GSB did not relate to breeding strategies, privately held pedigrees did play a role in artificial selection. They were used to prove that no inbreeding, an anathema to horse breeders by the late eighteenth century, had been practiced.[6] In fact the relative rarity of Arabians in the horse population of Britain virtually compelled some form of record keeping, in order to avoid inbreeding. While William Marshall argued in the eighteenth century that inbreeding methods developed the evolving Thoroughbred (he even stated that the practice originated at Newmarket, the centre of the racing Thoroughbred), it seems clear that he was mistaken. A famous breeder of farm animals claimed that horse breeders had always been reluctant to use the method, and he criticized them for not doing so.[7] Modern research on early Thoroughbred horse breeding records confirms that inbreeding did not play a significant role in the making of the breed.[8] Studies show that early horse breeders kept private records to guarantee the absence of inbreeding.[9] But public records carried more authority in this matter than the private records of the breeders, and the GSB came in to take over the guarantee role in breeding matters that breeder records had held originally. In 1815 the European horse breeder J. Justinus explained the fundamental thinking behind Thoroughbred horse breeding philosophy, which connected pedigree recording with quality: good horses resulted from an unsullied lineage of preserved purity, which was synonymous with quality. Selection should be made on the basis of genealogy, because that proved purity (and therefore quality), and maintained vigour and strength.[10] Inbreeding, still anathema, was not connected to these notions of purity and genealogy. The legacy of Thoroughbred theory, and in some cases, even breeding approaches, has been profound. The evolution of the Thoroughbred horse introduced the idea of a public registry system and the theory of purity. The legacy of the Thoroughbred for all future breeding would be an allegiance to recording pedigrees and the belief that pedigrees labelled animals as pure. Neither, ultimately, related to actual breeding methods.

Thoroughbred horse breeding also introduced certain cultural tenets to the breeding of animals. Better horses could be related to better people, specifically the nobility. Reverence for genealogy lay at the heart of the matter. This linkage of genealogy with quality would ultimately play a role in the rise of eugenics, when animal breeding served as an example for human breeding. Incipient eugenic thinking, however, was not fundamental to selection methods as far as these early breeders were concerned. The Thoroughbred horse breeders might have seen similarities between "noble" people and "noble" horses, but their system for breeding horses was based on what seemed to them to be principles governing heredity in livestock. The resulting cultural overlay evolved from following selection methods that worked with such principles.

The Selection Theories of Robert Bakewell and Sir John Sebright

By the second half of the eighteenth century, Thoroughbred horse breeding was based on a clear set of premises designed to shape artificial selection practices and to work with the process of heredity. Certain other theoretical approaches had been put forward by this time, and these contradicted the stance of the horse breeders. By the mid-eighteenth century a number of men, particularly in Britain, addressed artificial selection in attempts to improve farm animals.[11] Breeders in that country, influenced by the empirical thrust of the Enlightenment, focused on how to use artificial selection to improve farm animals for agricultural production. It should be pointed out, however, that many of the conclusions and approaches to methodology put forward at this time had been known to breeders – and practiced by the best of them – since at least as far back as Roman times.[12] It was the formalization of that knowledge into a practical structure, and the more extensive use of it, that were new. The best known of the Enlightenment British breeders, and the one who gets the lion's share of credit historically for this work, was a tenant farmer, Robert Bakewell of Dishley.[13] While Bakewell left little written material about his methodology, letters between George Culley and him between 1787 and 1792 provide information on what came to be known as the Bakewellian system of breeding. Another contemporary who wrote extensively about English farming, Arthur Young, added to our knowledge of Bakewell's breeding theories. To begin with, Bakewell advised careful selection of both males and females.[14] Next, he emphasized the use of inbreeding (that is the mating of animals that were related) and the avoidance of any outcrosses. Arthur Young described what was often called the "in-and-in" breeding theory of the Bakewellian method: "In breeding his bulls and cows (and it is the same with his sheep) he entirely set at naught the old idea of the necessity of variation from crosses; on the contrary, the sons cover the dams, and the sires their daughters."[15] Another important principle in Bakewell's system was the use of the progeny test as the basis for selection of stock to be used for breeding purposes. Animals were only good if they could produce good offspring. Bakewell tended to focus on males for progeny testing because of the greater ability to quantify results compared to females, not because he believed the input of males to be greater than that of females. In order to judge the breeding ability of his males more comprehensively, he lent out bulls, rams, and stallions. Under these conditions, he was able to evaluate the progeny of these males in various situations and in considerable numbers. He could, in other words, quantify results of a progeny test as far as was possible in eighteenth century circumstances, and to some degree, neutralize the effects of environment. Males whose progeny he found acceptable returned to his farm to breed females, who would be ever more closely related to such males.

It is worth pointing out that Bakewell's apparent emphasis on males in breeding was based on sound principles, not gender bias. He knew female input was as great as male. But he looked at the problem of inheritance in terms of populations rather than individuals, and this approach effectively forced him to focus on males. He could shift populations more effectively because of their greater capacity for reproducing in volume. It has been argued that he and the Enlightenment farm breeders were the first to emphasize populations rather than individuals in breeding strategies.[16] In effect, he worked out what could be defined as an eighteenth century form of quantitative genetics: quantifying progeny results and inbreeding to intensify the inheritance of quantitative traits. It is almost certain that Bakewell was aware of the effects of crossing lines emanating from different hereditary sources, namely increased vigour and also the inability to breed true to that increased vigour. But he was interested in stamping type and creating lines that bred truly – that is, in the creation of "breeds" – so he avoided any form of outcrossing.

These were innovative ideas for the times. To begin with, most breeders rejected the idea that female hereditary input was important to breeding programs, thereby following the more dominant gender bias of the times concerning the value of males generally. Breeders had also traditionally avoided all forms of inbreeding because experience taught them that such practices could have deleterious results, such as lowered fertility. Bakewell argued that only through the mating of related stock could animals be made to breed truly (meaning a likeness would continue generation after generation), and that careful selection could counteract negative effects of inbreeding. Eventually he made inbreeding acceptable when one was trying to fix type.[17] Bakewell's system provided a way to create "breeds." Resistance to inbreeding, however, did not end after Bakewell's time (he died in 1795). Its injurious effects continued to scare off many breeders. The idea that the progeny test should take such precedence in selection was also troublesome for breeders, who tended to look at an individual's excellence on the basis of its ancestors rather than its offspring. Population groups did not command as much attention when it came to breeding as did an emphasis on the value of particular individuals. Breeding by ancestry and individual worth were both far more common than breeding by progeny test and via population results. Breeder deviation from Bakewellian principles would have important ramifications on livestock breeding, especially in relation to the historical interaction of livestock genetics with farm animal breeding over the twentieth century.

It is worth taking a thorough look at the opinions of the British breeder, Sir John Sebright as described in his short book *The Art of Improving the Breeds of Domestic Animals*, published in 1809, because he explained eighteenth and early nineteenth century breeding in detail. Sebright, like Bakewell before him, accepted the theory that laws drove inheritance. He, also like Bakewell, did not believe it was critical to know what those laws were. It was only important to learn which selection method made the laws work in such a way as to bring desired results in the progeny. Sebright argued: "Many experiments must be tried, to establish a single fact; for Nature is sometimes capricious in her production, that the most accurate observer will be frequently deceived; if he draws any inference from a single experiment."[18] Experimentation would tell the breeder which methods of selection produced the best results.

Sebright supported many of Bakewell's views. Females, he agreed, were important to breeding programs. He elaborated on this point, though, by adding that choosing good females and good males would not provide for good progeny, "Were I to define what is called the art of breeding, I should say, that it consists in the selection of males and females, intended to breed together, in reference to each other's merits and defects. It is not always putting the best male to the best female, that the best produce will be obtained."[19] Inbreeding was at the heart of the matter for Sebright who pointed out that "Mr. Bakewell had certainly the merit of destroying the absurd prejudices which formerly prevailed against breeding from animals, between whom there was any degree of relationship," and added, "had this opinion been universally acted upon, no one could have been said to be possessed of a particular breed, good or bad; for the produce of one year would have been dissimilar to that of another, and we would have availed ourselves but little of an animal of superior merit, that we might have had the good fortune to possess."[20] In other words, before Bakewell's time, the best breeders, perhaps few in number, had always practiced inbreeding.

Sebright realized, however, that the bad as well as the good could be strengthened from such a procedure: "By breeding *in-and-in*, [a] defect, however small it may be at first, will increase in every succeeding generation; and will, at last predominate to such a degree, as to render the breed of little value."[21] He elaborated on inbreeding in the following words: "If the original male and female were of different families, by breeding from the mother and the son, and again from the male produce and the mother, and from the father and daughter in the same way. Two families sufficiently distinct might be obtained, for the son is only half the father's blood, and the produce from the mother and son will be six parts of the mother and two of the father."[22] He fully understood the fact that by crossing two breeds or strains, the progeny of the first cross might be good, but they would not breed truly.[23] Sebright turned away from perhaps the most important aspect of the Bakewellian system – namely, selection on the basis of the progeny test. Ancestry breeding took precedence over the progeny test, as far as Sebright was concerned. "We should not breed from an animal, however excellent, unless we can ascertain it to be what is called *well bred*; that is descended from a race of ancestors, who have, through several generations, possessed, in high degree, the properties which it is our object to obtain."[24] His views concerning ancestry breeding indicate that subtle changes in artificial selection theory were underway shortly after Bakewell's death. Thoroughbred breeding culture had begun to infiltrate general animal breeding.

Naturalist Attitudes to Artificial Selection

At the same time that Bakewell and the livestock breeders studied breeding strategies, another group of people became interested in the dynamics of artificial selection. By the early eighteenth century – before the horse breeders and Bakewell altered the animal picture – varying types of cattle, horses, and dogs clearly existed, a fact that was noted by naturalists. Why did artificial selection lead to distinctly different types within a single species? Certain underlying laws seemed to be at work, and those interested in natural history wanted to understand the laws themselves. They undertook breeding experiments in order to see what results occurred. The outcome of artificial selection strategies was only useful to them insofar as it could explain the process of heredity. Their focus, unlike the breeders, was on the laws that explained the results. In 1745 the French mathematician and astronomer Pierre-Louis de Maupertuis (1698–1759), for example, observed that artificial selection had altered traits, and he wondered if natural selection could do so even more effectively.[25] Could speciation result simply from such random selection?

By the second half of the eighteenth century, naturalists more commonly undertook breeding experiments using plants, with the hopes of finding underlying rules that governed how heredity worked. There were a number of people in various European countries who together might be called the founders of an approach that led to a hardening of an attitude towards correct experimental breeding procedures; initially in relation to plant breeding, but subsequently animal breeding as well. Experimental breeders decided to cross lines that differed from each other. The deviation in type of the parent lines was important to them, because of potential large variation in the resulting progeny. It was found that such a breeding system brought out interesting features that would not have been otherwise observed in the next generation. If one crossed two species that would interbreed, for example, would one get a third species from that mating? Could race/line-crossing explain how hereditary laws work? Did hybridization, then, explain speciation and hereditary laws? Could one successfully establish completely new varieties through hybridizing? The experimenters labelled their breeding method as hybridizing whether it was applied within or across species.

In 1751 Carl Linnaeus published a systematic discussion of plant hybrids.[26] The first to try experimenting with hybridizing, and a person who has been described as the founder of breeding by hybridizing,[27] was the German naturalist J.G. Kölreuter (1733–1806). He began hybridizing experiments in 1759, and published results from this work in 1761. Kölreuter argued that that new species could not arise out of hybridization.[28] He hybridized plants to establish the existence of sexual reproduction in plants and the fixity, or the constancy of "pure" species.[29] Kölreuter was not concerned with better crops or farm animals, but rather with the genetic laws behind differences between "varieties" and "species."[30] The British plant breeder, T.A. Knight (1759–1838), practiced inbreeding (it has been argued that he was the first to apply Bakewellian principles to the inbreeding of plants[31]) and the crossing of those inbred lines.[32] It did not take long for the experimenters to see that inbreeding could reduce the vigour and fertility of the plant, and that under these conditions crossing inbred lines brought a return of both traits. The progeny of the two inbred lines were also often better than either parent line. It soon became evident that the progeny would not breed truly to its improved state. These features of inbreeding and crossbreeding (or outcrossing) were familiar to animal breeders like Bakewell and Sebright.

After 1800 a few naturalists began experimenting with animal breeding. They did so along the hybridizing lines of the experimental plant breeders and for much the same reason: in an attempt to understand the process of heredity. In 1819 a Hungarian, Count Festetics, undertook inbreeding/crossing experiments with animals, and subsequently wrote "Hereditary Laws of Nature." He believed species inherited characteristics and did not acquire them, traits found in past generations could reappear, offspring could show variation from parent stock, and the dangers of inbreeding could be avoided with careful selection.[33] By this time, plant-breeding experimenters had begun to suggest that their work should be applicable to agricultural production. Knight, for example, argued as early as 1799 that experimental breeding practices could be used to help the betterment of farm plants.[34]

The European hybridizers started from a theoretical stance that differed in a particularly important way from that of farm animal breeders. Animal breeders worked from the premise that stock should be able to breed true to certain characteristics. If a cow had been bred to provide more milk than average, her heifer calves were expected to excel in the same way. Unlike what the hybridizers wanted, the progeny was not to deviate significantly: it was first, to replicate the parents, and second, to improve (if possible) on their best characteristics. This cleavage, namely breeding for true lines versus hybrid breeding that resulted in non true-breeding progeny, would act as a critical wedge between naturalists (later scientists) and practical breeders for a considerable length of time when it came to artificial selection methods to be used on farm animals. A hardening of attitudes was clearly in place as early as the late eighteenth century: true line breeding was the practical way to breed livestock, while hybrid breeding was the experimental way to breed. The contentious question that arose was: could or should the experimental way to breed be applied to livestock breeding?

A fusion of the ideas behind the practical approach and that of philosopher/naturalist in breeding was rare in the first part of the nineteenth century. One particularly significant instance of the application of principles emanating from both took place in Brno, Moravia (present

Czech Republic) in the 1830s, where sheep breeding for wool production was important to the local economy. By that time Bakewell's
method had had a profound effect on breeding practices in Moravia
and became enmeshed with attitudes towards proto-scientific questions. In the late 1790s a teacher and writer of science and economics,
C.C. Andre, introduced Bakewell's theories to Moravia.[35] These theories dovetailed well with existing sheep breeding practices in that area,
including those of Baron Ferdinand Geisslern, for example. Geisslern,
a sheep breeder of imported Merino sheep on a farm just northeast of
Brno, left no records, but Andre publicized his practice of developing
different strains within a single sheep breed.[36] It is clear, additionally,
that careful crossing, inbreeding, and progeny testing were practiced in
Moravia on sheep as early as 1797, a unique situation in central Europe.
Many Moravian breeders understood the dangers of inbreeding, which
compelled record keeping of the progeny test in order to keep it under
control. After 1800 (when information on his methods was published
in Brno) Bakewell's system was more widely applied by sheep breeders in Moravia than it had been in the 1790s.[37] Sheep breeding, under
Bakewellian principles, was designed to produce true breeding lines
via careful inbreeding.

Attitudes from experimenting naturalists soon became infiltrated
with those from the Bakewellian system, partially as a result of university professor J.K. Nestler's lectures that were based on the work of
Martin Köller (1779–1838). Köller understood the need to keep pedigree
and production records in order to properly progeny test. He believed
selection on the basis of one trait to be dangerous, and he outlined the
principles of individual and family selection. Nestler, though, while
promoting the theories of Köller, also discussed the hybridizing work
of the naturalists, and the natural phenomena of variation. He stressed
the importance and effects of inbreeding in relation to hybridizing and
to improvement in livestock breeding.[38] For Nestler, hybridizing, that is
the breeding of improved animals that would not replicate themselves,
had a role to play in livestock improvement. In other words, Nestler
used the thinking of the naturalists and practical breeders in his efforts
to develop a workable artificial selection strategy. Count Festetics' ideas
also became known to Moravian sheep breeders, thereby bringing in
another prevailing attitude towards the theory of hybridizing. While
the experimental breeders remained ambiguous, or perhaps at the very
least, non-committal, concerning the question of true breeding lines or
hybridizing when it came to livestock improvement, they tended to

focus on hereditary laws in a way that most practical animal breeders did not. The Moravian breeders were interested in both the underlying laws and the results.

The question of heredity itself, within the framework of artificial selection methodology under Bakewellian and naturalist principles, fully commanded the sheep breeders' attention in Moravia by the early 1830s. Over 1836–7, Prelate Napp, Nestler, and members of the Sheep Breeders' Society (an organization formed in Brno in 1814) discussed in detail the problem of heredity. The men were concerned with the principles behind the transfer of traits. As Napp put it: "We are not dealing with the theory of the process of breeding, but the question should be: what is inherited and how?"[39] (Prelate Napp later ran the monastery where Gregor Mendel did his experiments on peas. Napp actively supported Mendel's research, which, in 1865, resulted in the earliest known laws of genetics.)[40]

The Rise of Purebred Breeding

Attitudes of the practical farm animal breeders underwent significant change over the mid-nineteenth century, and in doing so made the possible meshing of practical breeding approaches with those of the naturalists (evident in Moravia in the early part of the nineteenth century) less likely to happen in the future. Patterns in breeding that were evident as early as Sebright's time took centre stage, and altered the structure of practical breeding dramatically during this period. Fundamental premises of Thoroughbred horse breeding theory modified Bakewellian principles. The eighteenth century saw European breeders introducing the idea that inbreeding stamped type, and that the way to improve stock was to progeny test. These principles were at the heart of the Bakewellian method and differed profoundly from those behind Thoroughbred horse breeding, where any sanguineous mating was anathema, and genealogy, as much or even more than the progeny test, served as the basis for selection. Something new therefore entered the breeding world early in the nineteenth century – purebred breeding – when aspects of horse breeding culture fused with the Bakewellian system within the world of British beef Shorthorn cattle.

Robert and Charles Colling created the Shorthorn in the 1790s by using Bakewell's principles in their cattle breeding operations. True to Bakewell's teaching, the Collings did not acquire foundation animals on the basis of genealogical background. In 1784 Charles Colling

bought a cow named Duchess from a tenant farmer and in his opinion
she remained the best cow he ever saw. He hoped to perpetuate her
through inbreeding, not improve on her. Meanwhile, Robert Colling
found a good bull grazing by the roadside, named Hubback, and purchased him from a local bricklayer, even though the bull's ancestry was
unknown. Colling then inbred intensely to him.[41] The Collings even
used cattle outside existing Shorthorn varieties in breeding programs
designed to create a new, uniform Shorthorn. The brothers introduced
Galloway blood, for example. Colling Shorthorns were recognized as
good beefing stock and soon replaced Bakewell's Longhorns both in
popularity and monetary value. The excellent qualities of Shorthorns
quickly attracted attention, and the animals subsequently became well
known internationally. By 1817 Shorthorns had been exported to Kentucky and Virginia in the United States.

In 1800 Thomas Bates began buying Shorthorns from the Collings,
and he acquired stock on the basis of genealogy.[42] Bates's decision to
choose foundation animals based on genealogical background (instead
of proven worth) represented a departure from the Bakewellian principles that the Collings had used, though it might have been only part
of an evolving promotional plan that Bates had in mind. He began to
practice inbreeding with his new animals, but apparently for different reasons than Bakewell's or the Collings'. Bates claimed it was the
inbreeding itself that was important to him because it preserved what
he defined as purity, which he saw as synonymous with quality. Neither Bakewell nor the Collings had connected inbreeding to notions
of purity. Bates maintained a constant level of inbreeding throughout
his breeding career, a fact that suggests he monitored his inbreeding
carefully.[43]

In 1822 George Coates established a public herd book for Shorthorns.
The breed achieved greater popularity as a result of this form of recording, and public pedigrees also allowed for the early geographic expansion of Shorthorn territory. Bates actively supported George Coates's
new public registry system, and some scholars suggest that Coates
began his work largely at the instigation of Bates.[44] Publicly recorded
pedigrees under Bates would become just as important as inbreeding.
He contended as well that pedigrees demonstrated different degrees
of purity, and even described some of the renowned stock of Charles
Colling as "mongrel" - a concept that would have been totally foreign to
that breeder.[45] Here we see the influence of Thoroughbred horse breeding
culture, rather than methodology, in Bates' thinking. Inbreeding became

attached to the idea of maintaining purity, as evidenced in pedigrees, and, in turn, purity came to define the meaning of quality. Genealogy, as found in public pedigrees, and not the progeny test, became the means of selection for breeding. Bates's attitude to pedigrees resulted from his absorption of aspects of the Thoroughbred horse breeding methodology and culture. He used both to shape and promote his Shorthorn breeding. Everyone at the time recognized that the background of foundation animals, admired for their perceived quality and used to fix a breed close to their type, was simply not known from a genealogical point of view.[46] Colling's bull Hubback serves as an example of this phenomenon.

When he connected pedigree breeding with inbreeding (even if only as part of his promotional plans), Bates laid the foundations for purebred breeding. Public registries worked meaningfully with inbreeding practices that no longer reflected the selective process espoused by Bakewell. Inbreeding by pedigree (certified in a public registry) preserved purity, which in turn meant quality - vigour with no degeneration. Pedigrees recorded in such registries explained to everyone the level of inbreeding, which in turn could be defined as purity to breed type in individual animals. Breeders could also control or manipulate the inbreeding technique by maintaining certain levels of inbreeding, thereby enhancing purity. The less the use of an outcross breeding technique the greater the line's ability to breed truly, and the level of the ability to breed truly came to be seen as a better indication of the degree of purity within a breed. Publicly recorded pedigrees could be related simultaneously to purity and inbreeding, thereby creating a basic philosophy behind the "purebred" method. Animals were believed to be "pure" to breed type, and therefore "purebred," when they carried public pedigrees that certified them to be so. The "purity" of such animals within breed type, however, differed and varied in relation to the quality of their pedigrees. Ideas about breed, the meaning of purity within breed, and the role of pedigrees in breeding became entangled with each other in a complicated way through this process. Methodology and principles of heredity became interconnected with cultural attitudes. Certain social evaluations could be attached to breeding results. Purity might be the key to breeding for consistency, but it carried with it implications that had nothing to do with that desired end. Purity was often seen as an end in itself, especially by poor breeders.

An animal was "pure" to type, that is purebred, if it was acceptable by genealogy for recording in a public registry. Sanguineous breeding had initially been part of the guarantee of purity, and proof of that breeding could be seen in pedigrees. For some time over the nineteenth century, breeding outlook demanded that purebred registered stock be inbred. Bates, for example, marketed his Duchess line of Shorthorn cattle on the basis of purity because of intense inbreeding as recorded in pedigrees. No outcross had "contaminated" them.[47] With the better fixing of type, however, any form of deliberate inbreeding within the restricted inbred pool of animals defined as purebred became less significant. The purebred system in its mature form simply meant that an animal could carry a pedigree if both its parents had been recorded in the stud book. Pedigrees guaranteed consistency of type because of the restricted hereditary background that had been established by earlier sanguineous breeding. Any method of breeding was acceptable within that framework – inbreeding and outcrossing, selection by the progeny test, or ancestry. The philosophy behind Thoroughbred horse breeding continued to shape attitudes to purebred stock, regardless of breeding methods. Purity was linked to genealogy, but not to any particular breeding method or selection process used to generate that genealogy. The work of Bates, then, ultimately affected everyone's perception about what it meant to define any animal as either improved or "purebred." It even led to shifts in Thoroughbred horse breeding. "Thoroughbred" thinking might have shaped the ethos of purebred thinking, but purebred breeding, in turn, infused the breeding of the Thoroughbred horse.

As early as the late eighteenth century, a horse could be labelled a Thoroughbred if both its parents were recorded as Thoroughbreds in the stud book, not just if the animal resulted from a cross to an Arabian stallion. (Arabians continued to be registered in the GSB until the 1960s, a hangover from the early days when the Thoroughbred breed was developing.) Over the mid-nineteenth century, breeders came to rely on the GSB as a breeding regulation, adhering to purebred principles. The Thoroughbred had become a "purebred" horse. Sanguineous breeding within that GSB-limiting horse population continued to be culturally unacceptable for some time, but it was up to the breeder if he wanted to use the method. Inbreeding did in fact become more common over the nineteenth century. The idea of "pureness" in Thoroughbreds reached a new height in 1913 when Lord Jersey managed to convince horse breeders in Britain to use only animals registered in the British GSB for breeding purposes. For over thirty years, racehorses from the United States and France of outstanding quality were not eligible for GSB pedigrees (because their sires and dams did not have GSB papers) and therefore could not be used in Britain for breeding.[48] Man o' War, or Big Red, the famous American racehorse, and therefore his progeny, did not pass the Jersey test. Purity in this case related to marketing issues that had nothing to do with breeding methods.

The linkage of pedigree keeping to Bakewell's methodology also led to a return in the allegiance of animal-breeding philosophy to the idea of race constancy. Scholars have commented on this retrograde attitude to animal breeding and argue that, by 1848, race constancy, a tenet of Thoroughbred breeding that was also supported by the pedigree concept, again held sway across Northern Europe.[49] Curiously, under these conditions, many contemporary breeders in North America no longer seemed to recognize what constituted purebred breeding, or the fundamental role that Thoroughbred horse breeding culture had played in it. Often Bakewell's methods were held to be synonymous with purebred breeding.

By the late nineteenth century, for example, while the North American farm press might consistently tout purebred breeding as *the* method to breed by, actual methodology describing breeding in the press outlined either the Bakewellian method, or the strategies of the Collings.[50] The methods of breeding taught in North American agricultural colleges in the late nineteenth century focused on how these eighteenth-century breeders practiced artificial selection, but defined these methods as purebred breeding.[51] Neither the press nor the agricultural experts in North America seemed to recognize the fact that the purebred system had nothing to do with actual breeding and did not reflect either Bakewell's method or the breeding strategies of the Collings. The idea that practical breeding (especially purebred breeding) related to science also attracted notice. While breeders often referred to purebred breeding as a science, not all accepted that idea. Some began to argue that it was difficult to define exactly what was scientific about the method, or about any animal breeding method known at the time. As early as 1876 an American Shorthorn breeder stated that science had done little to help in the breeding of better stock.[52] In 1896 a Canadian farm journal reported on "The Science of Farming" and noted: "In spite of the great advances that have been made by breeders of late years in the science of breeding, there are still many things either but little understood by them or totally beyond their comprehension."[53]

Standardbred Breeding as an Artificial Selection Tool

While purebred breeding would ultimately dominate farm animal breeding in both Europe and North America, another system carried considerable sway for some time. European methods of creating breeds

and defining them as such over the late eighteenth century and particularly over the nineteenth century differed from the Thoroughbred
system and purebred breeding. Artificial selection, under these conditions, was designed to produce animals that met certain set standards,
and therefore was labelled as standardbred breeding. This method was
most commonly applied to horses, but it came to play a significant role
in North American chicken breeding, as well as the dog fancy. Methods that regulated the development of the European Warmblood horses
serve as an example of how standardbred breeding worked. Generally
speaking, warmbloods resulted from the crossing of the so-called cold
blood of any heavy draft horse with the so-called hot blood of an Arabian or Thoroughbred. Various warmbloods were produced through
this type of crossing – the Trakehner, the Hanoverian, and the Dutch
Warmblood (subdivided into the Gelderlander and Groninger). Warmbloods received accreditation as a "breed" on the basis of type assessed
by inspection and testing for performance, not by genealogy as recorded
in pedigrees. The horses were bred to a standard. Methodology used
in breeding mattered even less than their genetic background.[54] Even
though warmbloods were created under a standardbred structure in
the nineteenth century, the use of pedigrees for breeding purposes
began to infiltrate the system. With time, moreover, purebred breeding
philosophy increasingly governed how warmbloods could be papered.
While inspection and testing for performance continued, those animals
seeking admission to the stud book increasingly came from the genealogical background of accepted, pedigreed horses. Performance testing,
however, counteracted the effects of purity philosophy that so perme-
ated the purebred and Thoroughbred systems. Warmbloods were (and
are) viewed as capable horses, not horses of "pure" breeding. Under
these conditions, breeding methods used to create them remained as
flexible as they had always been.

One of the best-known animals created under a standardbred system
was the standardbred trotting/pacing horse. Different North American
lines of trotting and pacing animals could be registered in a stud book
as early as 1839, but recording was done on the basis of their speed. The
horses were bred to a speed standard, and thus labelled as Standardbred. Speed standards were more formally organized and defined later
in the nineteenth century. By the late 1870s any stallion, mare, or geld-
ing (castrated stallion) that could trot or pace at the rate of a mile in two
and half minutes under set racing trials organized by the National Trotting Association could be entered in the Breeders' Trotting Stud Book.

An emphasis on breeding to a standard appeared to some to be more effective than purebred or Thoroughbred breeding. It was noted that the speed of Thoroughbreds increased only 8 per cent over a twenty-five-year period (1875-1908), while Standardbred ran faster at the much higher rate of 27 per cent over a twenty-five-year period (1877-1913). By breeding to a standard the trotting/pacing horses had increased their capacity to run by half a minute over less than fifty years.[55]

Even so, purebred breeding ethos tended to infiltrate many standard-bred systems over the late nineteenth century, and did so with respect to this trotting/pacing horse. By the 1880s, entry into the book was possible on the basis of genealogy that indicated potential speed. A horse sired by a registered Standardbred and from a mare sired by a registered Standardbred qualified for entry into the stud book. Any mare that was sired by a Standardbred stallion – or herself had produced a tested Standardbred – also qualified. By late 1898 entry into the stud book had become purely an issue of genealogy, not one of performance testing. An animal whose sire and dam were registered as a Standardbred qualified for recording.[56] Standardbred breeders could select under any method they chose to in order to breed. The only restriction was that they had to breed from a registered stallion and a registered mare.

The fundamental theoretical stance of the Thoroughbred, purebred, and standardbred systems became thus entangled in North American breeding circles. Even the words "Thoroughbred," "purebred," and "standardbred" were erroneously used interchangeably to describe the same thing. In reality, all described purebred breeding, which held hegemony over the other two by the mid-nineteenth century. Regardless, the standardbred system was an organizing rather than breeding system.

Purebred Breeding and Marketing: The Story of the Percheron Horse Trade

The gradual move to organize breeding under the purebred system resulted largely because of the enhanced marketing potential it provided. An example of how significant purebred breeding could be to markets and marketing can be seen in the American Percheron horse trade over the late nineteenth century. Breeders in the United States imported draft horses from France as early as 1876 and they began registering them in a stud book under the name Norman Percheron.[57] By 1882, over 1,800 horses had been recorded in this registry.[58] No French draft horse was called a Norman, however, and the Percheron type came in several styles – two facts that Americans at first did not recognize. Demands by importers for "purity," combined with conformation to specific characteristics, soon made Americans realize that the word Norman meant nothing to French breeders and that various types of horses qualified as Percherons. Breeders in the United States set out to change the situation. First, the Norman Percheron stud book dropped the name Norman.[59] Next, Americans managed by 1883 to force French breeders to set up a stud book to register Percherons. The *Breeders' Gazette* pointed out that this action would undermine a clique in Illinois that claimed total control and knowledge of French draft horses.[60] In other words, the move to define purity in French draft horses was as much about control of the American importing business as about purity. American importers also encouraged French breeders of Percherons to select for one type. The American desire for greater size and weight in draft horses made the French producers breed for those characteristics between 1879 and 1889. French breeders abandoned the lighter, trotting Percheron type, which they preferred, because of the marketability of the heavier but also awkward and bad-tempered horse that Americans wanted. (A return to lighter Percheron breeding began in France only after 1897, when the American market had collapsed.)[61]

French breeders, however, did not always comply with American wishes as much as United States importers wanted. The *Breeders' Gazette* complained in 1885 that some French peasants (as the editor of the journal described the breeders) were unbelievably stupid because many still did not see any point in recording their mares if these animals were not sold. However, they were slowly getting the message that Americans would not buy foals from unregistered dams: "But in some cases where the importance of registration of the brood mares cannot be beaten into the heads of such peasants as own really good mares, the stallioners [owners of the breeding stallions used on such mares] are paying the fee and thus securing the registration."[62] Meanwhile, back in the United States, breeders who had clung to the name Norman dropped out of the new Percheron association. They decided to call their horses French Draft, and formed the French Draft-Horse Breeders' Association. The American French Draft-Horse Breeders' Association asked the French government to set up another stud book, which it did in 1886 by establishing clear regulations that accepted horses under separate categories: draft of no distinct breed; or crosses of two dominant breeds, Boulonnais and Nivernaise; or Percherons (which dominated three quarters of the recording).[63] American French Draft breeders claimed that this situation allowed them to pedigree many of their imports as Percherons.[64]

The issue of French Draft (formerly Norman) and Percheron recording finally made the Illinois Board of Agriculture establish a committee to look into the matter of breeds in France. Were all French horses in reality Percherons, or did distinct breeds exist in France? The committee concluded that the essence of the problem lay in the fact that pure Percherons were expensive to buy in France, and American importers found that if they could prevail upon men back home to purchase miscellaneous types and breeds from France as purebred Percherons, they made a great deal of money. French authorities were called in and asked what the word "breed" meant in France. These officials argued that, in France, the word was understood to mean the same thing it did in the United States – distinct types that bred truly. The claim by the French Draft-Horse Breeders' Association that it recorded miscellaneous horses as Percherons was proven to be invalid.[65] It was a distinctly valuable victory for the Percheron breeders' association. Notably, attention to actual quality in horse types played no role in the debate.

Corruption continued, however, in the importing business. "There still remain those," the *Breeders' Gazette* stated in 1888, "who buy disreputable specimens of horseflesh at merely nominal prices, and with them prey upon the legitimate business of the importer."[66] Registered Percheron stallions found to be non-breeders were purchased by dealers from Paris. The dealers wanted the certificate, not the horse, and if they found an American who would buy any particular animal, the certificate would be attached to that horse. It seemed that the same practice was followed in the United States as early as 1887.[67] The horse depression of the 1890s took its toll on the lucrative Percheron/French Draft-Horse business. The Percheron Society effectively went bankrupt and collapsed in 1893.

Charles Darwin and Artificial Selection

Those concerned with natural history continued to be aware of the work of practical breeders, and also of the marketing power of purebred breeding. The subtle shifts in Bakewellianism that resulted from the rise of purebred breeding seemed masked by the stunning effect the system had on market values, and, as a result, confused the very people who had earlier been influenced by it in its original form. Charles Darwin, for example, apparently failed to recognize that changes in animal breeding practices had taken place after public recording was linked to Bakewellianism. In his *Variation of Animals and Plants under Domestication*, written in 1868, he commented on purebred breeding: "Why have pedigrees been scrupulously kept and published of the Shorthorn cattle, and more recently of the Hereford breed? Is it an illusion that these recently improved animals safely transmit their excellent qualities even when crossed with other breeds? Have the Shorthorns, without good reason, been purchased at immense prices and exported to almost every quarter of the globe?" Monetary value, Darwin suggested, guaranteed genetic quality. "Hard cash down," he said, "over and over again, is an excellent test of inherited superiority."[68] In these confused words, Darwin expressed the new approach to animal breeding prevalent after the union of public recording and inbreeding had taken place. Monetary issues reflect the cultural evaluation of the power of pedigree, though, and do not explain how heredity works. Darwin seemed to have had a better grasp of the idea of genetics from reading material on animal breeding much earlier. In *Origin of Species* he wrote that "the laws governing inheritance are quite unknown."[69]

Darwin might have been confused about the actual workings of heredity, but he spent a great deal of time reading about the way artificial selection was practiced by breeders.[70] In his attempts to explain the process of natural selection, Darwin studied, for example, the hybridizing work of plant breeders.[71] He also became a fancy pigeon breeder in 1856 and undertook to learn how pigeon breeders altered characteristics of the birds. He corresponded with pigeon breeding experts like John M. Eaton.[72] He focused on intensive selection practices of the pigeon breeders that brought about minor changes in the birds. Pigeon breeders avoided both inbreeding and most particularly crossbreeding. Purity was important to them, and they could cling to purity in breeding because pigeon breeds had long been established by the mid-nineteenth century. The use of crossbreeding to mould type and inbreeding to fix type was no longer necessary. Darwin came to see the practical breeding of the pigeon breeders as representing standard artificial selection methodology. It suited his analogy of artificial selection with natural selection where inbreeding and/or crossbreeding would only occur in a much weaker way and in conjunction with any other naturally occurring mating scheme. It has been argued that he failed to see that breeding systems that created breeds and vastly different types within domestic species virtually always reflected contrasting systems of inbreeding with outcrossing or crossbreeding.[73] After 1860 and the publication of *Origin*, however, it is clear that Darwin read extensively about inbreeding and corresponded with breeders who practiced it: Herman von Nathusius in Germany and his work in pig breeding being a prime example.[74]

Over the second half of the nineteenth century, much of biological research focused on explaining Darwin's evolutionary theory via experimental breeding. A myriad of academic theories (such as Neo-Lamarkism) concerning speciation and/or the ability of a species to change evolved as well.[75] The world of biology was exceedingly complex by the late nineteenth century, and Darwinism played a role in that complexity. It linked natural selection with artificial selection more tightly than ever. Associating natural selection with artificial selection seemed to encourage a greater focus on experimental hybrid breeding. Hybrid breeding came often to be seen as a viable tool for both the study of evolution and the improvement of agricultural plants in particular. The Dutch plant physiologist, Hugo de Vries (1848–1935), trained under the German scientist Julius von Sachs, was especially important to the trend of linking hybrid breeding with experimentation designed to bring about better farming and to explain the dynamics of evolution.[76] Experimentation in evolutionary biology should be done at agricultural experiment stations, he argued, because there were basic similarities between natural and artificial selection.[77] For biologists like de Vries, it was impossible to look at either component separately, and therefore any attempt to study one resulted in a study of the other. In a subtle way, biologists had moved themselves into the world of practical breeding, particularly in relation to plants. Under these conditions, the experimental methodology of the plant hybridizers came increasingly to be seen as the scientific way to practice artificial selection, and plant hybridizing would dominate much of agricultural plant research at state-run institutions across North America by the late nineteenth century.[78] The same was true in Germany after the 1880s.[79] Farm-animal breeding did not attract as much scientific attention as did plant breeding, but some hybridizing work was done on livestock.[80]

Practical (versus experimental) animal breeders, especially those involved with the larger farm animals, continued to follow purebred breeding principles, meaning they bred for lines that reproduced truly and did not focus on hybridizing. Regardless of this fundamentally different approach to breeding, practical animal breeders likewise tended to see artificial selection within the framework of Darwinism. A massive report released by the British Horse Commission in 1890, for example, revealed that many British horse breeders believed Darwin had clarified how heredity worked in artificial selection, even if they were unsure how that was the case.[81] A German professor of animal breeding at the Berlin Agricultural College, Herman Settegast, firmly believed that Darwin's work was important to animal breeders.[82] Not everyone involved in breeding, however, saw Darwinism as relating to artificial selection or as it being helpful for methodologies to improve livestock breeding: Nathusius, an experimental animal breeder and also a hybridizer, being a prime example.[83] Darwinism, laws of heredity, and artificial and natural selection had together developed interlocking – but increasingly confusing – links by the end of the nineteenth century.

The Rise of Biometry

Darwinism established a fundamental concept for science: the genetic architecture of animal populations could be altered by selection. This concept as such was, of course, basic to Bakewell's thinking, and therefore we cannot credit Darwin with originating the idea. But Darwin's inspiration brought the conception squarely into the realm of science. One outcome of this development was the rise of a scientific/statistical approach to variation in species – namely, biometry. It would be some time before some of the theoretical approaches of biometry became applicable to artificial selection in animals for farm improvement, but few developments would be as important to future genetic theoretical artificial selection theory. Understanding its early characteristics, then, is critical to an appreciation of why biometry remained detached from theoretical artificial selection for as long as it did; it is also important to consider it alongside the basic contributions it offered to artificial selection theory. The founder of the new methodology designed to study inheritance was the British geographer and mathematician, Francis Galton.

Born in 1822 and a cousin of Darwin, Galton studied medicine and mathematics at Cambridge University. Interested in geography (specifically meteorology and weather) in his early life, later he focused on the problem of inheritance. Consistent throughout his career, however, was his underlying fascination with quantification in patterns he studied. His idea that quantification could be applied to heredity initiated what would be known as biometry. He was particularly interested in applying biometry to human inheritance, and in 1883 he coined the word eugenics to describe a new "science" that studied human inheritance by way of mathematics. When he came to assessing differences in people, Galton looked at features that demonstrated quantitative inheritance; that is, characteristics that everyone inherited but in a quantitative way (for example, more or less of features such as intelligence, height, etc.). The quantitative approach to inheritance would be very significant to animal breeders in the future. Animal breeders were primarily interested in quantitative traits: how much or how little of a characteristic a farm animal would inherit as a result of certain breeding programs. Biometry also shared a particularly important outlook with the eighteenth- and nineteenth-century practical breeders: the notion that inheritance could be studied without focusing on concepts concerning genetic laws. The genetic architecture behind inherited characteristics was not essential for understanding the effects of either natural or artificial selection on hereditary patterns. This stance was radically different from that of the eighteenth- and nineteenth-century experimental hybridizers.

Galton amassed an immense amount of data in collected human pedigrees and on dogs, horses, and cattle, which he analysed by statistical means. In applying mathematics to biology in this fashion, he in effect invented biomathematics.[84] Biometry, a statistical way of looking at biological inheritance, seemed to elucidate the process of evolution under Darwinian terms, and the journal *Biometrika* was founded to pursue that study. In its first issue, Galton stated: "The primary object of Biometry is to afford material that shall be exact enough for the discovery of incipient changes in evolution which are too small to be otherwise apparent."[85] Biometry remained focused on natural instead of artificial selection and, as such, did not attempt to predict the outcome of any breeding strategy.

The work of Galton, while it established certain principles that were important to animal breeders, also resulted in the rise of eugenics, a form of thinking that did not dovetail well with the outlook of animal breeders. Eugenic theory would be based largely on principles found in animal breeding circles (devotion to purity/genealogy/quality), but there is little evidence that animal breeders viewed their work as relating to human breeding. A study of the Canadian farm press in the early twentieth century reveals that breeders saw no linkage between human breeding and animal breeding.[86] Early scientists concurred, bemoaning the fact that breeders showed no interest in eugenic work.[87] Breeders might have noted similarities between human and animal heredity patterns, but they distained strident eugenic views. Eugenics increasingly dominated any form of genetic research into the 1920s, and the disenchantment of animal breeders with science's approach to breeding escalated accordingly.[88] Because it was fundamental to the thinking of biologists/geneticists, eugenics played a role in retarding the infiltration of scientific approaches to animal breeding in the practical world before 1930.

Galton's methodology of studying eugenics and heredity in animals attracted the attention of two important scholars after the 1889 publication of his second book, *Natural Inheritance*. (One writer today believes that this important book has not generated the recognition it deserves, possibly because Galton's writing was somewhat muddied.[89]) The great statistician, Karl Pearson, and a marine biologist, W.F.R. Weldon, were both impressed with Galton's idea of using a statistical analysis to study biology – specifically, patterns of inheritance. In the 1890s Pearson took over the crude statistical tools that Galton had developed and refined them to look in many ways like modern statistics.[90] Pearson published extensively on correlation and curve-fitting methods of analysing metric inheritable traits.[91] "One of the most frequent tasks of the statistician," he wrote, "is to represent a series of observations or measurements by a concise and suitable formula." The formula should "enable us to represent by a few well selected constants a wide range of experimental or observational data."[92] In 1904 Pearson published the first mathematical model to explain how multi-trait inheritance could work in a continuous way.[93] Weldon used Galton and Pearson's tools to analyse data he had collected on shrimp and crabs.[94]

The work of Pearson in particular made biometry appear to be purely mathematics. The conundrum of mathematics versus biology was a problem that concerned scientists, especially biologists. While the use of mathematics in a study of biology made sense to them, it was a question of how much emphasis should be put on mathematics in application to biological issues. Pearson, the mathematician, battled the British biologist William Bateson over the issue. When Bateson attacked biometry in a paper published by the Royal Society, Pearson responded with some force concerning Bateson's poor command of mathematics, "It is ... with considerable sense of the gravity of the contest that I take up the gauntlet thrown down by Bateson, but it seems necessary to do so for the sake of our infant science." He continued: "I should have been content for the present to continue my own work, leaving the old school of biologists rigidly alone. It is Mr. Bateson who has forced the controversy by a brilliant and logomachic [incorrect use of words] attack. He does not attempt to meet biometric conclusions by new measurements, he

appeals to the significance of words, and to what he holds to be fundamental biological conceptions."[95] How does one talk to a person about
the problem when he has no knowledge of this mathematical subject?
Pearson asked. "When an opponent has not even a preliminary training in biometry, and the other fails to attach any clear ideas to the terms
by his antagonist, used apparently as if they had universally accepted
weight, it seems very hard to find a common ground for discussion,"
Pearson pointed out.[96] Bateson, a biologist, was not equipped to understand the mathematical sophistication that lay in Pearson's work, and
Pearson was not a trained biologist in spite of his interest in various biological themes.[97] (He studied the coat-colour patterns of cattle, focusing
on pure Shorthorns and Shorthorn crosses, for example.)[98]

The value of biometry to animal breeding remained hidden for some
time. Several reasons are provided here, but more on the subject appears
in the next chapter. To begin with, biometry tended to focus on human
inheritance, and it also created and promoted eugenics – a problematic
way of thinking for animal breeders. Second, biometry's emphasis on
a study of Darwinism (thus the effects of natural selection on genetic
architecture) increased a sense that it lacked relevance to artificial selection strategies. And third, biometry appeared primarily to be a child of
mathematics, a discipline that in its pure form seemed alien to animal
breeding. The biology/mathematic linkage, in fact, worried many scientists over the first half of the twentieth century.[99] Although influenced
by Galton and the biometricians, W. Johannsen, for example, argued
that biometricians saw heredity as mathematics, not mathematics as
being in service of understanding genetics. "We must pursue the science of heredity *with* but not *as* mathematics," he said.[100] The fact that
Pearson and other early biometricians claimed evolution should be seen
as a statistical problem more than a biological one only aggravated the
situation.[101] Future livestock genetics (and ultimately livestock breeders) would inherit the dilemma: were breeding predictions based on
science merely mathematical problems or were they biological ones?
The biology/mathematics dissension continued for some time, making it unclear how livestock genetic theory, often founded on statistics,
could be applied in the farm-breeding world.

Biometry also laid the foundations for a future contentious concept that
came to be described as "black box thinking." The assumption that hereditary results of certain breeding programs could be understood and predicted without any knowledge of the underlying genetic structure behind
the phenomenon was fundamental to practical livestock breeding theory,

and would also be foundational to quantitative genetics, biometry's child. With the rise of molecular genetics, such an approach became increasingly subject to criticism.[102] It seemed to some geneticists concerned with farm animal breeding by the end of the twentieth century that breeding strategies based upon biometric theory treated genes and gene interaction "as if occurring inside a black box" only knowable through its observable output.[103] R.C. Lewontin described the application of quantitative genetics to livestock breeding as "an attempt to produce knowledge by a systemization of ignorance." He further stated, "We use quantitative genetics to predict results of certain selection methods, but we do not know anything about the genetic architecture which will dictate the effectiveness of any such method."[104]

With the rediscovery of Mendel's laws in 1900, some of the basic principles underlying biometry went into relative eclipse, thereby changing the direction of hereditary science for a number of years. That story and the synthesis of their differences are the subjects of the next chapter.

Chapter Two

Early Developments in Genetics

The advent of Mendelism after 1900 initiated new directions in attitudes to heredity, but added both complexity and confusion to the hereditary science situation as well. Three scholars independently rediscovered Mendel's 1865 paper on the breeding of peas. Mendel had used the hybridizing inbreeding/crossing method of breeding and concluded much of what had been observed by the earlier hybridizers.[1] In modern terminology Mendel established two laws: when the gametes (or reproductive cells) form, the gene pairs separate (each unit of the pair is either recessive or dominant); and genes are both immutable and act independently (a view that subsequently has become somewhat modified). Mendel's laws established a new science that competed with biometry as the accepted way to view the inheritance process in evolution. The misunderstandings between them and the ultimate acceptance of the idea that Mendelism was compatible with Darwinian theory shaped how science's theoretical attitudes to farm animal breeding techniques would develop. In this chapter, I discuss the various approaches to artificial selection and heredity that emanated out of science and fused to form livestock genetics. The roots of theory are important if one is to understand how theoretical artificial selection could and/or did work eventually with practice, and also whether it differed in any fundamental way from the approaches of practical breeders.

Different stances of Mendelism and biometry towards evolutionary biology explained why early Mendelians did not develop better breeding strategies, and why they potentially alienated breeders from science generally. Mendalians had a tendency to see traits as being inherited in a qualitative way, meaning that any variation in traits that arose over generations resulted from mutation. The genetic architecture of a species did not change. New species evolved only through mutation; evolution occurred, therefore, through discontinuous inheritance. With regard to artificial selection, one would be forced to conclude under this theory that hens could not be bred for overall change in egg laying capacity, and cows would always give the same average amount of milk unless some mutation factor altered the situation. Any approach to inheritance that implied hens could not be made to lay more eggs over time via breeding strategies, and that the mean milk yield of cows would never change without mutation would make no sense to breeders. Artificial selection practices had clearly advanced the average output of hens and cows over generations, and there was no real evidence that these changes resulted from mutation. Mendelists also suggested that traits were inherited in a simplistic way; namely, as a result of the action of one gene. Known as unit character theory, inheritance of milking ability, for example, was thought to result from the action of one particular gene. Such an approach presented a stumbling block for practical breeders whose experience indicated that inheritance of such quantitative traits was anything but simple.

The attitude of biometry to the problem corresponded better with breeder-outlook. The biometricians believed traits were quantitative and that all variation in traits was inherited. Evolution occurred through continuous inheritance. Changes in genetic architecture could be powerful enough to result in the rise of new species. All traits demonstrated themselves in offspring quantitatively – that is, more or less in terms of measurement.[2] Under this theory, artificial selection could make hens inherit the capacity to give more eggs, and do so generation after generation. Similarly, cows could be bred to give more milk on a consistent basis. If breeders had been asked which attitude seemed correct, they would have said that egg-laying and milk yields should clearly be described as quantitative issues, and that inheritance was clearly continuous. As long as Mendelism/mutation played a major role in theory concerning evolution, hereditary science remained largely nonapplicable to animal breeders. Darwinism (and therefore biometry) had, in a fundamental way, a more natural affinity to breeder-thought than mutation (and therefore Mendelism), because breeders saw small changes over generations in the animals they bred.

This rift in science (which had in fact been initiated before the advent of Mendelism), while primarily a conflict between the British biologist William Bateson and the British mathematician Karl Pearson, tended to involve many other British biometricians and Mendelists.[3] Weldon seriously questioned the validity of Mendelian theory because it seemed to suggest discontinuous inheritance. He stated: "The fundamental mistake which vitiates all work based upon Mendel's method is the neglect of ancestry, and the attempt to regard the whole effect upon offspring, produced by a particular parent, as due to the existence in the parent of particular structural characters; while the contradictory results obtained by those who have observed the offspring of parents apparently identical in certain characters [learn] clearly enough that not only the parents themselves, but their race, that is their ancestry, must be taken into account before the result of pairing them can be predicted."[4] It is perhaps ironic that the founder of biometry, Francis Galton, sided with the Mendelians in the matter of discontinuous/continuous inheritance. Pearson and Weldon could only conclude that Galton had not understood his own methodology well enough to see that biometry supported Darwinism; namely, continuous inheritance.[5] Bateson, a Mendelian and the man who coined the word genetics in 1906, was the supreme champion of discontinuous inheritance. Bateson thought that the discontinuous nature of inheritance he observed in nature indicated that speciation occurred by mutation. Mendelism, it seemed to Bateson, supported the idea of mutation, and by doing so also denied the validity of a Darwinism that argued that small changes under natural selection could give rise to new species.

While the simple fact that the Mendelist/biometric debate helped cloud breeder understanding of any potential applicability of hereditary science to their needs, it is important to point out that the disagreement was not as strong in the United States (in fact it was largely confined to Britain[6]), a country that encouraged the use of biometric strategies in Mendelian experiments.[7] The fact that biometry was never completely cut off from Mendelism in the United States would be important to the future development of livestock genetics. This early elasticity laid the groundwork for taking a statistical approach to the continuous inheritance of quantitative traits in populations that arose in the United States in the 1930s, and formally initiated the birth of international livestock genetics.

Government Funding and Agricultural Research

The Mendelian/biometric divide in research should be seen within the framework of government funding for scientific research that developed over the late nineteenth and early twentieth centuries in North

American and European countries. Growth of the phenomena is exemplified here by a brief discussion of the American situation. In the 1850s only three state agricultural colleges existed in the United States.[8] The Morrill Land-Grant Act of 1862 resulted in the rapid rise of more publicly funded, American state agricultural colleges: the first being the Massachusetts Agricultural College, established in 1862, followed by the Connecticut Agricultural Experiment Station, founded in 1876. Others followed and increasingly these began to undertake various investigations in both the field and the laboratory.[9] The Hatch Act of 1887 stimulated scientific research in the United States by allowing federal funds to support experiment stations that were attached to state agricultural colleges. By the early twentieth century there were forty-eight such institutions doing agricultural research in the United States. Mendelism led to a hugely enlarged emphasis on experimental plant breeding, and also to the passing of the Adams Act in 1906, which increased funding for scientific research, often genetic in nature, at experiment stations. Similar institutions arose in Canadian provinces and in various European countries (which will be evident from information arising later in the book) over the same period. How applicable any of the research undertaken at these centres actually was to practical agriculture, however, is another question.

Problems of Practical Application: Clashes between Science and Practice, 1906–1920

Practical breeders knew about developments in Mendelism, if not the Mendelist/Darwinian debate concerning discontinuous/continuous inheritance. The potential of Mendel's laws for animal breeding excited them, in spite of its apparent impracticality to their work. Non-scientists and non-breeders were also fascinated by the potential implications of Mendelism for animal breeding. Since Mendelism meant hereditary science, and animal breeding meant working with heredity, many argued that science should drive breeding practices. People of this mindset thought any disagreement of thinking between the two should end. They would be aware, however, that there were difficulties in bringing about this happy union, even if they did not understand how significant the continuous/discontinuous dilemma was to the division. British W.H. Heape, for example, hoped to bring the scientist and the practical breeder closer together as early as 1906, but he recognized that the breeders saw no value in the scientist's work from an application point of view, and that scientists had little knowledge of or respect for the work done by breeders.

Heape was acutely aware of what appeared to be a growing antagonism between the two groups. "At present the breeder, in our march of progress, is neglected," he stated. But, to some degree this was a problem brought about by the breeders themselves. "Practical breeders are very apt to look about purely scientific aspects of the study of breeding as quite inadequate for their purpose," Heape explained, thereby marginalizing them from advances in science. But that situation was not helpful to science either. "[T]he man of science is apt to overlook the immense value of the store of knowledge which the breeder has at his disposal, he is apt to consider it of very little use to him," he wrote. Scientists, as a result, tended to see practical breeders as working with superstition and untrustworthy ideas. This type of dismissal led scientists to underestimate the practical breeder's understanding of the dynamics of hereditary laws.[10] Heape went on to elaborate on what the "art" of breeding meant, and here we see clearly the embedded confusion in thinking that was so typical of the times: which often dismissed practical breeding as being merely "art," and which separated science from practice on the basis of that art. Heape's words reflect the prevailing outlook that linked practical breeding simply with purebred breeding. Thoroughbred horse culture, with its deeply embedded sense of beauty in breeding and in the creation of beauty via breeding, had become part of purebred culture.

> So far I have referred only to the science of breeding, I have not yet mentioned the 'art' of breeding. I will do so now. … It may be defined as a special attribute of eye and hand which is born with a man, and in its highest development it is a very remarkable gift. I say gift advisedly, it cannot be wholly taught or learned, many have it to some extent, but only those how possess it in perfection have been, and are, the great breeders of the past and the present day. It has been called falsely science, and those who possess it are said to be scientific breeders; but it is *not* science, it is pure art, and of the greatest value; by means of it the infinitely minute variations in animals are recognized and seized upon, and by its exercise successful mating is divined. But, it may be said, if breeding is an art what need is there for science? The answer is plain, the scientific knowledge of anatomy is no less necessary to the born sculptor than is science to the born breeder. The condition of the breeding industry now is no whit better than was the art of sculpture before anatomy became a science, and by the application

of science the breeder will gain no whit less than the sculptor gains by the knowledge of anatomy.[11]

Heape proved to be overly optimistic: a division between the two groups remained in place. Often conflict and confusion followed, resulting partially from the fact that neither side seemed to see that they were hopelessly splintered over the continuous/discontinuous approach to heredity. An example of the complicated confusion, arising from intertwining of various ideas and misunderstandings, can be found in the exchange between a scientist who undertook experimental chicken breeding and a renowned chicken breeder. While working at the Maine Experiment station between 1907 and 1916 (after being hired under the Adams Act), Raymond Pearl looked at the breeding of chickens with better egg production in mind. Trained in zoology and a biometrician as well, Pearl used Mendelism in his experiments, thereby providing another example of elasticity in the United States with regard to the Mendelism/biometry divide. Pearl wanted to identify the way egg-laying qualities were inherited. He concluded that mass selection on the basis of either individual worth or ancestry – namely, choosing high-producing hens and the sons of high-producing hens to breed for the next generation of layers – did not work. This selection method had not increased the egg-laying averages of the flock at the Maine station over a ten-year period. He believed too that the progeny test was the best means of selection.[12] He thereby reintroduced that basic Bakewellian principle to the problem of artificial selection.

Breeding and dissecting work led him to conclude that egg-laying ability seemed to be passed on through males. By 1911 his impressions concerning fecundity and male transmission of the trait had convinced a number of agricultural experts that breeding efforts aimed at egg production in a flock should concentrate more on the breeding of cocks, rather than hens.[13] But his approach was unconvincing to many important chicken breeders, partially because Pearl lacked clarity of thinking in his written comments. He was unclear, for example, as to whether females transmitted egg-laying ability at all. Pearl did not explain adequately, either, how the process of male inheritance of fecundity worked. Because good breeders had always appreciated the value of the male in egg production (via his female ancestry), Pearl's theories were not eye-opening for many practical breeders. R.R. Slocum, who worked for the Bureau of Animal Industry and who published a book on the breeding and mating of chickens in 1920, stated that Pearl's advice on how to increase egg production through an emphasis on males matched what any good breeder would already be doing. Breeders knew sons of good layers were important for egg production, Slocum pointed out.[14] Equally critical to ensuing misunderstandings between the scientist and the general poultry public was the fact that many failed to realize that Pearl's conclusions should be seen within the framework of the progeny test. Practical breeding had moved away from the principle of the progeny test as early as 1809, as Sebright's thoughts made clear. Purebred breeding principles had subsequently deviated from that basic Bakewellian approach increasingly over the nineteenth century, promoting instead selection on the basis of the individual in relation to its ancestry.

Pearl's comments in Bulletin 305 about the Maine Experiment Station provoked critical response from an articulate and prominent chicken breeder, and thereafter revealed something of the dilemma of defining what divided "science" from "practical" in early twentieth-century breeding, or precisely what was new in a scientific approach. Pearl seemed to be saying that the standard way of breeding did not work, at the same time that he advised breeding under what really amounted to the old method. He stated: "Selection to the breeder means really a system of breeding. Like produces like, and breed the best to the best; these epitomize the selection doctrine of breeding. It is the simplest system conceivable. But its success as a system depends upon the existence of an equal simplicity of the phenomena of inheritance."[15] If, for example, a breeder mates an individual that is larger than average to another individual larger than average and always gets offspring larger than average, then breeding "the best to the best" would, as Pearl put it, "offer a royal road to riches."[16] But if, he continued, "a character is not inherited in accordance with this beautifully and childishly simple scheme, but instead inherited in accordance with an absolutely different plan, which is of such a nature that the application of the simple selection system of breeding could not possibly have any direct effect, it would seem idle to continue to insist that the prolonged application of that system is bound to result in improvement."[17]

H.H. Stoddard, a man who had been concerned with breeding for better egg production for over forty years and who had written extensively on the subject in various poultry journals, responded to Pearl's views. In his articles for the *American Poultry Journal*, Stoddard challenged Pearl's conclusions, questioned the innovativeness of his suggestions, and took issue with some of the inflammatory language found in the Bulletin. Stoddard stated that:

> The fact is the bulletin is wrong. High fecundity and low, too, may descend from either sex to either sex or it may not descend at all directly from either to either. There will sometimes [be] great irregularity, and scattering every which way, and reversion to remote ancestor types, especially if there has been a cross of strains considerably diverse. Selection for the purpose of breeding from the best to get the best, even if it is 'childishly simple,' will continue to be the only way to fix characteristics, and among many misses there will be some hits. Breed 'the best to the best' and though you may find that some of the progeny may not be as good as the average of their parents, yet some may be as good and some decidedly better.[18]

Stoddard pointed out that this breeding method – namely, mass selection on the basis of individual worth and ancestry – had brought the original egg production of wild fowl from six or eight eggs a year to at least and often more than fifty in domestic chickens.[19] He also argued that fecundity could be inherited from hens by their female progeny: "I do not deny the influence of the male bird in helping to build up a strain of great laying. Neither do I know of anyone who would … What I do deny is that dams have no finger in the pie of hereditary fecundity. They have a great deal to say about it."[20] Stoddard assumed that Pearl must mean, even if he did not articulate it, that the selection of males proceed on the basis of their mothers: "What Dr. Pearl really teaches us is that fecundity is transmitted equally by both sexes, and by alteration," Stoddard argued, "but in his summary (misleading because incomplete) has laid a trap for his readers; for, although it tells what the daughters inherit, and from which parent, it is silent as to what the sons inherit and from whence."[21] Pearl actually intended that males be chosen via the progeny test – on the basis of their daughters. It was a point that Stoddard failed to pick up.

Stoddard concluded that Pearl had needlessly confused the existing situation by advocating what breeders had always done – namely, practice mass selection on the basis of individual worth and ancestry for both males and females. Under "Pearlite" theory, as Stoddard called it, a heavy layer would be mated to a cock whose dam was also a heavy layer. Isn't that the same as the best to the best, he queried; that is, both males and females from families in which the female members were good egg layers? Pearl just called it something different. Stoddard summarized his impressions concerning Pearl's approach to breeding in the following words: "If the 'childishly simple scheme' or 'breeding from the best to the best,' which 'is the simplest system conceivable,' was so totally and disgustingly fruitless in the past, will the identical practice result differently because of masquerading under a new name?"[22] Stoddard stumbled on a main point when it came to comparing craft with scientific breeding: the same breeding method could be utilized by both groups.[23] Explanations set out by scientists like Pearl implied little innovation, and even seemed to indicate how seamless practice and science were in breeding approaches. What was science and what was not? How did it differ from practical breeding in place for centuries? The scientists appeared to be no more able to answer those questions than the practical breeders. The fact that scientists who were Mendelian in their thinking tended to believe that all variation in traits was not inheritable, while breeder experience clearly suggested otherwise, did not seem to be widely appreciated as an underlying problem in how one group could relate to the other. Pearl himself would eventually, like so many Mendelists, support the theory of discontinuous inheritance with respect to natural selection.[24] His stance on discontinuous inheritance would be completely at odds with Stoddard's experience in artificial selection.

Mendelism generally was nothing more than practical breeding under a different guise, Stoddard thought. "Mendelism, or the new genetics,[25] or whatever it may be called," Stoddard noted, "offers at its present stage not new practical instructions for mating and breeding either the lower animals or humans. The professors who say that the old rule of 'breeding the best to the best,' is no good; turn right around and prescribe methods that amount to the same thing."[26] Stoddard believed, though, that Mendelism ultimately would offer practical breeders aid. He concluded: "The whole problem offered by Mendel's discovery, one of the most important as well as wonderful, in the annals of science, is such a complicated one that it will take generations to solve it, and at present the breeders of domestic animals … can derive little benefit or none at all from all that Mendelism can offer – in its present stage of development."[27] Perhaps Stoddard instinctively knew that acceptance of Darwinism – especially of the continuous inheritance of variation – had the capacity to change how Mendelian theory might be applied in breeding situations. The assumption that traits could not be inherited in a quantitative way was, quite simply, antagonistic to any artificial selection methodology.

J.G. Morman, another chicken breeder, grasped the underlying assumption of discontinuous inheritance of variation in Pearl's approach, apparently more effectively than Stoddard had. Morman believed that one could only conclude from Pearl's work that fluctuation in egg-laying capacity was not an inheritable trait; rather, it was discontinuous. Pearlism, if viewed that way, was a dangerous trend for the future of animal breeding strategies, as far as Morman was concerned. The famous English biometrician, Karl Pearson (with whom Pearl had studied), was already convinced and had "relegate[d] the problem of inheritance of egg-laying power in fowls to oblivion." Morman's experiments in 1912 on the problem convinced him that egg-laying was an inheritable characteristic, while also causing him to wonder what good science was if the experimenters could be so misguided as to believe that chickens do not inherit egg-laying?[28] As far as breeders like Stoddard and Morman were concerned, efforts to substitute "scientific" advice for breeding practices appeared confusing, non-innovative, and/or ill-supported by evidence.

The apparent conundrum that both scientists and practical breeders were concerned with the same problems – yet did not work together – continued to puzzle people. When reviewing the effects of genetics (effectively Mendelism) on agriculture in 1918, two American scientists, E.B. Babcock and R.E. Clausen, tried to explain what was different and what was the same about "the art of the breeders' craft" and the approach of "the geneticist with all his knowledge of natural law and principle." They had some difficulty in expressing how one interacted with the other – partially for the fundamental reason that they did not see how incompatible the view that inheritance of variation was to any principle driving artificial selection theory. While admitting that practical breeders had been successful in breeding better livestock, they implied obscurely that scientists were equipped to do a better job in the future. Breeders had learned methods that worked through experience. Scientists, when interested in livestock breeding, might seek similar results but they started from a theoretical base rather than one acquired through experience. Babcock and Clausen fundamentally believed that Mendelism would explain why the art of breeding worked, and they opaquely suggested that this knowledge would eventually yield even better results.[29] They offered no explanations as to how this could be the case. Some scientists, however, began to argue that practical breeders knew nothing about proper ways to breed. As early as 1911 W. Johannsen discounted even the historical input of practical breeders to the issue of heredity. In fact Johannsen argued that Darwin himself had over-valued the importance of breeders to questions of genetics.[30] While it might have seemed that before 1915 and the chromosome work of T.H. Morgan, livestock breeders commanded as much respect as hereditary scientists in the matter of heredity knowledge (they were equally important to the American Breeders' Association), a schism between the two groups was apparent. The virtual takeover of that association by scientists heavily influenced by eugenic thinking did nothing to encourage the affiliation of breeders with the organization.[31] It would be some time before it was apparent to scientists/geneticists/eugenicists that the principles of practical breeding were an essential part of livestock genetics.

Roots of Livestock Genetics: Hybridizing Ideas for Plant Improvement in the United States, 1908–1920

At the same time that the Mendelist/biometrician debate fractured the thinking of many academic scientists, the theoretical foundations of livestock genetics were being defined. Patterns evolving in plant breeding research between 1908 and 1920 would be important to future livestock genetics. Biologists specifically interested in plant experimentation continued to follow the breeding strategies of the earlier hybridists, but now with a focus on the dominant/recessive characteristics of certain traits. Within this framework it was work in maize genetics that was especially significant. As early as 1908, the American scientist G.H. Shull had wondered how Mendelism explained the improved fertility and vigour resulted from the crossing of inbred parental lines of corn, and why such progeny could not sustain that superiority when bred with each other for the next generation.[32] It was clear to him that lack of overall heterozygosis (or variation in genetic make-up), which resulted from inbreeding, meant deterioration. It seemed to Shull, though, that a simple return to genetic variation resulting from the cross of inbreds did not fully account for the return to vigour. He became increasingly interested in a Mendelist explanation for hybrid vigour, and in 1914 named the phenomenon "heterosis."[33] Fundamentally, Shull wanted to understand the genetic architecture behind the well-known effects of crossbreeding. While breeders like Bakewell and Sebright cared more about the outcome of crossbreeding – namely better vigour, but also the inability of the progeny to breed truly in the next generation – Shull was interested in the dynamics that brought about those results.

In 1917 another American, D.F. Jones, reasoned that dominant/recessive characteristics of gene pairs played a role in heterosis. The expression of dominance in gene pairs resulted in hybrid vigour because the most dangerous and debilitating genetic influences came only from recessive expression in gene pairs. "If, for the most part," he wrote, "these favourable characters are dominant over the unfavourable (if normalities are dominant over abnormalities) it is not necessary to assume complete dominance in order to have a reasonable explanation for the increased development of [the first generation] over the parents or any subsequent generation."[34] Jones believed the stunted growth of inbred corn resulted from the presence of such abnormalities, "There is abundant evidence to show that many abnormal characters exist in a naturally cross-pollinated species [such as corn], and that they are recessive to the normal condition."[35] It was only inbreeding that revealed such "lethal" recessives, a pattern geneticists labelled as inbreeding depression.

Jones also turned his attention to the use of hybrid breeding for maize production. He hypothesized that a double-cross hybrid method of breeding would overcome the weakness of single inbred lines, which often lost sufficient vigour so that few seeds could be collected for cross-breeding. By inbreeding four strains, next crossing the four to produce two lines, and then finally crossing those two lines, he postulated that he could restore the lines to the original fertility by the first cross, and bring out superior hybrid vigour by the second cross.[36] In other words he planned to take two inbred lines, A and B, and cross them to produce a first generation of AB line. Two other inbred lines, C and D, would be crossed for another hybrid line, CD. Next AB and CD would be crossed to produce hybrid vigour. In order to maintain that level of hybrid vigour over succeeding generations, seeds from the final cross would not be used for breeding. New generations demonstrating heterosis would always be regenerated by stock belonging to the parent and grandparent generations. Jones had invented the double-cross hybrid corn breeding method. It would take some years and considerable effort and expense, however, before the method was made to work more effectively than traditional plant-breeding methods that selected for lines that reproduced truly, in spite of propaganda that suggested otherwise.[37] Jones' double-cross hybrid breeding system used the old eighteenth century experimental breeding methodology followed by naturalists in developing an artificial selection to be used in farm production. In doing so he introduced a fundamentally different concept to agricultural breeding – that is, abandoning the breeding of lines that reproduce truly. His work attracted the interest of corn breeding companies that funded research devoted to making the method work. Corporate involvement in breeding introduced an entirely new element to the practice of breeding.

Roots of Livestock Genetics: Artificial Selection and Animal Breeding, the Work of W.E. Castle in the United States, 1905–1915

Over the same period, the American Mendelian, W.E. Castle, developed different theoretical approaches to artificial selection, which proved important to future livestock genetics. As a result of his animal breeding experiments, Castle ultimately rejected classic Batsonian mutation theory in favour of seeing inheritance of traits by animals as a continuous and quantitative issue. A short review of his work reveals why he did so. While primarily interested in studying Darwinism and its applicability (or lack thereof) to evolution, artificial selection strategies to better improve farm animals also concerned Castle. He wondered if the results of breeders' work – namely the rise of divergent breeds – could reveal something of the dynamics of hereditary laws, and whether knowing more about such dynamics would lead in turn to more effective selection practices. Unlike Bakewell and early breeders like Sebright, but in keeping with the plant hybridist tradition, Castle thought understanding hereditary laws could shape how breeding should proceed. "What we need to know is how, precisely, are new breeds formed … The successful practical breeder, the man who originates breeds, is a keen observer, a man of unusual intelligence and skill and of infinite patience. Yet if we ask him how, in general, he does his work, or how a particular result was obtained, we rarely get a satisfactory answer," Castle wrote in 1905.[38] In keeping with Bateson's point of view, it seemed initially to Castle that new breeds arose primarily as a result of mutation. Breeders worked with the progeny of a few mutated individuals to form breeds. Breeders could simply find a good individual made better by virtue of mutation, or could crossbreed and end up with superior mutated individuals with which to work. Castle concluded:

> On the whole, it appears that the formation of new breeds begins with the discovery of an exceptional individual, or with the production of such

an individual by means of cross-breeding. Such exceptional individuals are mutations. An examination of stock registers points in the same direction. The beginnings of new breeds are small. Pedigrees lead back to a few remarkable individuals or to a single one ... But given the exceptional individual, and a new breed is as good as formed. The few generations which the breeder usually employs in 'fixing' or establishing the breed and during which he practices close breeding serve principally to free the stock from undesirable alternative characters, not to modify the characters retained.[39]

Castle also argued, at this point in time, that selection alone could not create breeds: "Modification of characters by selection, when sharply alternative conditions (i.e., mutations) are *not* present in the stock, is an exceedingly difficult and slow process, and its results of questionable permanency. Even in the so-called 'improved' breeds, which are supposed to have been produced by this process, it is more probable that the result obtained represents the summation of a series of mutations rather than of a series of ordinary fluctuating variations. For mutations are permanent; variations transitory."[40]

Castle's tendency to reject inheritance as a quantitative and continuous process had aroused the ire of Pearson, who challenged him in 1904. Pearson argued that Castle's poor understanding of statistics made him unqualified to judge the validity of a biometric approach to patterns of heredity.[41] Castle's work with hooded rats, rather than Pearson's opinions, quickly changed his mind about how traits could be inherited. His altered approach represents another example of the American elasticity when it came to the Mendelian/biometric divide, this time in reverse to the experience of Pearl. By 1911 Castle had come to believe that mutation was not the only way to change the hereditary architecture of animals. Selection alone could do this, by quantitatively intensifying certain traits.[42] His thinking was now more in line with the biometricians. Castle continued to credit breeders with good work, but now believed the outcome of that work resulted not from mutation, but from the intensification of traits inherited quantitatively via inbreeding. Inbreeding, not mutation, proved to be the key. Castle stated that master breeders in the past had been able to create all modern improved breeds of livestock (lines that bred truly or homozygously) without the input of science by using some form of inbreeding that had intensified the presence of quantitative traits in animals.[43] He also realized that farmers regularly crossed breeds for greater productivity and knew such crosses should only be used for market purposes. They understood, in other words, that the crossed progeny should not be used for breeding purposes because the resulting next generation would not present either uniformity or increased vigour.[44] As far as respecting livestock breeders, Castle's problem continued to centre on the fact that they did not know why any of their results occurred, not on their lack of ability to breed.[45] It was this characteristic that divided practice/art from science in his view. Inbreeding was the key, and scientists needed to study its effects more closely.

Inbreeding experiments, it seemed to him, were fundamental to the study of improved agricultural production in livestock. Castle's own inbreeding experiments were designed to generate both self-perpetuating lines and hybridized lines. This approach harnessed the fundamental philosophies behind the strategies of practical animal breeders and plant hybridists. Castle's true-breeding-line work, which theoretically matched the aims of practical breeders, led to the development of specialized laboratory mice by his student, C.C. Little.[46] The potential of inbreeding and the crossing of such lines – following the hybridist tradition – also attracted Castle's attention.[47] His inbreeding and line-crossing research influenced Sewall Wright, who ultimately laid the groundwork for agricultural genetics applicable to livestock.

Dominant Patterns in Experimental Animal Breeding in the United States and Europe before 1930

Castle's interest in agriculture (and his biometric approach) tended to separate him from mainstream animal-breeding research. Concern with better farm animals and quantitative traits played only a very minor role in animal-breeding experiments undertaken in the United States and Europe before 1930. Chickens were a favourite species to work with, because of their cheapness and rapid reproductive life. Results from research were evident in a far shorter time span than would have been the case with cattle, for example. Shortly after 1900, Bateson, who had studied both plant and chicken breeding before the advent of Mendelism, began to experiment with chickens to show dominant and recessive traits that had nothing to do with prediction of better productivity from the birds.[48] Chicken-breeding research done at experiment stations in the United States between 1900 and the late 1920s illustrated that same pattern – specifically, concern for dominant/recessive traits and no interest in farm productivity. Biologists/Mendelians explored how features such as feather colouring, shape of comb, and skeletal defects were inherited from a dominant/recessive point of view. Examples of other characteristics studied were flightlessness, crooked neck, feathering, silkiness, ragged wings, feathered shanks, multiple and double spurs, blindness, and dwarfism. Inheritance of characteristics on the basis of sex also interested early Mendelians.[49] Because inbreeding and crossing of inbred lines – the hybridizing scientific approach to the practice of breeding – was the methodology used to explore these features, we see a continuation of experimental breeding patterns established in the eighteenth century. Mendelism did not introduce a new breeding technique.[50] The attempt by academic Mendelians to understand the process of heredity on the basis of hybridizing experiments resulted in the rise of what was known as unit-character theory, which led many to argue that inheritance worked via the transmission of single units. (This stance clearly matched Pearl's approach to egg-laying inheritance.)

Developments in Europe with respect to livestock genetics and its relationship to farm breeding were not more robust than those in the United States. Perhaps the Bateson/Pearson disagreements played a role in that pattern, with biometry being forced to the background in a way that was not true in the United States. Certainly the First World War stunted academic research in Europe. In 1927 L.C. Dunn toured some genetic centres in Europe and reported on the work being done at them. Dunn, who undertook a number of experiments for Connecticut research station at Storrs between 1920 and 1928, investigated the inheritance of plumage colour patterns by chickens (in the classic Batesonian tradition) in attempts to understand speciation and to link Mendelism to evolution. (He also looked into skeletal variations, the presence of lethal genes, egg-laying patterns of different poultry breeds, and colour of the leg shank.)[51] While he used animal breeding experiments in order to understand the process of speciation, Dunn was also concerned with the scientific breeding of agricultural animals for production reasons. His observations, then, are useful when trying to perceive, at least in overview, the scientific animal-breeding situation outside the United States, and how research in genetics related to practical breeding.

In England, Dunn found Cambridge to be the most important centre for animal genetics. He was ambiguous, however, in explaining how research at that university connected with practical farm animal breeding. While noting that agricultural animals were used in breeding experiments, Dunn stated that farm problems did not dictate research direction.[52] He commented on genetics in Scotland at Edinburgh University, which tended to focus on physiology of sex via experiments in laboratories using mice. (The genetics department at Edinburgh was founded in 1919. Named the Animal Breeding Research Department it had a director but no other staff and no building at that time.)[53] He added that "some good observational work with sheep [was] made possible by the cooperation of agricultural colleges and farmers." However, he noted that "scientific growth … [was] hindered by lack of contact with other scientific departments. This together with inadequate library and facilities for either breeding or laboratory experiments at present [were] the chief disadvantages for students going to Edinburgh for animal genetics in any capacity. At the other Scottish Universities there apparently [was] little activity in subjects related to genetics."[54] While Dunn concluded that animal breeding research in Britain was confined to Cambridge and Edinburgh, he warned readers not to overlook "the fact that there [was] a very generally distributed interest in the subject, as evidenced by the large membership of the Genetical Society, and the large number of biologists who [were] actually involved in work bearing indirectly on genetics … for in Britain one notes fewer specialists and possibly a wider diffusion of knowledge and interest in genetics among men who remain biologists."[55]

When assessing the robustness of animal genetic research and its potential relationship to farm breeding in continental Europe, Dunn commented on the situation in Sweden, Norway, Germany, and Russia at a centre near Moscow.[56] He spoke enthusiastically about the animal-breeding scientist, Christian Wriedt, who had left his native Norway for Sweden (where research on animal breeding and genetics existed), but had returned to Norway by 1927, hoping to open a privately funded research centre devoted to animal breeding. "Wriedt is a man of tremendous energy," Dunn stated, "and has been able to carry out important work with pigeons and farm animals only because he has been able to secure the cooperation of farmers and breeders all over Norway."[57] Wriedt was clearly interested in the potential of genetic research for the betterment of farm animals. Dunn found research facilities for genetics very fine in Berlin.[58] But he was more struck by the work done at a research station near Moscow, which focused on chickens, even though it had little to do with improving the birds for commercial use. The emphasis at the Russian station was on what chicken breeds could tell us about evolution, not about better agricultural production.[59] Dunn summarized the situation in Britain and continental Europe as follows: "The chief centres seem to me to be Cambridge, Edinburgh, Berlin,

and Moscow. It is difficult to rank these centres, but all things considered, Berlin seems at present the most important, with Moscow close
behind."[60] Dunn appeared to be evaluating the general rigour of genetic
research at these locations with these words, but not with any particular focus on the improvement of farm breeding. He overestimated,
for example, the importance of the Berlin Agricultural College for animal genetics in relation to agriculture at that time. Research on farm
breeding seemed to be in general decline in Germany after the end of
the nineteenth century, with animal breeding playing an increasingly
smaller role in any German genetic research.[61]

Dunn clearly did not provide information on centres devoted to genetics and animals in all European countries. He did not write about the
situation in France, for example. But since modern research suggested
that a sort of "genetic lag" prevailed in that country in this period, it is
possible that he would have had little to say about French research if he
had visited France. French biologists resisted Mendelism for quite some
time. The first chair of genetics at a French university was not established until after the Second World War and animal-breeding research
did not begin until the early 1950s at the Institut National Agronomique
in Paris.[62] There were important pockets of genetics research in relation
to farm animals at universities in Europe, however, that Dunn did not
seem to be aware of: the centre at a university in Louvain, Belgium, for
example, organized and run by Leopold Frateur from 1899 until 1936.
Frateur founded an animal husbandry institute at the university in
1908, and set out to study the relevance of Mendelism in the improved
breeding of Belgian cattle. Interested in both qualitative and quantitative traits, he conducted various breeding experiments on cattle; and,
after the First World War, he did the same with chickens and pigs. Frateur spoke at a number of international congresses on the subject of
animal breeding and science.[63] Another scientist deeply concerned with
livestock breeding at the time – and not mentioned by Dunn, either –
was A.L. Hagedoorn, who after 1924 lived in his native country, the
Netherlands. Hagedoorn tried to enlighten breeders on how to improve
their stock with better selection practices, and although he lived by private means and was not attached to a public research institution, he had
published material aimed at farm breeders from as early as 1912.[64] It is
unclear how much any of this European work affected actual breeding
on farms – likely very little. The situation did not seem that dissimilar
to the one in North America. Even though Dunn might not have given a
truly comprehensive report of the genetics research situation in Europe,

especially in relation to farm animal breeding, it seems clear that animal breeding research generally, and agriculture specifically, was no stronger in Europe than in the United States. The impact of academic animal breeding work on farms generally was probably negligible as well.

Groundwork for Livestock Genetics: The Work of R.A. Fisher and Sewall Wright, 1915–1930

Important to all future theoretical livestock genetics was the general healing of the rift between biometricians and Mendelists, or, in other words, the return of biometric principles that supported continuous inheritance and/or Darwinism to the study of heredity. No real attempts to develop artificial selection theory could progress until a synthesis of Mendelism and Darwinism had evolved. Some scientists had learned to overcome the fractionalization – Castle, for example, and even earlier the British mathematician G. Udny Yule[65] – but the split still hampered growth in the science of heredity, and certainly curtailed any widespread attempt to develop better systems of artificial selection, theoretically or practically speaking. The important 1918 paper by R.A. Fisher, the brilliant British mathematician and a biologist as well, has been credited with initiating a healing of the conflict within the international academic world and stimulating the rise of population genetics.[66] Fisher's work viewed biometric results in terms of Mendelian inheritance. He showed that Mendelism was not necessarily incompatible with Darwinism or continuous inheritance.[67] Fisher was primarily interested in eugenics. He was unusual in that unlike many scientists studying inheritance he was knowledgeable in both biology and statistics. Even so, he tended to see mathematics as the ultimate oeuvre when it came to the study of biology, thereby lending his support to the Pearsonian idea that statistics in effect *was* biology.[68] Fisher argued that quantitative traits could be inherited in a variable way because a large number of segregating Mendelian units acting together could explain the dynamics of that process.[69] This advancement would be known as the infinitesimal model of inheritance when it came to quantitative traits, and it refuted unit character theory. Fisher's concern with human inheritance meant that he assessed accumulated data that had arisen from what might be described as natural processes and therefore did not emanate out of planned breeding experiments. One geneticist would argue that this situation explained why Fisher did not put forward a theory of selection for metric traits, even though he provided the tools necessary to establish such a theory.[70] Fisher did not, therefore, attempt to predict how different practices of artificial selection would affect the inheritance of particular quantitative traits, and he was not concerned with animal breeding as such. The result, it seemed to one geneticist at a later period, was that much of the quantitative genetic work done with livestock over the next number of years merely repeated the theoretical work that Fisher had already demonstrated in 1918.[71] Fisher, then, may have established what would be known as quantitative genetics – and the fact that it was grounded in Mendelian principles. Livestock genetics, however, was not part of his legacy.

Theoretical approaches to artificial selection strategies for the betterment of livestock really began with the work of the American Sewall Wright. A student of Castle, Wright took Castle's conclusions regarding livestock breeding to new theoretical levels when he argued that populations could undergo genetic shift via inbreeding. Early scientists like Castle and Pearl did not believe that artificial selection could have such a wide-ranging effect.[72] Primarily an evolutionary geneticist, Wright studied livestock production (especially the historic breeding of Shorthorn cattle[73]) and also worked for the Bureau of Animal Industry. The effect of inbreeding on the genetic make-up of populations was theoretically important to both his interests. Wright explained as follows:

> My general project in the Animal Husbandry Division was clarification of the roles of inbreeding and selection in the breeding of livestock …
> Assessment of Shorthorn pedigrees helped me in this. From my studies of gene combinations … I recognized that an organism must never by looked upon as a mere mosaic of 'unit characters,' each determined by a single gene, but rather as a vast network of interaction systems … It was apparent from my studies of the breeding history of Shorthorn cattle … that their improvement had actually occurred essentially by the shifting balance process rather than by mere mass selection. There were always many herds at any given time, but only a few were generally perceived as distinctly superior; those of Charles and Robert Collins [sic, should read Collings] near the end of the 18th century, those of Thomas Bates and the Booths in the first half of the 19th century, and that of Amos Cruickshank later in the century. These herds successfully made over the whole breed by being principle sources of sires.[74]

By inbreeding a small population of animals, in other words, one could in fact shift the genetic make-up of larger populations by breeding that inbred group into a more general population. The implications of this phenomenon were important to understanding both the process of speciation and also how to practice artificial selection on livestock in order to achieve improvement. In this approach, Wright matched the thinking of Bakewell: one could establish "breeds" using this method, or extend the scope of existing breeds.

Another important thing Wright did for the future science was develop a way to quantify the effects of various inbreeding strategies.[75] As he himself put it, "I developed an inbreeding coefficient (the theoretical correlation between united gametes) that could be shown to measure the decrease of heterozygosis [genetic variability] from that in the foundation stock."[76] His work in effect theoretically quantified earlier breeder theory, and provided a more complicated method of controlling the level of inbreeding, something that Sebright had recognized as early as 1809, and Bakewell even earlier. Practical breeders had long believed that inbreeding had to be accompanied by careful selection, but their ability to quantify levels of inbreeding had remained primitive; they did not use any form of advanced statistics. Consequently, while the theoretical tools had been put in place, it would be some time before Wright's path coefficient had any effect on practical breeding programs.

Wright devised a way of calculating the level of shared genes that would result from different inbreeding systems between 1915 and 1922 – brother to sister, first cousins, double first cousins, half brother to half sister, and so on – using statistics obtained from Castle's inbreeding experiments on guinea pigs, along with his own experimental results.[77] Wright had established an accurate quantitative way to measure the level of homozygosis resulting from any type of inbreeding. His achievements were all the more remarkable considering the fact that he was not a mathematician. (Wright was not aware of Fisher's work when he developed his path coefficient and his ideas concerning genetic shifts as a result of inbreeding.)[78] Trained in Mendelism, Wright learned statistics by reading the work of Pearson.[79] (He serves as another American example of a scientist who combined Mendelism with biometry.) Wright argued that he could quantify the level of inbreeding and thereby reduce its intensity, thus avoiding some of the dangers it could incur. Curiously, only animal geneticists made use of the path analysis in breeding experiments for more than forty years, possibly because Wright himself left no textbook or instruction manual and only animal breeder scientists had any training in the method. Population geneticists discovered the method in 1948 from a textbook published in China.[80] This situation provides an interesting example of how the metropole interrelates with the periphery, or, how knowledge flows back and forth around the world. China might have seemed like the periphery when it imported Wright's theories, but, interestingly, it was China that brought about the reincorporation of Wright's thoughts into broader genetic perspectives.

Hybrid vigour also interested Wright, and he knew from his own experiments that crossing inbred lines often led to progeny superior to either parent.[81] He wrote, "By starting a large number of inbred lines, important hereditary differences ... are brought clearly to light and fixed. Crosses among these lines ought to give a full recovery of whatever vigour [had] been lost by inbreeding, and particular crosses may be safely expected to show a combination of desired characters distinctly superior to the original stock."[82] Wright was well aware of work being done on inbred corn, particularly by Shull and Jones. The path coefficient ultimately would make it easier to create vigorous lines resulting from matings of related animals that could, in turn, be used to cross lines via Jones' hybrid corn breeding method. (The coefficient could equally well be used to produce superior pure lines designed to breed truly.) Research geneticists began to explore how hybridizing could work within the controlled framework of Wright's path coefficient theory.[83]

By the late 1920s a fusion between the thinking behind Darwinism and Mendelism, through the work of J.S.B. Haldane, R.A. Fisher, and Sewall Wright, had led to the birth of population genetics (and its subset quantitative genetics).[84] This synthesis initiated the rise of theoretical livestock genetics, whose foundations were established by Sewall Wright and R.A. Fisher. Wright (with his work on inbreeding and population shifts, and his path coefficient) and Fisher (with his statistical approach to the inheritance of quantitative characteristics) provided quite distinct and different contributions. While both men were interested in heredity on the basis of populations, they held distinct views on the theoretical side of this problem.

Wright focused on inbreeding and the way it could shift the genetic make-up of a population. He addressed the problem by experimenting with breeding animals and by assessing records of past breeding as applied particularly to Shorthorn cattle. Fisher, because of his attention to eugenics, applied his knowledge of mathematics and statistics to the problem of variability in inheritance of certain traits within a human population. He demonstrated mathematically how variability could be inherited. Basically Wright assessed the distribution of gene frequency in a population by analysing inbreeding using his path coefficient, while Fisher looked at the problem by statistical methods suitable for the Analysis of Variance (ANOVA) in very large populations.[85] Wright also undertook experiments to test predictability of outcome. Fisher, alternatively, worked with naturally occurring phenomenon and analysed what existed, rather than experimenting to see what would happen. It should be pointed out that the primary interest of neither was agricultural production, or even the betterment of breeding strategies. The process of evolution and the application of Mendelian theory to Darwinism were their main concerns. In keeping with Darwin, both Wright and Fisher looked for patterns in artificial selection that might explain how natural selection worked. It would be for others to explore how artificial selection practices could be modified to change the outcome of breeding on farms. In order to achieve any alteration on farm breeding practices, however, an animal science that focused on theoretical artificial selection strategies had to be developed. It was primarily J.L. Lush who undertook this challenge.

Livestock Genetics: J.L. Lush and Artificial Selection Theory in the United States, 1920–1950

Jay Lush, at Iowa State University in the United States, took the Wright/Fisher concepts concerning heredity in populations and applied them to theoretical animal breeding. In the process he focused on developing breeding methods that brought about desired results in livestock. He was most influenced by the work of Wright, and therefore concerned with the effects of inbreeding. He, like Wright, believed that inbreeding should be practiced with crossing of inbred lines; thereby, hybridizing. Lush's concern with improvement via artificial selection focused on the theoretical ideas of Wright that lent themselves to fixing populations with a genetic shift from the average. The Lush school would identify itself in this way with population genetics.[86] Importantly, the first phase of livestock genetics, in spite of Wright/Fisher synthesis, tended to be concerned with the metric implication of inbreeding, and, as an extension to that, breeding for hybrid vigour. The implications of Fisherian theory became somewhat hidden within the framework of Wrightian inbreeding theory.

Lush came from a farm background and was trained at Kansas State Agricultural College in animal husbandry. Through a professor at the college who taught animal breeding and was a friend of Wright, Lush began corresponding with Wright as early as 1918. Between 1918 and 1922, while Wright developed his path coefficient theory of inbreeding and wrote about systems of mating, Lush kept up with the literature as it appeared. He quickly saw that much of it was applicable to the development of livestock breeding strategies that ultimately could be used on farms. He took courses on evolution from Wright, and found Wright's work on inbreeding in historic Shorthorn cattle of great interest.[87] Lush wanted to assess which artificial selection strategy worked best under various conditions. In 1923 he published one of the first papers on animal breeding that used statistics to study animal improvement. It dealt with the influence on the effects of individuality, age, and the season on weights of fleece that range sheep yielded.[88] Many of his papers between 1926 and 1930 devoted energy to developing more accurate ways to measure quantitative traits.[89] At Iowa State University, Lush assessed inbreeding levels practiced on other livestock species.[90] He was as concerned with the process of hybrid vigour as he was with inbreeding, a situation that made his work fit theoretically within either population genetics or quantitative genetics. (By the late 1940s new computation methods for understanding inbreeding levels overcame the unwieldiness of Wright's system when it was used in ever-larger breeding populations.)[91] He experimented with poultry breeding at Ames in 1945 in an attempt to produce chicks along the lines of hybrid corn breeding.[92]

The study of statistics in relation to agriculture at Iowa State in the 1930s also played a role in shaping Lush's theories. In 1913 G.W. Snedecor (1881–1974) was appointed to the mathematics department at Iowa and by the 1920s he had become interested in developing statistics to serve agriculture. His most significant work was *Statistical Methods Applied to Experiments in Agriculture and Biology*, published in 1937. He was also the director of the Statistical Laboratory, the first of its kind in the United States, founded in 1933. Snedecor invited Fisher to lecture at Iowa State in the summers of 1931 and 1936, which proved to be highly significant events for the development of Lush's work.[93] Snedecor effectively made statistics an integral part of agricultural genetics. While Snedecor would irredeemably link mathematical statistics and agricultural genetics with each other, this statistics/genetics fusion went against certain trends in general biological research in the United States at the time and until after the Second World War. The earlier concern of some American scientists with biometry seemed to have gone into decline by the 1930s, outside the world of agriculture. The interwar years saw a move away from theoretical mathematics in relation to the general study of biology in the United States.[94] American biologists interested in ecology in particular reacted against biomathematics in the 1930s. When a greater emphasis on biomathematics and statistics arose after 1945, it reintroduced an old dilemma, embedded since the advent of biometry, in any biological study: what in essence was mathematics and what in essence was biology? When did a biomathematical study cease to be biology and become simply one of mathematics? The question would attract the attention of important academic biologists who had nothing to do with livestock breeding or artificial selection theory in the first half of the twentieth century: Canadian entomologist W.R. Thompson, Italian scientist Vito Volterra (who was interested in theoretical mathematical physics as well as biomathematics), and British zoologist C.S. Elton being prime examples.[95] The work of Snedecor might have linked statistics to agricultural genetics in an irrevocable way, but some geneticists working in the field of livestock breeding continued to be troubled by the conflict. Quantitative geneticists specializing in livestock breeding questioned an overreliance on mathematics in the 1970s, as will be evident later in this book.

When R.A. Fisher lectured at Iowa State in 1931 and 1936, Lush was impressed with his statistical approach to quantitative inheritance. He absorbed information from Fisher, with its greater emphasis on the idea of variability as opposed to that of genetic shift in the make-up of the population. Lush synthesized the theories of Wright, the Fisherian approach to statistics, and what he could learn from Mendelians at the college to develop experimental animal breeding strategies, designed ultimately to be applicable for farm use.[96] He was unique in combining the thinking of Fisher and Wright.[97] At Iowa State he gathered graduate and postdoctoral students who would learn his theories and, by the 1950s take them literally all around the world.[98] Lush always felt indebted to Wright's work, even though Lush himself actually created a more usable and practically oriented set of theories designed to improve farm animals.[99] After 1940 Lush and his students concentrated on applying population genetic theory to data collected under commercial conditions instead of in the laboratory. Studies were done on beef and dairy cattle, swine, poultry, and even honeybees.[100] It remained difficult, however, to make experimental conditions (even if commercial operations played a part in experiments) match the variation that would be seen on farms in different environments. The workability of certain strategies might prove to be effective on some farms but not others.

In the 1930s and 1940s Lush wrote extensively about animal breeding methodology and the historical background to livestock breeding within that context. He argued, fundamentally, that the breeders had known the methods of selection that he advocated for centuries. The main innovation, for Lush, was the fact that Mendelian theory in conjunction with biometric approaches to continuous inheritance explained why certain results occurred when different selection methods were used, and perhaps more significant suggested ways of predicting what might occur when a certain method was followed. It would be possible, Lush believed, to learn which method of selection worked the best in a breeding program.[101] His most important paper was one published in 1947 on pig breeding.[102] In it, he used population genetic theory and biometric principles to assess how much attention ought to be paid to the merits and defects of littermates when choosing which boars and gilts (young females) to use for breeding.[103] Lush then extended his experimental problem to ask how much a population could be shifted on average in certain characteristics by selection of the basis of individual performance versus family merit alone, or by using both criteria as selection bases. Predictability, Lush believed, was important because it saved time and expense, both beyond the reach of farmers. Experiment stations should undertake research work emanating from his theoretical principles.[104] Lush fully subscribed to the idea that government and science should work together to set ways to promote better agricultural production. Improvement, at this stage, was unlikely to be generated by farmers themselves. While much of Lush's work was theoretically oriented, he hoped to affect practices on farms through the results of experimental work. How extensively Lush's work actually changed farm breeding over the 1930s and early 1940s is not known. By the late 1940s, however, practical breeders were increasingly aware of Lush's ideas and the move on farms to using results of breeding experiments done at experiment stations had begun.

Livestock Genetics in Europe and the Work of A.L. Hagedoorn, 1930–1950

Scientific interest in the whole area of animal breeding was much less robust in Europe than it was in the United States throughout the 1930s and 1940s, and Lush's work was not widely known. Many geneticists working with animals did not focus on farm needs. For example, animal-breeding research in Britain at Edinburgh was devoted to rabbits.[105] Attention to agricultural production influenced the work of some important geneticists, though. Chicken breeding for farms concerned the scientist R.C. Punnett, who had worked with Bateson at Cambridge to establish the study of genetics.[106] He is perhaps best known for what is called the Punnett square, a tool designed to predict the probability of possible genotypes in offspring for practical breeders. He also looked at auto-sexing of certain chicken breeds by feather colour, work that he began during the First World War.[107] Sexing baby chicks was important to British chicken producers who concentrated on egg production – because they wanted pullets as opposed to cockerels.[108] An important European interested in artificial selection for better farm animals in this period was Christian Wriedt, the well-known Norwegian that Dunn mentioned in 1927. Wriedt also wrote a book, *Heredity in Live Stock*, published in English in 1930, on animal breeding for the practical producer, in which he emphasized breeding for hybrid vigour and his work with chickens along those lines. In articles, he explained the effects of crossbreeding on milk production in cattle, and campaigned against purebred breeding's emphasis on such characteristics as colour coat.[109] While farm breeders might have read Wriedt's writings, it was not likely that many carefully practiced his strategies. Conditions on farms were rarely conducive to following the hybrid-breeding recommendations advanced by scientists. Inbreeding for crossing purposes was not possible given the volume of livestock on individual farms.

The Dutch geneticist A.L. Hagedoorn was the other important European livestock-breeding scientist at this time. Well known in international circles and the author of many papers, as well as two books on livestock breeding (*Animal Breeding*, published in English in 1939; and with G. Sykes, *Poultry Breeding: Theory and Practice*, published in English in 1953), Hagedoorn wrote about genetics and how to utilize Mendelian laws in artificial selection strategies for the practical breeder. He believed that scientific approaches to artificial selection had firm foundations in farm breeding knowledge. Trained by de Vries and a Mendelist, Hagedoorn argued that scientists of heredity – until at least 1940 – owed virtually all of their information on large livestock breeding to the accumulated knowledge of the farm breeders.[110] He worked closely with chicken breeders in both Holland and Britain until his death in the 1950s.[111] He advised sheep breeders in Australia, having been asked to tour the continent in the 1940s.[112] Hagedoorn clearly interacted with practical breeding operations.

A firm believer in the progeny test, Hagedoorn became embroiled in an argument with Dutch breeders of Black and White Friesian cattle in the 1940s. Breeders of the popular milk breed had begun to select bulls on the basis of beauty and competitions at shows, and not on the strength of daughter milk production.[113] Cattle breeders believed "prettier" cattle were in fact healthier, but they were also heavily influenced by the ethos of purebred breeding, which promoted a greater emphasis on ancestry breeding. Breeders often favoured a bull whose mother was a good milker, with no regard for the milking ability of his daughters. Dutch breeders reacted to Hagedoorn's attack by pointing out that no effective progeny testing was possible under farm conditions in Holland. Most farmers had only five to seven cows and might produce only three calves a year.[114] These numbers were far too small to effectively progeny test. Much of Hagedoorn's theoretical approaches to artificial selection were sound, but he lacked the mathematical ability to statistically test many of his hypotheses. At least one historian believed this deficiency explains why Hagedoorn did not leave a significant mark on livestock genetics in the wide-ranging international way that the Lush school (which evolved from Wrightian and Fisherian statistical approaches to testing breeding strategies) did.[115] While Wright did not have the mathematical sophistication of either Fisher or Pearson, he had enough mathematical competency to assess experimental results. Furthermore, Snedecor's work had cemented statistics to livestock genetics from an international point of view, making them basically inseparable.

Hagedoorn knew that animal-breeding science was more robust in the United States than in Europe. He argued that by 1930 there were not as many as twelve scientists, regardless of possible training in statistics, working in any form of agricultural animal breeding in the world outside that country.[116] Hagedoorn's first book, Wriedt's *Heredity in Live Stock* and Lush's *Animal Breeding Plans*, published in 1937, were the only major works on livestock-breeding science and artificial selection strategies to improve farm animals written before 1940 and with practical farmers in mind. While all three tended to emphasize inbreeding/ crossing for hybrid vigour and chicken breeding, they also looked at cattle and pig breeding within a Mendelian framework. While actual artificial selection practices advocated by the scientists often were not reasonable given farm conditions, their work explained Mendelism to breeders and showed why certain breeding strategies yielded certain results.

Agricultural Chicken Breeding and Genetics
in the United States, 1930–1950

A number of American geneticists (D.C. Warren, a poultry geneticist at the Kansas State University, for example) became highly focused on applying the hybrid corn-breeding method to chickens by the 1930s. Agricultural genetic research moved beyond attempts to illustrate the dynamics of simple Mendelism with this work. Instead, research was devoted to finding breeding methods that would improve production.[117] With hybrid corn breeding as their model, geneticists at the colleges and experiment stations undertook complicated inbreeding and outcrossing programs, designed to utilize hybrid vigour. Many scientists were discouraged by the results. Hybrid performance of chickens might have been better than that of the inbreds in the experiments, but it did not match that of existing superior, non-inbred stock. The expensive and time consuming experiments often ended up with stock that, after returning to normal vigour, reached a level of productivity that had existed before any breeding system had been applied to them. Many of the inbred crosses were done within a strain (thereby weakening the potential for hybrid vigour), and in most cases no attention was paid to the performance of the birds used to create the inbred lines.[118] Even so, while results from inbreeding/hybridizing experiments were still discouraging for scientists at research stations in the early 1930s, hope was on the horizon.[119]

It was, quite simply, possible to experiment quantitatively with chickens via inbreeding and hybridizing in a way that was not true of the large livestock. In contrast to large livestock, fowl were cheap individually, and poultry reproduced relatively quickly and in sufficient numbers to warrant quantification of results. As Lush explained some years later at a Fact Finding Conference of the Institute of American Poultry Industries, "With the larger animals, the volume of business, the time demands, the value of the individuals, is so great as to make one hesitate a lot toward [looking for large numbers of strains or numbers within a family]. You in poultry, I think, are pioneering in that direction as far as the animal field goes."[120] Lush recognized that the cost factor, related to time needed to apply breeding experiments and quantify results for the larger farm animals, made such work daunting for either small-scale breeders or the government.

The tendency to view hybrid corn breeding as the model for Mendelism – and consequentially as the scientific approach to breeding – made many argue that any other strategies used for breeding did not reflect Mendelian principles and therefore were innately lacking a scientific basis.[121] Under these conditions, genetics and genetic experiments could do little to alter the way larger farm animals were bred by the late 1930s because it was virtually impossible to use the hybrid corn breeding method on them. As the *Journal of Heredity* stated: "the application of genetics to animal breeding ... [had] been almost infinitesimal ... Most of the excellent breeds of fairly uniform animals [were] not the result of application of genetic knowledge, but of long continued empirical selection for the desired types."[122] Some agricultural experts believed in a vague way that eventually better results would emerge from genetic research,[123] but others argued that that would never happen via the hybrid-corn-breeding method. It could not be applied to large animals, they believed. Experiments resting on inbreeding and hybridizing were too costly due to the slow reproductive rate of the large animals, and the breeding, which might take place at the stations, registered the effects of too many environmental variables.

With the synthesis of Darwinism and Mendelism, which resulted from the rise of population genetics, one particular scientist, Wright, turned his attention to the breeding of animals and the affects of artificial selection on the genetic make-up of herds and flocks in his efforts to explain Darwinism. He focused on the effects of inbreeding, as had his mentor, Castle. Wright saw heredity in terms of groups or populations when working with inbreeding, using a hybridist tradition as well. At nearly the same time, Fisher established mathematical ways of assessing variables in naturally breeding populations; in this case, he often used humans. Both perspectives lent themselves to the understanding that Darwinism and Mendelian theory were congruous. Neither Wright nor Fisher were primarily interested in animal breeding or in changing practices on farms. It was Lush who more fully developed the ideas of both Wright and Fisher when it came to artificial selection theory to be used on farms. Lush approached Wright's theories concerning inbreeding within the context of studying variance under Fisherian terms. The 1930s saw the rise of statistics as a separate study yet still closely affiliated with theoretical genetics. Throughout this volatile period, plant experimentation had focused on hybridizing. Success of plant geneticists with superior farm maize production encouraged livestock experimentation within the hybridist tradition.

While scientists had moved, via experimentation, towards accepting what were fundamentally the principles of Bakewell in artificial selection, the practical breeders had moved away from that methodology in its pure form. Improving breeders were heavily influenced by purebred breeding, as well as by standardbred breeding, which often fell under the guise of a purebred ethos. All improved attempts at breeding carried an aura of Thoroughbred horse breeding culture that valued purity and beauty – characteristics foreign to scientific thinking. There were other factors that separated the scientific approach from being implemented, and which made academic theory impractical. Farms in both Europe and North America did not have the breeding numbers necessary to generate sufficient data to make results meaningful. Environmental conditions varied so extensively, that what might be applicable to one farm or area was not germane to another. Other factors masked the effects of artificial selection; namely, feed conditions, ability to control disease, and the nature of housing facilities. Chickens dramatically increased their productivity in the 1930s in North America without any input from breeding strategies, and largely due to the introduction of corn feeding. Milk yields of dairy cows in Canada dramatically increased over the 1890s, simply as a result of better feeding programs. Artificial selection was only one part of the picture when it came to improved livestock, therefore. Animal improvement could be dramatic as a result of husbandry practices.

Livestock genetics, still mainly theoretical in nature by the late 1930s, was characterized by certain cultural features, and had taken approaches to breeding that were not necessarily the only ones that could have been pursued. The very shaping of the science dictated how it would interface with practical breeding. And the very nature of certain livestock breeding industries would dictate how they would interface with genetics. It is no accident that chicken breeding in the United States came to represent the most advanced scientific breeding in the livestock world by the late 1940s, as the next chapter demonstrates.

Chapter Three

Practical Breeding via Theoretical Population Genetics

I hope it will [be] agreed on not to let a Ram to any person (live where he will) but will engage not to sell any Rams but what he shall see killed before they go out of his hands, or take any to market but what are disposed of for the season with such other regulations as shall be thought proper.

— Robert Bakewell[1]

So wrote Bakewell in a 1791 letter, which described how he hoped his Ram Society would protect the intellectual property embedded in the work of a group of sheep breeders. The attempt to control the distribution of animals, via some form of patenting, has an ancient history. By the late 1930s livestock genetics had adopted an approach – namely, inbreeding and crossing for hybrid vigour, which worked well with such business strategies. It provided a natural patent for breeders, and that fact quickly became apparent to men interested in investing in the business of breeding.

In animals, scientific breeding via inbreeding/crossing for hybrid vigour took hold most dramatically in chicken breeding, and the innate structure of the historic poultry industry does much to explain why that became possible. This chapter concentrates, therefore, on the world of chicken breeding in North America from roughly 1880 until the 1940s and assesses historic approaches to breeding, and critical divisions within the poultry world that affected the dynamics of breeding. It is then possible to see both how and why these patterns interconnected with the evolving American world of genetics. I turn next to reviewing attitudes to hybrid breeding in beef cattle and pigs within their respect breeding structures, in order to compare their situation with that of chickens.

Breeding Principles of Nineteenth-Century Chicken Breeders

All of these nineteenth-century chicken-breeding systems relied on general artificial selection principles established over the late eighteenth century; specifically, an overarching concern with the effects of inbreeding (that is, the mating of related stock in order to achieve uniformity) and outcrossing (that is, the breeding of unrelated stock in order to inject change and counterbalance the effects of inbreeding) on a population, and with hopes of balancing the two. A focus on populations, not individuals, was critical. Simple reliance on one individual alone was not part of any sophisticated breeding program. Individual worth and ancestry might play a role in how selection for breeding worked within this overall framework, but so did other approaches. Mass selection – that is, favouring various individuals according to some set criteria (such as looks or production history) – was the simplest and most common method. It had an ancient history that probably went back to domestication. Progeny testing was another way to select breeding stock. For good chicken breeders in the early twentieth century, it was often a question of balancing mass selection in particular (as well as progeny testing to some degree) with individual worth and/or ancestry breeding. A few examples of nineteenth-century chicken-breeding systems, which grew out of this overarching approach, follow.

H.H. Stoddard, a founding member of the American Poultry Association, an experienced American breeder, and a renowned publisher of poultry journals (and the man who responded to Raymond Pearl), wrote numerous articles on an inbreeding/outcrossing system for egg-laying hens. In order to create a working commercial flock, he advised the breeder to start with sixteen unrelated strains and combine the genetics of these over five years, by selecting only males from certain lines for breeding and only females from other lines for breeding. Selection of both males and females within the various strains could be done on the basis of mass selection, progeny testing, and/or individual worth and genealogical background. Culling should be done at every level and no inbreeding incurred. The final cross of purely unrelated stock resulted in a vigorous flock that would be inbred, brother to sister, for at least four years. The inbred lines were to be used for table egg production. The inbreeding after four years would weaken the stock, but by that time the breeder would have new birds, arising from his original sixteen strains, available for inbreeding and egg production.[2] Stoddard advised a well-planned balance between inbreeding and outcrossing in order to harvest the advantages that inbreeding could bring – uniformity in the progeny – without incurring its dangers; namely, a tendency to reduce fertility and vigour in the offspring. Stoddard's approach bore surprising similarities to the way future population genetics would approach artificial selection strategies.

Another method, developed about 1870 in Britain and used in North America almost immediately after that, focused on the development of a distinct male line and a distinct female line within two separate inbreeding/outcrossing systems that used various selection methods.[3] One line was created by mating females producing especially good pullets (young hens) to males producing particularly good pullets, while the other was made by mating females producing good cockerels (young males) to males producing good cockerels. It was the cross of the male line on the female line that brought about the desired final results. The males and females that came from the male/female line cross, however, were useless as breeding pairs because they would not reproduce truly. One had to access the parents or grandparents to achieve the same results. This situation created a biological lock and brought another meaning to the idea of mating the "best to the best." The breeding lines produced the "best," but members of one sex in each line were not the "best" in themselves. This double mating system provoked a lot of controversy late in the nineteenth century and into the early twentieth in North America.[4]

Breeders who opposed double mating questioned the desirability of forcing farmers back to breeders for replacements – thereby making farmers relinquish any role in breeding, a situation that the double mating system enforced. The philosophy that all farmers should be breeders for the good of ongoing livestock improvement was at the heart of the matter. The breeding and producing of stock should be a seamless operation and improvement overall was deemed to arise from the concerted work of the many. To put breeding into the hands of the few would seriously undermine results. A great deal of discussion about elitism in breeding, and restricting the occupation to a limited number of people, took place in the poultry press, usually in relation to the double mating system as applied to a significant utility breed, the Barred Plymouth Rock.[5] Certain developments within the general poultry industry would soon soften breeder resistance to such biological locks. The double cross mating system, controversial as it was, bore parallels to the hybrid corn breeding method that also ensured a biological lock.

In the 1870s the American master breeder, I.K. Felch, created a scheme that focused primarily on inbreeding. Felch advised controlling inbreeding, so that outcrossing would not be needed to avoid incurring the problems that inbreeding promoted. Inbreeding preserved and perpetuated desired characteristics, he argued, while following a percentage system of inbreeding could guard against the expression of undesirable features. Felch developed a complicated chart that showed how the progeny over generations of a foundation male and female could be mated with each other in order to intensify the genetic input of one over the other. For example, in the third generation, a male could be mated with a female in order to intensify the inheritance of either member the original pair. Ultimately, the breeder could produce stock whose genetic make-up showed a varying percentage between one-half and seven-eighths of each foundation member. By recombining the blood of a selected foundation breeding pair through different mating combinations over generations of their descendents, one could inbreed forever without experiencing seriously reduced vigour.[6] The heredity of the original pair would be improved by blending their input to varying degrees. Starting with non-related stock helped check problems with inbreeding, because one could revert back to the equal input of blood from each.[7] With careful selection, Felch could shift the general hereditary make-up of the flock towards both uniformity and improvement.[8] Felch's strategies are reminiscent of the general thinking behind Sewall Wright's theories concerning inbreeding and his path coefficient.

A focus on outcrossing without inbreeding was a common nineteenth century, North American way to produce meat birds.[9] The crossing of breeds (an extreme version of outcrossing) was known to result often in superior progeny that demonstrated what was called hybrid vigour. Mating systems based on crossing three breeds might be used rather than simply relying on two breeds.[10] An 1898 issue of the Canadian journal *Farming* noted the propensity to seek this type of breeding for increased vigour. The crossing of breeds of fowl worked well for the production of superior birds, the journal had to admit: "By crossing two … breeds that are very dissimilar, we secure an increase in hardiness in the first cross, as well as the special qualities in each breed to a high degree."[11] The inability of crossbred stock to breed truly for those good qualities in the next generation, however, made the breeding method subject to disapproval. The journals discouraged crossbreeding, even if it led to increased vigour and better productivity. "There is, it must be confessed, one great disadvantage attending to the rearing of cross-bred fowls – they are ... unsaleable as stock birds," the *Canadian Poultry Chronicle* warned as early as 1871.[12] Crossing of breeds also seemed to be out of line with good farming practices for the agricultural experts. Breeds that carried good quality and also bred truly for such quality had been established.[13] As the egg industry gathered increasing momentum early in the twentieth century, any attempt to focus on meat breeding (and concurrently on crossbreeding) decreased significantly.

Erosion of Breeding Skill: Organization of Chicken Breeding and the American Poultry Association

By the late nineteenth century, knowledge of these sophisticated breeding strategies had begun to decline in a widespread way, resulting in an erosion of breeding skills in the industry. The trend became increasingly evident over the early twentieth century. Many factors played a role in this development. The position taken on breeding by the organization that presided over chicken affairs in North America initiated the trend by introducing a serious dichotomy: was chicken breeding a matter of beauty or a matter of utility? The association's unclear position on the matter led to ongoing confusion within the chicken-breeding world, as a short review of its history and activities reveals.

Chicken-breeding organization in North America developed along patterns invented in Britain. Initiated in 1865–6, the British structure for organizing breeding reached North America almost immediately, and led to the 1873 formation of the American Poultry Association by poultrymen from both the United States and Canada. The following year the Association published a document that described all breeds and varieties, called the Standard of Excellence. It was renamed the Standard of Perfection in 1888.[14] The Standard made it possible to measure the quality of individuals within each breed for exhibition purposes. No rules were set for breeding methodology. All breeding aimed simply to produce birds that matched the standard as closely as possible. A basic dichotomy in chicken breeding was established from this beginning: were chickens to be bred for beauty or for utility? Were the two compatible? The importance of the show ring made it difficult to uncouple the two ideas. The same dichotomy arose over the way horse, cattle, and dog shows affected breeding in both Britain and North America.[15] Still, the beauty/utility conflict became most blatant within the poultry breeding/show system.

By the late 1880s the idea that beauty and utility could go together in breeding attracted open ridicule. In 1888, for example, *The Mark Lane express* in the United States addressed the issue squarely in the following words: "The fancier who minces the matter, preferring to allow the world to continue to believe that is exhibitions instruct and improve the people in a particular direction is insincere. In answer to the question, What has the poultry fancy done for profitable poultry? We must answer, clearly enough, nothing."[16] "Feathers count and *feathers only* [italics in the original],"[17] the *Farmer's Advocate* in Canada fumed in 1900. Look at the "football-haired Polish." Fanciers produced breeds that were "a curse to any farmer or practical poultryman," noted the *Advocate*.[18] State involvement in breeding, via supplying prize money for exhibition fowl clearly useless to farmers, bothered both Canadians and Americans.[19] Furthermore, the shows taught nothing about breeding. In 1892 the *Canadian Poultry Review* pointed out that while poultry exhibitions were designed to help farmers and not fanciers, farmers rarely went to shows and therefore learned nothing about the work of professional breeders.[20]

It should be pointed out, however, that not all efforts at breeding under the standardbred system were useless from the point of view of agricultural production, a fact that makes the whole beauty/utility issue so complicated. It was under this system that the best commercial breeds were created – breeds that would go on to form the backbone of all modern chicken genetics for both the table egg and meat industries. Breeders in the United States, working under the American Poultry Association's direction, developed the American Leghorn (a table egg breed), the White Plymouth Rock, the Rhode Island Red (both meat but potentially table egg breeds), and the American Cornish (a meat breed) over the late nineteenth century. It would be from these breeds that all modern producing lines for both poultry industries evolved.

The American Poultry Association considered standards based on productivity as early as 1903, when the committee in charge of revisions suggested that good utility should be recognized.[21] Pressure to do so grew after that time. At the 1907 meeting of the association, members resolved that "the American Standard of Perfection [gave] undue prominence to the beauty value of standard-bred fowls, to the detriment of the utility value of domestic poultry."[22] The association should organize standards for utility, just as it had for beauty, it was argued. Many thought the association should run egg-laying contests, a fashionable way to assess productivity by the early twentieth century in both Britain and North America. The first had taken place in Britain in 1897, when the Utility Poultry Club of England ran a competition, with only seven entries of four hens each.[23] By 1912 egg-laying competitions were common in the United States and Canada. All breeds took part in the contests that often degenerated into competitions between breeds.[24] Breeders used success in the egg-laying contests to promote both their breed and their own stock.[25] In the process an element of sport became attached to egg-laying contests.

Some breeders opposed egg-laying contests, and conflicts over their value would be ongoing within the American Poultry Association. Those who disapproved of egg-laying contests were often accused of being only interested in beauty. "At the present time there appears to be a tendency, on the part of a few, to revive the old time warring between the ultra-fancy and ultra-utility poultrymen," the *American Poultry Journal* noted in 1915.[26] Breeders who disliked the contests were not, in reality, necessarily focused on breeding for beauty. They thought competitions served no useful purpose because identifying heavy layers did not teach anyone how to breed them. As far as these breeders were concerned the competitions promoted poor breeding practices. The contests encouraged the idea that breeding should be guided by individual worth, they argued, and only on the basis of ancestry.[27] Relying on one hen, a winner of an egg-laying contest, to produce superior daughters was not, in their opinion, the proper way to breed.[28] They advocated following selection strategies developed by master chicken breeders over the late nineteenth century.

Erosion of Breeding Skill: The Record of Performance (ROP) and Breeding for Eggs

The authority of the state in breeding regulation reinforced the divisive factors evident in the American Poultry Association's position over breeding. North American governments created a structure to orchestrate breeding that combined the two main organizational tools established by the breeders; namely, exhibition standards of the American Poultry Association and the testing of egg-laying capacity. The idea of government involvement with poultry breeding for eggs began before the First World War in North America and was first initiated by William Graham of the Ontario Agricultural College in Canada, a country that had used governmental authority to regulate livestock breeding organizations in a way less likely to be seen in other Anglo-countries.[29]

By summer of 1919 regulations for a national Record of Performance, known as the ROP, had been established. Only birds of standard varieties, free from disqualifications outlined by the American Poultry Association in the Standard of Perfection, and capable of laying a determined number of eggs per year, were eligible for acceptance under ROP certification.[30] The ROP firmly linked beauty with utility. The state system fed into the beauty/use dichotomy by uniting beauty standards with production standards.

A move towards an ROP developed fairly quickly after the Canadian structure was established. The reasoning behind ROP regulations in the United States, however, was somewhat different. The issue of falsified egg-laying records seemed to plague American breeders more than Canadians, and that problem, more than attempts to direct breeding, drove support in the United States for an ROP.[31] While the American Association of Instructors and Investigators in Poultry Husbandry authorized the initiation of the national American Record of Performance Council as early as 1919, ROP regulation in the United States tended for some time to stay more regionally- or state-oriented.[32] (It was not until 1930 that sixteen states joined to form the United States Record of Performance Association, which recorded egg production, egg-weight, and body weight made on the breeder's premises. The regulations matched those of the Canadian ROP standards. Each bird had to be a good representative of its breed under the Standard of Perfection as well as meet a certain standard for egg laying.[33])

Poultry ROP did not command the same respect from utility breeders in the United States that it did in Canada.[34] The affect the Record of Performance had on breeding principles alienated American farm breeders who had opposed the contests for breeding reasons, even though the ROP was not based on the idea that egg-laying data should provide the basis of competitions. The ROP used the data to create a standard. Emphasizing a standard in this fashion, however, encouraged many of the undesirable breeding habits that the contests did: a reliance on individual hens and the promotion of ancestry breeding with little or no attention to the effects of balanced inbreeding and outcrossing.[35] The ROP also tied utility breeding to breeding for beauty. In order to qualify for the ROP a bird had to meet the Standard of Perfection. The ROP aggravated the inherent tensions in the traditional chicken-breeding world by dividing utility farm breeders into two camps (particularly in the United States): those who entered the ROP and those who did not.

Erosion of Breeding Skill: the Chicken Industry
and a Breeder/Producer Divide

A critical industry division in the structure of the North American poultry world had evolved over this period and came to play a deciding role in how future chicken breeding would be practiced. The division was a key factor in the erosion of breeding skill among poultrymen. A schism between the breeder and the producer/grower, detaching producers from the breeding activity, had developed. By the end of the 1920s most poultry producers took no part in breeding. Those with interests in poultry tended to fall into two separate camps: true breeders who created distinct lines of stock and producers/growers who simply multiplied and/or used the birds. While the professionalization of breeding and the masking of its methodology by such structures as the ROP distanced producers from the breeding activity, the hatchery industry played an even more critical role in that separation. In this case technology altered not just the structure of the poultry industry but also the process of breeding.

The hatchery industry's use of the superior artificial incubation methods in place by the early twentieth century[36] and the fact that baby chicks did not need to be fed for seventy-two hours after hatching (thus allowing them to be shipped considerable distances) encouraged a division between the breeder and producer/grower. The work of running huge incubators meant that increasingly by the 1920s men operating them were not involved in either the breeding or the producing side of the chicken business.[37] The concurrent growing pattern of buying day-old chicks from off the farm and from non-breeding hatcheries further reduced involvement of producer/growers with breeding. They had traditionally acquired stock from breeders and subsequently made decisions on how to multiply or reproduce those genetics. The move to relying on hatcheries instead of breeders for birds started a trend that would be ongoing; namely, the "deskilling" of producer/growers in breeding matters, and the reduction of their control over the genetics they used.

The cleavage between breeder and producer/grower was complicated by the gender of the people involved. Most breeders were men, and most producer/growers were women who raised the breeders' stock in order to produce marketable eggs and meat.[38] Ads for poultry feed and equipment also seemed to be aimed at women.[39] Articles referring to actual breeding strategies for farm poultry seemed almost entirely to have been submitted to the press by men (either fancy or utility-oriented breeders) or as having been written by press editors. Women wrote articles in the farm and poultry press, but rarely on the subject of breeding (although preference for breed was frequently stated).[40]

The hegemony of the American Poultry Association (an organization that set standards but did not stipulate breeding methodology), the standards of the ROP (a structure that also set standards but did not provide information how to breed), and perhaps more importantly the development of an industry divide between the breeder and the producer resulted in the reduction of a wide-spread understanding in both the United States and Canada of breeding methodologies promoted by men like Felch and Stoddard. Increasingly, it was unclear to the average breeder how breeding should proceed. As early as 1914 C.D. Cleveland (secretary of the New York Poultry and Pigeon Club, breeder, judge, and writer of many articles on chicken husbandry) commented on the poverty of material concerning selective breeding methods of breeders: "I have never been able to understand why it was that the breeder was so loath to give away anything in regard to the essentials of the way he breeds his varieties ... Breeders ought not to hold back their so-called breeding secrets." The editor of the *American Poultry Journal* thoroughly agreed, stating: "Mr. Cleveland's remarks summed up the situation very nicely. We have been trying for years to get articles [on breeding] we want, and believe should be published, on the how and why of mating and breeding, but we can't get the information ... Perhaps some breeders may not be able to tell how they get results."[41] Articles devoted to breeding methods appeared less frequently in the poultry press, a pattern quite evident by 1930. Virtually no articles on the subject were printed after 1930.

Chicken Breeder/Scientist Interaction, 1900–1920

In spite of the apparent limited innovative value of early Mendelians, some interaction or communication between breeders (particularly in the United States) and scientists took place on occasions. It is difficult, however, to assess if or how much one influenced the other in the years before and immediately after the rise of Mendelism, because parallel thinking often dominated much of the artificial selection strategy used by both sectors. Some biologists did not follow the hybridizing practices so common to much of the academic breeding research. Not all scientists emphasized progeny testing either, and not all breeders relied primarily on mass selection. Similarities existed between the actual breeding methods of many early biologists and practical breeders to such a degree that even when it is known that breeders communicated with them, it was not apparent how much the scientists affected breeders. Often breeders seemed to follow what experience, not science, had taught them. It is not evident, for example, how much (or even if) the large Texas breeder, M. Johnson, used scientific input when he abandoned the breeding of exhibition poultry in favour of utility features.[42] American D. Tancred discussed breeding with G.M. Gowell (who preceded Pearl) at the Maine Experiment Station and relied on sibling or progeny testing.[43] But whether he did so via his own initiative or as a result of Gowell's advice is not clear. Tancred was also inclined to quantify breeding results, but there is no proof that he followed that practice on the advice of Gowell. Another breeder who conversed with a poultry scientist at an experiment station was American J.A. Hanson.[44] His system resembled the future breeding of geneticists, but it had an affinity to inbreeding and crossing systems designed by Stoddard. One American breeder, E. Parmelee Prentice, believing in the future value of Mendelism, relied entirely on scientists. A Chicago lawyer, he bought a country estate, Mount Hope, in Massachusetts in 1910 and became interested in the potential of Mendelism for better farming.[45] Prentice hired H.D. Goodale, poultry scientist at the Massachusetts Experiment Station, to run cattle and chicken-breeding operations at Mount Hope on a full-time basis. A specialized poultry staff worked under Goodale.

As the above examples make evident, it cannot be said that utility breeders uniformly dismissed the work of scientists as being non-innovative even if it was difficult to assess how Mendelism could provide useful strategies for breeding. It certainly cannot be argued that breeders were prepared to ignore early Mendelian activities. The scientists might, after all, have something to offer in the future. Many breeders hoped Mendelism could ultimately evaluate known selection methods and the structure of breeding programs in useful ways. With time, the effectiveness of mass selection, for example, could be tested against the effectiveness of progeny testing. Even Stoddard was prepared to admit that Mendelism might ultimately be of revolutionary value to chicken breeding.[46] Increasing breeder-receptiveness to scientist inputs, however, exacerbated the ongoing conflicts that existed in the chicken-breeding world by enflaming die-hards of the American Poultry Association.

Old American fanciers believed that the ongoing reduced significance of standardbred breeding under these conditions resulted not from a confrontation between beauty and utility interests or divisiveness among utility breeders, but rather from what often appeared to be clashing approaches to breeding between practice and science. At the 1924 meeting of the American Poultry Association in Toronto, Canada for example, American E.B. Thompson claimed that the press and the agricultural colleges had assassinated the standardbred system for poultry breeding, both of which ridiculed craft/farm breeding generally and, at the same time, supported genetic research in breeding. Canadian breeders did not see the divisive tension as positioning science against traditional breeding in the way many Americans did. For Canadians the issue was clearly a utility-versus-beauty dilemma, but the problem related only to how standardbred breeding was practiced and what science could teach within that framework. It was not a question of standardbred breeding versus science. Research concerning the process of heredity was far less robust in Canada than it was in the United States (thereby perhaps making it less threatening), a fact that might explain why attitudes in Canada differed from those in the United States.[47]

Hybrid Corn Breeding, Corporate Involvement, and Scientific Chicken Breeding for Eggs

By the 1930s the American corn breeding companies had succeeded in making the hybrid breeding method work; more importantly, conceivably, they had convinced farmers to buy seeds from them every year rather than breed next year's crop. The success of the hybrid-corn-breeding method as applied to plant breeding made the companies want to explore the idea of using the same system for the production of egg-laying chickens. Under these conditions, Lush's (and his school's) inbreeding and, more importantly, hybrid work with poultry attracted their attention. It would be the beginning of crucial new developments in relation to chicken breeding; namely, the entrance of corporate involvement and an emphasis on breeding for lines that did not produce truly. Unlike many scientists in the United States working at research stations in the 1930s, the managers of the corn companies believed the hybrid-corn-breeding system could be successfully applied to chickens. They were willing to finance expensive experimental programs in order to find a way to achieve this target. The Wallace family – that is, Henry

A. Wallace who developed the hybrid seed company Hi-Bred Corn Company in 1926 (renamed Pioneer Hi-Bred Corn Company in 1935), and his son Henry B. Wallace – initiated this effort to produce commercial hybrid chicks in 1936. By 1942 the Wallace family was selling hybrid egg-laying Leghorns under the name of Hy-Line.[48] How these hybrids chicks were generated was kept secret, even if inbreeding and line-crossing were central to the process.

There were a variety of ways for geneticists to produce the four inbred pure lines that went ultimately into the final terminal cross. One could set out to establish many inbred pure lines, designed to breed truly, by performing the closest inbreeding possible (brother to sister), and through time letting the ones excessively weakened by inbreeding die out naturally. Everyone knew that certain lines tolerated inbreeding better than others, but it was only through actual breeding that one could one know which ones stood up.

Another, perhaps more productive way, was to carry fewer inbred lines under less intense inbreeding practices. That way the geneticist was not forced to rely on a line simply because it could tolerate intense inbreeding.[49] Large numbers of birds were needed in order to quantify varying results.[50] The hybrid-corn-breeding method, as applied to chickens, needed a huge number of birds in order to operate properly because many breeding experiments – and the stock used in them – would have to be discarded. No ordinary breeder had the numbers required to carry out any breeding system of this nature. Flocks on farms (breeding or otherwise) rarely numbered above a few hundred at this time, and the average per farm was in fact a great deal lower and had not changed in size much since the nineteenth century. I.K. Felch estimated that in the 1870s American flock size varied from twelve to fifty. Larger ones were extremely rare.[51] By 1913 the flocks on some American farms had risen to between 100 to over 200 hens.[52] As late as 1930 the average flock in the United States remained about the same as Felch's estimate for the late nineteenth century. Over half of farms reporting chicken-keeping in 1930 stated they had less than fifty hens, and the average number of hens per farm was estimated to be no higher than twenty-three. The vast majority of commercial flocks until after the 1950s continued to number below 200.[53] The hybrid-corn-breeding method as applied to chickens, which was immensely expensive and also wasteful, needed the kind of corporate enterprise that the Wallace family could provide.

By the early 1940s American geneticists working with breeding companies had succeeded in breeding hybrid hens with increased egg-laying capacity via the hybrid-corn- breeding method, and companies of egg-laying birds began to franchise hatcheries, which were themselves independent, in the United States and Canada. Both Hy-Line and the DeKalb Hybrid Corn Company marketed chicks through franchised Canadian hatcheries by the 1950s.[54] Hatcheries advertised what they had been franchised to sell.[55] Canadian hatcherymen were forced to confront the fact that producer/growers liked egg-laying birds that resulted from the crossing of breeds or of strains within a breed, even if these emanated out of the United States.[56] As early as 1941 one Ontario hatcheryman who ran a large operation with incubators in a number of locations, A. Seiling, stated that 65 per cent of his sales had been hybrids, and he prophesized that that number would increase in the next year. Once a producer had experienced hybrids, Seiling explained to other hatcherymen, it was impossible to sell him/her anything else.[57] ROP birds were not wanted, and few if any hybrids were available from Canadian breeders. The result was increased importation of American hybrid chicks into Canada.

Hybrid Breeding and the Rise of the Broiler Industry: Breeding for Meat

Meanwhile, dramatic changes in chicken breeding for meat were underway in the United States. By the 1930s a renewed interest in meat breeding resulted in concerted efforts by some American breeders to supply the hatcheries with crossbred chicks to be raised for the rapidly growing broiler market in large eastern seaboard US cities. (No similarly sized market existed in Canada at this time.) Breeders, such as the Hall brothers, returned to traditional ideas of crossbreeding for better meat production.[58] Crossbred stock, of course, implied terminal stock when it came to breeding purposes. Broiler growers of Hall crossbreds would be forced to return to supplier hatcheries for replacements – these in turn bought from multipliers of Hall-bred birds. Crossbreeding for hybrid vigour brought with it a biological lock. Since chicken producers had largely abdicated breeding, and since the crossbred stock grew so well (making them economically attractive to farmers), poultry producers/growers readily accepted the crossbred hatchery chicks, and the concurrent biological lock that they possessed.

The most important event for the poultry meat breeding industry in the twentieth century was the Chicken-of-Tomorrow contest, run by the Atlantic and Pacific Tea Company (A & P) between 1948 and 1951 in the United States. A & P agreed to sponsor and underwrite a long-range project designed to improve meat-type birds by teaching breeders what consumers wanted in a meat bird. The Chicken-of-Tomorrow contests established what might be described as the golden cross for meat production; namely, a Cornish/Plymouth Rock/Rhode Island Red cross (or a Cornish/New Hampshire cross, the New Hampshire being comprised of a Rock/Red cross) male and a White Plymouth Rock based female line. The contests did not teach how to breed, but rather showed breeders what to breed for. Furthermore, the dressed poultry meat contests gave breeders a chance to compare their work to that of others, a situation that stirred up a sense of competition and instinct of improvement. Several breeders who were fundamentally unknown outside their local communities skyrocketed into national fame, particularly the two winners of the national contests.[59] The contests paved the way for complete acceptance of crossbreeding for hybrid vigour (and not the breeding of lines that reproduced truly) in broiler production, first within the United States and then in Canada. Such an approach to breeding ensured the presence of biological locks, and fit neatly with the entrenched geneticist attitude to breeding for agricultural improvement. The Chicken-of-Tomorrow contest, then, encouraged a hybrid geneticist approach to breeding, which in turn was attractive to corporate enterprise. American breeders who had been successful in the contests found themselves at the head of companies that functioned increasingly with the aid of geneticists.

Corporate breeding also initiated a change in the way genetic research applied to chicken breeding would reach the American public. Traditionally the dominant centres for poultry breeding research had been found at agricultural colleges, experimental stations, government agencies like the Bureau of Animal Industry under the USDA, and universities where aspects of evolution theory, chromosome theory, and other genetic questions were studied. Even if much of this work done on poultry before 1930 was of little use to farmers, the philosophy behind public support for such undertakings initiated at agricultural colleges and experiment stations arose from the conviction that American farmers were the expert breeders, and any information available should be aimed at them in order to improve their breeding methods for egg-producing hens. After 1915, if not before, geneticists clearly hoped ultimately to reach breeding farmers directly through the research they did at government-funded agencies.[60] Increasingly by the late 1940s, though, geneticists produced information of more interest to the developing new private companies that then sold the resulting product to farmers who earlier might have been breeders, or to farmers' wives who had always been producer/growers. Breeder companies did not disseminate information to these people on how to breed. They were perceived to be consumers of expertise practiced by someone else, not learners of it. Under these conditions, American university geneticists served the interests of the companies and not farmers. Increasingly men like Lush acted as advisors to geneticists working in companies or alternatively trained people who eventually would work for these organizations. By 1950 the demand for trained geneticists by the breeder companies was so great that it outstripped supply. By the mid 1960s, with fewer breeders and the consolidation of breeder companies into smaller numbers, the need for trained geneticists declined.[61]

Reasons for the Scientific Revolution in Chicken Breeding

Chicken breeding had been a sophisticated endeavour from at least the mid-nineteenth century, and many of the attitudes of the best breeders were very similar to those of Mendelians working with artificial selection after 1900. Breeders knew that various artificial selection strategies brought certain results in the progeny, just as the scientists did. Felch and Stoddard clearly comprehended the effects of inbreeding, and appreciated the percentage inbreeding ideas put forward much later by Wright. Hybrid breeding and crossbreeding were fully understood by breeders long before heterosis interested scientists. By the 1920s tension between practical breeders and geneticists that had developed after the rise of Mendelism showed signs of dissolving, perhaps as some appreciation of commonalities in practical/scientist thinking became evident, and in spite of ongoing critical but confusing beauty/utility dichotomies in the poultry breeding world. Significant American breeders spearheaded what could have become a more widespread interest in science, and in the distinct approaches being developed by population geneticists towards artificial selection by the early 1930s; specifically, a clearer and stronger emphasis on the progeny test, selection on the basis of population groups, and sophisticated quantification of results – all of which were designed to produce good true breeding lines.

Why was it, then, that quite a different genetic breeding revolution, one with a focus on hybrid breeding or non-true breeding lines, took place? There were three fundamental reasons. First, North American chicken breeding was hopelessly divided in outlook by the 1920s, thereby presenting a vacuum that the corporate breeders could take advantage of. Traditional organizations that supported chicken breeding had introduced a particularly serious and entrenched dichotomy: should birds be bred for beauty or for utility, and did beauty mean utility? The tenacity of the American Poultry Association to abiding by the Standard of Perfection and beauty breeding had forced a confrontation between commercial breeders and fancy breeders. The linkage of programs like the ROP with fancy did not help that situation. Increasingly, too, traditional breeders had lost a clear understanding of what breeding methodology actually meant, because they had come to see either the ROP or the Standard of Perfection as methodology. The cleavage of the breeding world had become ever more pronounced: fancy breeders who created both beautiful and hopefully useful birds, breeders who worked within science/government structures, and breeders who followed their own instincts often with the aid of biologists at agricultural institutions. A small nucleus of breeders, intent on change and receptive to ideas potentially emanating out of genetics by the 1920s, could not shift or heal the deeply embedded and ideologically divisive tensions in the chicken-breeding world.

The second reason (which also further encouraged an erosion of breeding knowledge) lays in the structure of the general North American poultry industry and the divide that existed between the breeder and producer/grower. To begin with, the divide helped to mask the fact that chicken-breeding strategies had been based on sound artificial selection principles, a situation that encouraged the conviction that chicken breeding had been primitive before the advent of genetics. The divide also provided the rationale for a critical change in the philosophic outlook behind chicken-breeding strategies. The idea that breeding should result in true producing lines was no longer important if the producer was not part of the breeding structure. Once the producer/grower abdicated any role in breeding, one of the main advantages that true breeding lines offered – that is, the possible inclusion of producing farmers in the breeding process – was no longer critical to the breeding structure. Hybridizing, clearly understood but considered to be a questionable way to breed in the nineteenth century, took centre stage under these conditions. The producer/grower began to demand hybrid stock when it became available. The producer/grower might consume the product of the breeder, but the opinions of the former directed what breeding would be acceptable and therefore how breeding would evolve.

Finally we come to the third reason: the entrance of corporate enterprise, which was essential for the funding needed to create good hybrid lines. The cheapness of the individual birds and their fast reproductive life all lent itself to the potential establishment of many breeding lines, but breeding operations of this magnitude required capital investment to make them work. Success of the American corn companies with inbreeding and crossing for hybrid vigour triggered the support of private enterprise regarding financing the expensive experiments. The funding of such capital was attractive because the biological lock indigenous to hybrid breeding created a natural patent, thereby protecting the investment of the company. Corporate hybrid chicken breeding could only work if poultry farmers agreed to buy the chicks and, in the process, abdicate any role in breeding, a situation in place by the late 1920s. The buyer of company hybrid stock could not use that stock for breeding and would, therefore, be a return customer for every new generation. Since the producer/growers were the main buyers, and because they did not care to breed, they had no problem accepting hybrid stock. The triumph of crossbreeding over lines that breed truly was in place, and with this came biological locks that in turn supported the idea of private enterprise because they provided a protective patent.

Beef Cattle Breeding and Hybrid Strategies

Breeding for hybrid vigour via inbreeding and crossing seemed to make sense within the beef cattle world, if for no other reason than simple crossbreeding was a common way to produce beef stock for market. For centuries, cattle producers had selected bulls in order to achieve hybrid vigour. Classic crosses for hybrid vigour resulted from the breeding of Shorthorn cows to Angus bulls or Shorthorns cows to Galloway bulls. The progeny from such crosses were known as "blue roans." Angus were also crossed on Herefords and the progeny were called "black baldies" – cattle that were black in colour instead of Hereford-red, but with the white faces of the Hereford. But the historical structure of beef cattle breeding played a role in how selecting for hybrid vigour would work. The production of beef resulted from a classic division between breeder and producer, but the boundaries between the two were not as sharply delineated as in the chicken industry. The beef cattle world has been described as having two sectors: the breeder and the feeder. The division is ancient, dating back to the Middle Ages in Europe when cattle were walked from their breeding grounds to feeder farmers who, living nearer to consuming urban markets, brought the animals to maturity. The division became increasingly marked over the eighteenth century, especially in Britain, the largest beef-consuming nation in the world by the nineteenth century, but also in North America where the Spanish system of ranching dovetailed with a feeder structure.[62] By that time, the situation had become more complicated with respect to the designation of breeder and feeder. The breeder arm had itself become divided into two sections. It was made up of the elite purebred breeders and those who might be called the multipliers who ran what is still described as cow/calf operations. The feeder farmers acquired the calves from the cow/calf operators and fed the stock until it was ready for slaughter. The whole structure was tied together by the fact that the feeders decided what stock they wanted to buy from the cow/calf operators, and the cow/calf operators decided how much, how little, or if they wanted to use the purebred breeds in their programs. At least as far back as the early nineteenth century, cow/calf operators had tried to utilize hybrid vigour by crossing bulls of a certain breed or type on cows with a different genetic background. Beef cattle production, then, might have been structured on a pyramid of purebred breeding, but the dominance of terminal crossbreeding meant that breeding in a more structured way for hybrid vigour should have seemed attractive.

One successful chicken breeder thought it made a lot of sense. The founder of Shaver Breeding Farms, who had produced the successful egg-laying Leghorn known as the Starcross 288, Canadian D. McQueen Shaver, decided in the 1950s that he could extend the culture of his chicken breeding operations to cattle. His efforts reveal that attempts to create biological locks via inbreeding and crossing in beef cattle production did not work. Shaver aimed his efforts at cow/calf operators and ultimately, through them, the feeders, not at the purebred cattle breeders. Shaver started with breeds that were uncommon in North America. He created a synthetic line, or composite breed (that he named Shaver Blend) based on crosses of the Lincoln Red, North Devon Red, and Maine Anjou breeds (he later added the genetics of five other breeds). He intended that bulls of this composite line be crossed in cow/calf operations on the different prevailing genetics in North American herds – namely, Holsteins, Angus, and Herefords. It appeared that he hoped to maintain markets for his bulls by instituting a biological lock, thereby following standard poultry breeding culture of the time. (Shaver never sold cows –"they are gold to us," his son explained. The holding back of females provided a biological lock.)[63] Cow/calf operators did not flock to Shaver Blend bulls, but not because they opposed the idea of working with hybrid vigour. It could be argued that it was a resistance to biological locks, which effectively would remove cow/calf operators from deciding what market stock they wanted to produce, that underlay their attitudes to Shaver Blend bulls. Cow/calf operators formed a critical part of the breeding arm of the beef industry and resisted biological locks, which hindered their control over breeding for feeder demand. The use of crossing inbred lines to protect the intellectual property invested in breeding would not work when confronted with a breeding sector within a livestock industry that would not accept the lock. It would only work when the stock went to the final producer and when that producer had relinquished any control over the genetics that went into the stock.

By the late 1960s cattlemen – that is, cow/calf operators and feeders – favoured crossing the newly imported breeds like Charolais and Limousin on traditionally-bred Angus/Herford "black baldies." This three-way cross designed to increase hybrid vigour, and a reinvigorated interest in structured crossbreeding in beef cattle, made Shaver believe that cow/calf operators would welcome bulls bred specifically to provide for hybrid vigour on traditional cows. All earlier crosses had relied on the continued existence of purebred breeds that the purebred breeders generated. Cow/calf operators had been linked with the feeders, cattle producers who used their stock, but also the purebred breeders. No such structure had existed in the chicken world. To some degree the very nature of the beast – namely, the relative expense of females, the length of time to raise stock, and the space needed to do so – dictated how the cattle breeders, cow/calf operators, and multipliers interacted with feeders as cattle producers.[64] Feeders might have been simply producers, but the cow/calf operators interacted with the elite breeders and even acted as breeders in their multiplication work. All sectors were affiliated in a way that was not true in the chicken world.

Failure to attract cattlemen to Shaver Blend bulls led to the development of the Shaver Blend as a "purebred" breed of cattle. By the 1990s both cows and bulls were sold to breeders working together under Shaver guidance. The cattle were pedigreed under a private registry with established regulations for entry, the Shaver Beef Blend International Registry. Shaver Blend breeders formed an association of their own that continued to emphasize the values of hybrid vigour their bulls offered when used on commercial cows. "Pedigreed Bulls are available only from D.McQ. Shaver Beef Breeding Farms or from Licensed Shaver

Herd owners," promotional literature stated.[65] "All Licensed herds of Shaver Beef blend have access to Shaver's breeding recommendation which ensure the most effective matings to maximize the benefits of heterosis," cattlemen were told.[66]

Pig Breeding and Hybrid Strategies

Hybrid breeding of pigs interested livestock geneticists from the time of Lush. Among the larger livestock species, pigs reproduce relatively quickly and in sufficient numbers compared to cattle and horses, thereby making the species appear to be a good candidate for hybrid breeding. Generally speaking, it was assumed that features of economic importance in pigs were controlled by many genes each having small effects – specifically, that inheritance functioned under Fisher's infinitesimal model. Inbreeding could collect these accumulative changes in a population; then, by crossbreeding, characteristics that could not be added accumulatively (that is, non-additive variance) could be achieved. This was the formula for hybrid breeding. In the case of pigs it became apparent that growth and carcass traits were moderately inherited, but they responded poorly to crossbreeding for hybrid vigour. At the same time, reproductive traits showed low heritabilities accompanied by high hybrid vigour resulting from crossbreeding. Selection was therefore aimed at lean growth traits, while crossbreeding was used to boost sow productivity. Different attributes were looked for in the sire and dam lines, as well.

The tendency in pig breeding was to work within breeds and subsequently cross them for terminal results, rather than creating synthetic lines for crossing. The low levels of hybrid vigour compounded the detrimental effects that resulted from the inbreeding used to establish such lines. The expense of creating inbred synthetic lines did not make sense. In the 1950s, within breed selection was done by the progeny test, but performance testing superseded this practice.[67] Quantitative genetics had, by the 1980s, provided accurate methods of estimating breeding value, which in turn allowed for a more nuanced approach to hybrid breeding. Pig breeding, then, even though it capitalized on terminal crosses, did not follow the formula of the hybrid corn breeding method. Any form of hybrid breeding, however, provided the biological locks necessary for a working form of patent, and that feature led to the privatization of pig breeding by companies.

International pig-breeding companies gained increasing shares in the pig-breeding industry in the later part of the twentieth century. Even in this new climate, purebred breeding remained central to any hybrid operation. The companies maintained superior lines of purebred animals that they used in crossbreeding programs. They also bought stock from purebred breeders to maintain those lines. That fact meant that independent purebred breeders continued to have a role in the whole structure. Selling gilts (young sows) was as important to the companies as selling boars, but they did not rely on packages the way the chicken-breeding industry did. The genetics of gilts and boars could be bought separately by a commercial pig farmer who then made his own choice as how to work with the hybrid vigour embedded in either sexed line. The pig-breeding companies also relied on multipliers to increase the volume of stock before it went on to commercial farms where the final product was generated, but the commercial farms still had some choice in how they would utilize the hybrid vigour from a pig-breeding company, especially if they did not contract with a processor. They remained, in other words, part of the chain of breeding, carrying forward the breeder's work and manipulating it as they thought best. As late as 1989 the pig-breeding companies only sold 28 per cent of commercial boars and 14 per cent of gilts bought by American farmers. Company dominance of the breeding industry has rapidly increased since that time.[68]

The stories of chicken, beef cattle, and pig breeding, as well as the entrance of genetics into the industries, illustrates that it is not always advancements in genetic knowledge that are critical to the animal-breeding process. Biology certainly plays a role – rate of reproduction, for example. The state of genetics at a particular point in time – in relation to the dynamics of a breeding industry at the same time – is vital. Hybrid breeding, from the classic geneticist point of view, did not work with beef cattle; significantly, this was due to the cultural structure of the breeding industry. Hybrid breeding did eventually come to pig breeding, but the culture within that industry helped shape the way it would work. Biology accounted for some of the continued reliance on purebred breeding as the foundation of the system, but the historical shape of the breeder/producer sectors discouraged the complete removal of pig producers from breeding decisions. It also seems clear that any concerted effort at hybrid breeding required the financial input of corporate organizations.

The success of geneticists in the hybrid breeding of chickens encouraged them to denigrate breeding systems that remained impervious to the method. They disparaged any reliance on purebred breeding and aimed remarks at beef cattle breeders in particular. Born in the nineteenth century, the purebred system had become the anathema of livestock geneticists, who believed by the 1940s that hybrid breeding was simply the scientific way to breed for improvement. As I.M. Lerner (a geneticist from the University of California) and H. Donald (co-director of Animal Breeding and Animal Genetics at Edinburgh University) wrote in the 1960s:

> The business of breeding [purebred] stock for sale is not just a matter of heredity, perhaps not even predominantly so. The devoted grooming, feeding and fitting, the propaganda about pedigrees and wins at fairs and shows, the dramatics of the auction ring, the trivialities of breed characteristics, and the good company of fellow breeders, constitute a vocation, not a genetic exercise ... In the larger classes of livestock, the biologically untrained men are still in charge and the differences in background, outlook, and in sophistication between them and the academic or research biologists, needless to say, leads to mutual antagonisms ... It is possible that geneticists underestimate many practical difficulties in their simplified theoretical concepts, but it is certain that many breeders refuse to understand and to credit geneticists having advanced their cause in any way.[69]

Traditional ways of breeding livestock, which were fundamentally related to purebred breeding, remained entrenched in other industries, partially because their industry structures were so different from that of the chicken industry. That situation would not change until fundamental shifts in the thinking of geneticists had taken place. The importance of livestock industry culture to the way genetics will interact with its breeding arm will become even more evident when later developments in genetics are matched against the organization of dairy cattle breeding.

Chapter Four

New Directions: Artificial Insemination Technology and Quantitative Genetics

One of the major developments within the story of artificial selection was the rise of artificial insemination, a technology that allowed for a concentrated focus on the productivity of breeding males in a vastly more comprehensive way than had been possible in earlier times. Of all the species that used artificial insemination (AI) cattle were particularly important. Because of overwhelming dairymen acceptance of AI, and the marginalized use of it in the beef cattle industry (as of 2003 less than 5 per cent of the world's beef cattle were artificially inseminated, largely as a result of the different structure of that industry), AI had been most significant for dairy bulls. I concentrate, therefore on the interconnection of AI with dairying from a practical and scientific breeder point of view. Dairy-farmer use of bulls, livestock geneticists, and AI organizations together would virtually revolutionize the way artificial selection operated in the production of dairy cows on a world scale.[1] One particularly important feature of dairy breeding in conjunction with AI and quantitative genetics, however, was the importance of purebred breeding within the process.

I begin the chapter by reviewing historical regulation put in place for breeding males, assessment of female productivity as a way to evaluate males, and the role of the state in data collecting structures – all of which played a role in shaping the modern AI industry. After a brief look at early developments in the AI industry before 1960, I describe attempts at cross/hybrid breeding in dairy cattle. I then turn to discuss how AI worked with dairymen, the purebred industry, and livestock geneticists in the breeding of dairy cows. I briefly examine the beef cattle and pig situations in order to compare them to patterns in the dairy industry.

**Background to the AI Industry: Early State Regulation
of Breeding Males**

The AI industry developed in conjunction with purebred breeding, but
the industry's structure evolved from a complicated background, part
of which reflected historical attitudes towards the public regulation of
breeding males. A move by the state to control the public service of
males began in Europe in the seventeenth century. The original focus
was on stallions, in efforts to improve the horse population for use by
the army. The first organized attempt by the state to regulate what stallions would breed was initiated by Louis XIV of France in 1665.[2] Peter
the Great encouraged the use of particular stallions in Russia as early
as 1680 by creating state studs, and Frederick the Great began a similar
approach in East Prussia in 1732.[3] By the nineteenth century, Italy and
the Austrian Empire also ran government studs that collected breeding stallions.[4] The situation in Britain contrasted sharply with activities
in continental Europe. The state did not establish government studs,
but tried to encourage the use of certain stallions, again primarily for
military purposes, via a different method. Over the late nineteenth
century a system of promoting privately owned stallions for general
stud service developed. In 1887 a Commission on Horse Breeding was
founded to encourage the better breeding of horses across the nation.
Good breeding stallions should be supported as such through remuneration provided by the state, the commissioners thought. The question
was, though: how to identify these animals? Purebred breeding clearly
defined quality, as far as the commissioners were concerned, but that
was only part of the story. How, in effect, did the mysterious principles
or forces within the animal cause weak legs and poor breathing? Were
these problems more important than purebred breeding?[5] The commission decided to investigate the question of heredity and horse breeding
generally, and in 1889 asked the Royal College of Veterinary Surgeons
to define defects that resulted from inheritance.[6] What the college wrote
back provided the foundations for what would be accepted as hereditary equine unsoundness for at least the next thirty years, not just in
Britain, but also in all of the United States and Canada.

Of the 2,000 questionnaires sent out to vets, only 300 responded
with information. The endorsement of those who replied was nebu-
lous. Many did not believe the profession had a right to an opinion on
the subject of heredity in horses.[7] Veterinarians tended to believe that
the only way to understand heredity was through experience based

on observation. Such conflicting attitudes from veterinarians did not stop the college from dealing with the Horse Commission's query. The council of the college replied by sending the commission a list of defects that qualified as being hereditary in nature. The commission presented the list, dealing with the defects separately, to each witness – all of whom were either veterinarians or horse breeders – at the 1889 hearings. The confusion that arose from the complicated discussions that took place at the 1889 hearings[8] did not deter the faith of horsemen everywhere, even after the release of an enormous report in 1890 that in fact concluded nothing, that experts had answered the problem of heredity and unsoundness. By the early twentieth century better horse breeding came to be more closely linked with non-military equine uses, but the importance of both purebred breeding and hereditary soundness stayed in place. When a law was finally passed in Britain in 1918 requiring stallions to be inspected for soundness before being certified by a license, horses were judged on the basis of the list presented in the 1890 report.[9] The same was true of all stallion legislation passed in both the United States and Canada in the early twentieth century, where the need for better horses also related to agricultural and industrial uses. Quality, purebred breeding, and the ability to detect the genetic inheritance of certain characteristics were firmly linked by laws that licensed stallions to stand at public stud.[10] A very complicated entwining of attitudes to breeding practices and the assessment of breeding potential was embedded in all stallion legislation.

The idea that genetic improvement of other farm animal species could be promoted by state certification of privately owned breeding males quickly took hold in Britain at about the same time. Attempts to link improvement directly to hereditary laws, as was the case in stallion legislation, did not take place. Instead agricultural experts and government officials defined improvement simply as purebred breeding and hoped government schemes would promote the spread of purebred influence on the general herds and flocks. The Livestock Improvement Scheme launched by the British government in 1914, for example, was designed to help farmers have access to purebred (believed to be better) bulls. By the 1920s purebred breed associations actively called for moderations in the Scheme to make it match the 1918 stallion act, even without a report like the 1890 British Horse Commission's to fall back on, which stipulated that animals at public stud had to be licensed.[11] The Improvement of Livestock (Licensing of Bulls) Act was passed on the recommendation of the purebred breeders' organization, the National

Cattle Breeders' Association, and the at least tacit agreement of the National Farmers' Union, a body representing the ordinary farmer, but not until 1931. The bull licensing system stated that farmers had to have bulls over ten months of age inspected. Disqualification meant the bulls had to be slaughtered or castrated.[12]

The regulation of cattle breeding under the Scheme laid the groundwork for government efforts at orchestrating breeding in Britain and, while it promoted the interests of the purebred breeders and labelled purebred stock as superior, it acclimatized the farm public to state intervention in the breeding of livestock.[13] It also encouraged the idea that a more generalized use of particular males led to better artificial selection practices. It did not, however, lead to farmer organizations acquiring bulls, a factor that would be significant within the framework of AI structure in that country. No cooperative groups were formed in England and Wales to collectively own bulls, as was the case in some European countries and North America. The fact that no cooperative-type bull organizations existed in Britain before the advent of AI meant that the shape of the British AI industry would differ from the industry in North America and in other European countries.[14]

State encouragement, rather than legislative action, promoted the use of purebred males in the same period in North America. The movement never had the teeth that stallion legislation did in North America, however, and encouragement rather than force (as in the British case) dominated North American cattle improvement activities. An American campaign against "scrub" bulls was initially triggered by concerns about quality of beef cattle in relation to stock being raised in Argentina.[15] In this case attempts to shift breeding practices within the United States were triggered by the dynamics of an industry in a distant land. In effect, the periphery directed activities in the metropole. American farmers were encouraged to use purebred bulls through a number of financial incentives. Cooperative bull associations, in which farmers owned a bull jointly, also developed, with the first being established in Michigan in 1906.[16] In Canada, a system of loaning purebred bulls to farmer associations across the country was in place by 1909.[17] Regional activity was also strong. In 1920 the purebred breeder organizations in Ontario, Canada, under the Ontario Cattle Breeders' Association and with the financial aid of the Dominion and provincial governments, set out to remove bulls deemed poor (unusually of grade quality) in certain counties under its Better Bull Campaign. This move could only be accomplished, though, with the cooperation of farmers owning so-called scrub bulls.[18] No legislation was advocated (or indeed put in place) to force farmers to abandon their non-purebred bulls. Increased federal government financial support for the provincial plan came in the form of the Sire Loaning Policy and the Sire Purchasing Policy, both designed to help farmers access purebred superior bulls inexpensively.[19]

Assessment of Males via Female Production: Early Data Collection on Milking Cows and Artificial Selection

Structures developed to evaluate females in the dairy industry proved to be just as critical to the functioning of the AI industry as attitudes to the regulation of breeding dairy males. In fact understanding female productivity would ultimately be the primary way of assessing the quality of breeding males. Understanding how historical data collection on milk yields of dairy cows developed, then, is critical to appreciating how the AI industry evolved in relation to the dairy breeding industry. By the latter half of the nineteenth century, attempts to assess the hereditary worth of cows on the basis of their milking ability became increasingly common in Europe. Interest in cows for dairy purposes, as opposed to the beefing quality of cows, was especially strong in northern European countries like the Netherlands and Switzerland, and far stronger than in North America.[20] Only a small dedicated group of North American dairymen were as interested in the hereditary aspects of superior dairy cows as were their north European counterparts, and virtually all these were purebred breeders.[21] By the 1880s, testing cows for milk yields was done in both Europe and North America when animals at shows were milked in competitions.[22] The relationship of purebred breeding to milk testing was intimate, and when American breeder associations began requiring milk tests for animals entered in the herdbooks, that linkage became increasingly strong. In 1880, even before a Holstein breeder association existed in the United States, Soloman Hoxie developed the idea of an Advanced Registry for the Holstein breed, with inspection for type and testing for milk production requirements.[23] By 1886 testing was needed for entry into the Holstein book, in 1900 to the Guernsey book, in 1902 to the Ayrshire and Brown Swiss books, and in 1903 to the Jersey book.[24] The Holstein Friesian Association of Canada established a system of Advanced Registry in 1901 for cows with superior milk yields with its Record of Merit (ROM). In 1905, at the request of several dairy breed associations, the Canadian department of agriculture initiated a state-endorsed test. It ran ten or twelve months of the year (one of the earliest tests in the world to run this length) and was known as the Record of Performance (ROP).[25]

Milk testing continued to be an integral part of North American pedigree registration structure and breed promotion, and played an increasingly prominent role in the breeding and marketing of purebred dairy animals. In 1926 the American Ayrshire Association created a herd test known as the Herd Improvement Registry (HIR); it required the enrollment of all registered cows within a herd. However, the stipulation that low records could be canceled or concealed if the animal was dropped off registry in the herdbook weakened the usefulness of the test, by distorting results.[26] Other dairy breed associations adopted the structure within a few years, each establishing its own rules for the program. In 1929 the Holstein association established a herd classification standard, by which the cows could be assessed for their physical type by a trained person. Jerseys adopted a system in 1932. From the beginning, breeders were just as interested in conformation as milk production. Classification standards had become increasingly similar across breeds.[27] In 1940 the purebred dairy breeders formed the Purebred Dairy Cattle Association to further coordinate their efforts.[28]

The invention of the Babcock test in 1890 for fat production in milk greatly facilitated the keeping of milk records of cows, because before about 1915, butter fat went into cheese and butter, the most important products of milk. The Babcock test revolutionized milk testing by introducing a way to collect data outside the structures directed and run by the purebred breed associations. Organized testing for milk yields by cow testing associations, run by dairymen outside breed organization, originated in Denmark.[29] Such organizations spread rapidly to other countries. The system arrived Germany in 1895, for example, when the first incorporated society to test cows was founded by thirteen breeders in northern Germany.[30] The Netherlands adopted milk testing along Danish lines in the 1890s as well. Friesland initiated the movement, and practiced milk recording over the years more than any other Dutch province.[31] In 1906 a Danish immigrant, Helmer Rabild, started the first cow-testing association using the Danish model in Michigan. The US Department of Agriculture became interested and, in 1909, hired Rabild to promote cow-testing associations federally. The movement soon gained state support after its initial start. Dairymen formed Cow Testing Associations, which became known as Dairy Herd Improvement Associations (DHIA) after 1927. By 1929 cow-testing associations had sprung up in all forty-eight states and were supervised under state agricultural departments.[32] Ordinary dairymen, regardless of whether they used purebred cows, were interested in the results. The move of the general dairy industry producers into the process of milk data collection ultimately ushered them into the process of breeding, for they would already be providing the volume of documentation needed to quantify results of breeding bulls. It should be pointed out, though, that the vast majority of dairymen were not part of any strategy to collect data on milk yields at this point in time. The purebred breeders only represented a tiny fraction of dairy farmers in the United States, so their input was small. They also tended to use the data they collected to promote sales, not progeny test bulls.

The idea that milk data could be used to identify good bulls had been theorized by Nils Hansson of Norway as early as 1913. He suggested that, when the production of the daughters of a sire were compared with that of their dams, it might be possible to see how much of a daughter's milking ability had been inherited from their sire. A bull index works on the assumption that the level of inheritance of a daughter is halfway between that of her sire and her dam, and that by knowing the milk producing level of dam and daughter, the sire's transmitting ability could be calculated.[33] By the mid-1920s, it had become apparent that the milk data records were more useful in identifying good bulls for breeding than identifying cows. In the United States, DHIA records were used in an effort to prove the value of sires as early as 1925, generally by looking at the records of both daughters and dams. Purebred breeders followed suit. Progeny testing of bulls was largely in place in the HIR structures by 1930, as a result of the infiltration of biometric approaches to data collection.[34] A national sire-proving program, in which daughters of each sire were compared with their dams (known as bull-indexing), was initiated in 1936.[35] Milk data was firmly linked to sire progeny testing by this time, although its effectiveness would be much improved on over the next seventy years.

Dairy Cattle Breeding Experiments: Crossbreeding versus Purebred Breeding

The ongoing importance of purebred breeding in the dairy industry did not mean that crossbreeding in dairy cattle received no attention over the years. A general concern with hybridizing that so marked the early years of Mendelian experiments, and later had such an explosive effect on chicken breeding, made some people in both Europe and the

United States wonder if crossbreeding, and even inbreeding and cross-ing inbred lines, would improve the productivity of dairy cows. In 1906 a crossbreeding experiment involving the Jersey and Red Danish milk breeds was undertaken on a large scale on Count Ahlefeldt Laurvigen's estates on the island of Langeland in Denmark in order to study the inheritance of fat in milk, not to look for hybrid vigour. This experi-ment, which ran for twenty years, provided little conclusive evidence. In the United States T.J. Bowlker began a crossbreeding experiment in 1911 using Holsteins and Guernseys at his farm in Massachusetts. While he died in 1917, his wife continued running the work until 1919 when the herd was sold to the University of Illinois. Unsurprisingly, it was found that the first cross was superior to the second. Raymond Pearl at the Maine Station began a crossbreeding experiment in 1913, attempting to establish whether volume of milk or milk fat dominated patterns of inheritance. In 1912 L.J. Cole's experiment, which ran from 1912 until 1933, involved crossing of Jerseys and Holsteins with Angus cattle, and the data suggested negative hybrid effects. The US Bureau of Dairy Industry ran a crossbreeding experiment in 1939 and in 1946. The data showed that the breed crosses exceeded their purebred dams in both milk and fat yields, a fact that caused considerable discomfort in the purebred dairy-breeding world. A much smaller experiment in crossbreeding began at the South Carolina Experiment station at Clem-son in the early 1940s, which indicated increased milk yield in cross-breds over purebreds.

A major experiment, which began in 1949 and ran until 1969, took place in Illinois. It was designed to assess how much crossbreeding could account for milk yield and components, and to compare vari-ous economically important features in crossbreds and purebreds. The experiment showed conclusively that, on the basis of income per cow per lactation, crossbreds exceeded purebreds by 15 per cent. On the basis of income produced per cow per year, crossbreds also exceeded pure-breds.[36] Other experiments confirmed a real potential for crossbreeding in dairy cattle.[37] The inbreeding/line-crossing method was also tested. Intense inbreeding of lines within a breed of dairy cattle and cross-ing those inbred lines was done to see if effects were similar to those found in chickens bred that way. Results proved to be disappointing. An assessment by Lush of the records of the Dairy Herd Improvement Association indicated that there was no evidence of sufficient hybrid vigour to warrant inbreeding/crossing programs. A prolonged inbreed-ing/line-crossing experiment using Holsteins, conducted in California and running from 1928 into the 1950s, suggested some hybrid vigour response.[38] A review done in 1965 on the value of inbreeding/line-crossing did not support that conclusion.[39] Another study, started in 1948 and reported on in 1979, assessed the performance of six inbred lines of Holsteins and the effects of crossing those inbred lines. The results revealed that there was no increase in milk-yield traits, but that there was some evidence of better reproductive performance.[40]

In the practical world of North American dairy cattle, breeders emphasized true line breeding and demonstrated little interest in pursuing the hybrid vigour made evident by crossbreeding experiments, or in inbreeding/line-crossing for hybrid vigour. Breeding proceeded via purebred breeding for true breeding lines, and remained impervious to attitudes emanating out of science that promoted breeding for terminal crosses. Eventually, purebred breeding led to cattle superior to those created by any crossbreeding program. It was evident by 1980, for example, that crossbreeding using Holsteins showed no improved production over pure Holsteins. They exceeded crossbreds for general performance by 10 per cent.[41]

The Purebred Industry and Cow Improvement in North America before Artificial Insemination

Advances in milk data collection and involvement of general dairymen in the procedure via DHIA structures did not undermine the strength of the purebred breeders in the dairy industry. Neither did the influence of government and agriculture colleges on dairying affairs, both of which had developed considerably after the late nineteenth century. In 1906 dairy instructors and investigators (of college rank) from the agricultural colleges, experiment stations, and dairy divisions of the United States and Canada (H.H. Dean of Canada was one of the primary members) founded the American Dairy Science Association. The American Dairy Science Association tried to coordinate the work of the dairy breed associations in both countries with that of the agricultural colleges and experimental stations. The association adopted uniform production testing procedures for all official tests run at state institutions such as experiment stations. In 1917 it began publishing the *Journal of Dairy Science*.[42] The purebred breed associations continued to have a strong voice throughout these years, using data collection as promotional material. Increasingly scientists, the United States Department of Agriculture (USDA), and the Department of Agriculture in Canada chafed over the control the breeders had over the functioning of the industry. A willingness of breeders to collaborate with advancing science at least to some degree, however, characterized the purebred dairy industry in these early years; the same was not true of the standardbred/purebred chicken industry. The American Poultry Association did not work closely with the Poultry Science Association (founded in 1908 by American and Canadian poultrymen). Still, by the 1930s many remained deeply concerned over the continuing power the dairy purebred industry had over dairying generally.

In 1936 the USDA published an exhaustive survey of farm-animal breeding in the United States, as well as comments on dairy breeding under the various milk testing schemes and sire testing in relation to them, which provided a good summary of the U.S. dairying data collection situation before the advent of AI. The report emphasized the relative failure of record keeping for dairy cattle, stating that in spite of the work done by educational organizations very few farmers collected data, and when they did it was only done to single out poor-producing cows that could be culled. Dairy herd-improvement associations, the report continued, argued that only one-third of cows produced enough milk to make them profitable to farmers, while another third provided enough to break even, and the last third were so poor that they cost their owners money. It should be remembered too that less than 2 per cent of the nation's dairy cows were listed in any herd improvement association, and that these were probably the best producers, the associations stated. The USDA report then turned to the quality of breeding males in the dairy cattle population, saying that:

Of the 4,302 sires that have been analyzed in the 707 herds, 48 percent had too few daughters with records to permit an appraisal of their inheritance; 7.9 percent were rated excellent; 8.4 percent were rated as good; 10.6 percent were rated as fair; and 3.5 percent whose daughters were very nearly but not quite as good as their dams were rated as undeterminable; and 21.5 percent were rated as poor. Thus, under the practice of selecting sires that has been followed in the better herds, only about one-half of the sires were used extensively enough to determine with any accuracy what inheritance they possessed, and only about one in four of the total number possessed an inheritance that could be gauged positively as good enough for improving the germ plasma of the herds in which they were tested.[43]

The report emphasized the powerful role of the breed associations in data collection on milk yields – and on sire testing via those milk yields. The publishing of good milk yields was a primary way of promoting breeds and herds within breeds, and breed associations pressed their members to collect data with this in mind. But only good results were made public.[44] They enhanced pedigrees and therefore promoted sales. Breed associations played into the problem by only reporting animals that met a minimum requirement for entry into the herd book, "preferring to ignore the failures." The USDA report concluded that "This shortcoming proved to be a severe limitation on the use of records for breeding studies. If a half or a third of the progeny of a given sire appeared in the test lists, one could only speculate as to whether the other died, were sold, failed to meet the requirements, or did not have an opportunity to prove their worth."[45] The advent of AI, used in a widespread way, would solve a lot of these problems. But the AI industry relied on the already existing structures to collect data, and to generate breeding males. It simply made the whole system work better, and also allowed for the utilization of quantitative genetic theory.

The Rise of the Artificial Insemination Technology and the AI Industry

Interest in AI has an ancient history. Practiced by the Arabs on horses as early as the fourteenth century, modern research on the technology of AI was initiated in Russia in 1899 when E.I. Ivanov managed to get better conception rates with horses on government breeding farms. In 1909 the Russian ministry of agriculture set up a laboratory to study the physiology of reproduction and to train technicians in techniques of AI.[46] AI research continued in Russia, and was also carried on in Britain and the United States after that time. While the potential of AI to extend the influence of superior males in a breeding population was obvious to all in this early period, it would be impractical until artificial service conception rates could at least match those resulting from natural service, which did not occur until the 1930s.[47] The first AI cooperative association for cattle breeding was established in Denmark in 1936 by Edward Sorensen of the Royal Agriculture and Veterinary College of Copenhagen and others.[48] In 1938 the first American farmer-owned cooperative AI breeding association was established in New Jersey, after E.J. Perry of Rutgers University had visited Denmark and saw how it could be done.[49] By 1946 some eighty-four associations dealing in bull semen existed in the United States. Often bull cooperatives developed into AI cooperatives. Farmer-owned AI cooperatives gradually replaced the older bull associations, and were largely dominated by dairymen. (Purebred beef breeders took a dim view of AI, believing in pasture breeding but probably more influenced by the idea that AI might reduce their bull sales.[50] Purebred dairy breeders also feared AI for bull-market reasons, but the more developed system in place to progeny test dairy bulls, and the rising pattern of joint ownership of bulls, already undermined their ability to sell bulls in the volume that beef breeders could.) The number of AI organizations peaked in 1950 at ninety-seven.[51] Purebred dairy associations agreed together to set uniform requirements for the artificial insemination of dairy cattle, and in 1946, along with other purebred breeders of livestock, formed the National Association of Animal (Artificial) Breeders (NAAB). As early as 1947 the Canadian organization supporting artificial breeders worked closely with NAAB. Antibiotics were added to AI semen in 1946 and 1949.[52]

In 1952 a freezing technique for semen was invented in Britain. Trials with frozen cattle semen commenced in North America immediately.[53] Cattle biology lent itself to frozen semen in a way that was not true of other livestock: chicken semen freezes so poorly that only fresh semen is used. The freezing of horse, pig, sheep, and goat semen works only to a limited degree.[54] This factor alone encouraged the dominance of cattle in AI organizations. The first large AI organization to use only frozen semen was the Waterloo Cattle Breeding Association in Ontario, Canada, in 1954. Use of frozen semen had a pronounced effect on the AI programs, which increasingly focused on dairy bulls, by facilitating shipping over long distances; most AI operations became nationwide after 1955. The cost of converting to frozen semen put many small dairy farmer-owned cooperatives out of business. They were forced to merge or consolidate in order to survive.[55] By 1955 some 30 per cent of registered American dairy cows were artificially inseminated with frozen semen.[56] (By 1965 almost all cattle semen in the U.S. was frozen.)[57] The importance of frozen semen cannot be overemphasized. It allowed for the use of dairy bulls under enormously different conditions, which meant non-genetic factors (such as environmental influences) could be neutralized when evaluating the breeding potential of the animals.

Frozen semen did not impact the AI industry in other countries as rapidly as it did in North America. This situation would play a role in how extensively newer ideas in livestock genetics would be applied to dairy cattle breeding. The Netherlands' experience is a particularly interesting example because of the importance of the Friesian breed (which evolved into the Holstein) to the world's dairy industry. The first Dutch AI organization was established in 1939 to help small farmers fight venereal disease. After 1945 the number of AI organizations began to grow, but the breeders of Friesians generally opposed the movement because AI cut into bull sales. In reality in Friesland itself, AI bull calves of the breed were denied registration in the herd book until 1946. By 1951 the number of AI organizations had reached its peak in the Netherlands – at 160. Numbers began to contract, with 131 functioning in 1960, and by 1969 some 93. Contraction did not seem to result from – or encourage the use of – frozen semen, as was the case in North America where size proved to be a factor in conversion. Frozen semen use was marginal in the Netherlands for a considerable length of time. By 1964 only 1 per cent of inseminations in the country resulted from frozen semen, and by the end of the 1960s only 20 per cent of cows were inseminated that way. AI companies bought champion show bulls of a young age, which hampered proper progeny testing until the late 1960s, and were known to reject well-tested bulls with good progeny if their type was not right. In 1960 progeny-tested bulls sired only 35 per cent of registered dairy cows. By 1965 some 55 per cent of cows were sired by tested bulls. Many still did not think a good bull was necessarily one that performed well on the progeny test.[58]

Before the advent of frozen cattle semen, the selection of dairy bulls in North America to be used for AI varied little from pre-AI days (similar to the situation in the Netherlands until the late 1960s), and AI's main advantage, as in the Netherlands, was control of the spread of venereal disease.[59] Bulls were, for one thing, used only locally because the semen could not be moved any substantial distance. Committees comprised of dairy farmers made the decision of bull selection, and they usually chose purebred bulls from renowned breeders, often animals that had been tested under breed association standards. Bulls normally only had a few daughters and usually daughters only within one herd. Such bulls were often old. Local breeders, who sometimes were directors or patrons of the AI cooperative, put pressure on farmers to buy their own bulls for the unit. The AI organizations used the same criteria to choose breeding bulls that the breeders themselves did: limited progeny testing, pedigrees, and show results. Bulls chosen in this fashion might work well in herds where good management prevailed, but the results were often less than satisfactory in herds where management was poor, thereby masking how much genetics had contributed to the good results.[60] Evidently the restricted testing of elite bulls had masked their genetic worth.[61]

Artificial Insemination and New Directions in Livestock Genetics

In the mid 1940s geneticists in both Europe and North America began to discuss the implications of AI. Because it extended a bull's breeding territory, in 1944 the American geneticists, G.E. Dickerson and L.N. Hazel, wondered about its potential to make progeny testing more effective and comprehensive.

The same idea attracted the interest of geneticists at Edinburgh University in Scotland. The AI situation in Britain provided them with particularly good, unbiased data because of the centralization of the industry. The British industry was coordinated under the Milk Marketing Board, which fundamentally ran all AI units in the country after 1945. Bulls were moved around the country to ensure some rotational breeding and to reduce dangers of inbreeding.[62] By 1951 the British AI industry was based on a single structure, making it the largest in the world. The lack of farm cooperative AI organizations, and the coordinated running of the AI industry by the Milk Marketing Board, provided the geneticists with data collected across the country, a situation that allowed for the neutralization of inbuilt variable factors and, by extension, for national progeny testing.[63] Geneticists could progeny test bulls on the basis of daughters in a wide range of circumstances and emanating out of vastly different herds.

It was no accident either that work within Britain using the data from AI would be initiated at Edinburgh, which had become an important genetics centre with respect to livestock after the Second World War. Britain had fallen far behind the United States with respect to livestock genetics over the 1930s, in spite of the importance of R.A. Fisher's work to the development of that branch of science, and his presence in a genetics department at Cambridge. Lush's ideas were virtually unknown. Genetics generally languished in Britain before the war. As the geneticist W.C. Hill reminisced, "There were few geneticists of any sort, and they tended to be regarded as eccentrics pursuing an incomprehensive subject."[64] There were only three genetics departments: one at University College in London under J.B.S. Haldane, at Cambridge under R.A. Fisher, and in Edinburgh under A.E. Crew. That situation changed rapidly after the war. In 1945 the Agricultural Research Council (ARC) arm of the government had set up the Animal Breeding and Genetics Research Organization (ABGRO), designed to make research work in agricultural genetics more robust. By 1947 Edinburgh, where an animal genetics department had existed since 1919, had become a major centre for livestock breeding research.

The Work of Alan Robertson: Quantitative Genetics Applied to Dairy Cattle

Alan Robertson was one of the first to join the new and enlarged Edinburgh genetics team, and he quickly became acquainted with trends in livestock genetics that had developed in the United States. Robertson spent nine months with Sewall Wright and J.L. Lush. In 1948 I.M. Lerner (a geneticist who had first done research in British Columbia, Canada, and subsequently moved to California) visited Edinburgh, and he further enlightened Robertson and other British scientists on the meaning of Wright's path coefficient. Wright himself spent a sabbatical leave at Edinburgh in 1949.[65] Even with the input of American ideas concerning livestock genetics, the Fisher approach would ultimately influence the direction livestock genetics took at Edinburgh with respect to artificial selection strategy more than that of Wright/Lush. And the Fisher thrust fundamentally emphasized the study of metric quantitative traits within a population; it did not approach the effects that hybrid breeding had on such metric traits. Fisher's theories fit well with breeding in lines that produce truly as opposed to inbreeding/line crossing for hybrid vigour. Lush's ideas emanated out of Wright thinking, but Robertson's ideas emanated out of Fisher thinking – even though there was an underlying intersection of Wright/Fisher theories in the approaches of both Lush and Robertson. The new emphasis on Fisherian attitudes would change the course of livestock genetics.

Robertson and J. Rendel began to study the artificial selection implications of AI in dairy cattle improvement in 1946. They postulated that AI would help them pinpoint genetically superior bulls that could, in turn, be used to increase the breeding quality of bulls in the next generation. Bulls could be selected to sire the next generation of breeding bulls that would be progeny tested just as their fathers had been – namely, from daughters across different herds.[66] As F.W. Nicholas put it: "If a cycle could be established in which selected bulls were then used to breed the next generation of young bulls, which in turn would be progeny tested in the same way as their fathers had been, then continued genetic progress would result.

This realization gave rise to the first scientifically-based proposal for the use of AI in an animal improvement program, and to the development of the contemporary comparison method of progeny testing."[67]

Robertson went on to quantitatively study artificial selection strategies for improvement over generations along four separate paths: bulls to breed better bulls, bulls to breed better cows, cows to breed better bulls, and cows to breed better cows.[68] Together with Rendel in their important 1950 paper, he demonstrated quantitatively the theoretical merits of progeny testing.[69] Robertson and Rendel were aware that widespread use of AI could potentially spread deleterious recessive genes through inbreeding, which would result from overuse of popular sires that had done well in progeny tests. AI could, however, offer a wider variety of bulls throughout any location, thereby counteracting the spread of dangerous recessives and reducing the likelihood of inbreeding. Robertson and Rendel suggested that AI units should run their own progeny testing, and that before a bull was widely used in service he be mated to at least twenty of his own daughters to see if he carried deleterious recessives.[70] Ultimately, all AI companies in the world would use the progeny testing system they designed.

The work of Robertson made it clear that data should be as unbiased as possible, and the collection of it should be as wide as possible and under similar circumstances. Breed associations in the United States began to work more effectively with centralized government organizations after the input of geneticists made it clear how important data on as large a scale as possible was. The breed-association-testing methods soon became intertwined with the DHIA. Most of the herds enrolled in an HIR program had also been part of a Dairy Herd Improvement Association, but the dual system of recording had gone on for a number of years because the breed associations wanted to maintain a separate form of record keeping. The work of Robertson and others convinced the North American associations to drop their record keeping, and between 1955 and 1965 milk-yield recording was gradually taken over completely by the Dairy Herd Improvement Associations.[71]

Quantitative Genetics and Statistics: The Work of L.N. Hazel and C.R. Henderson

Statistics used to estimate potential breeding value of an individual animal were by nature complex. Parents pass on their genes, but not their own genotype, which, because of segregation, is freshly recreated anew in the next generation. It is the average effects of a parent's genes that will determine the genotype of the progeny. The breeding value of any animal has to be estimated by the mean value of the progeny resulting from its use in breeding.[72] It was the ability to predict the breeding value of males – specifically, methods of sire indexing – that attracted particular attention within that framework. The difficulty in devising the statistics to make sire indexing work was evident as early as the 1930s. Significant innovations with respect to the problem took place in the United States at Iowa State, with the work of L.N. Hazel in the 1940s, and that of C.R. Henderson from the late 1940s into the 1960s.

In 1943 L.N. Hazel recognized that he faced a daunting question and one that had been recognized by breeders since the beginning of recorded breeding history. He stated:

> The idea of a yardstick or selection index for measuring the net merit of breeding animals is probably almost as old as the art of animal breeding itself. In practice several or many traits influence an animal's practical value, although they do so in varying degrees. The information regarding different traits may vary widely, some coming from an animal's relatives and some from the animal's own performance for traits which are expressed once or repeatedly during its lifetime ... These factors make selection a complicated and uncertain procedure; in addition fluctuating, vague, and sometimes erroneous ideals often cause the improvement resulting to be much less than could be achieved if these obstacles were overcome ... From the studies of heritability which have been made for economic traits in different farm animals, it seems that the best indexes which can be constructed will be far from perfect. The confusing effects of environment ... in masking genotypes cause the progress in the present case to be less than half of what might be made if genotypes could be recognized precisely.[73]

Hazel has been credited with being the first to clarify the issue of multiple trait selection, by defining genetic and environmental correlations and suggesting how they could be estimated. He was, however, working with linear and matrix algebra, areas of mathematics that could only be dealt with by very specialized statisticians. Sire indexing was not yet possible using Hazel's work.[74]

Selection indexing clearly had potential – as Hazel made evident – but there were still many complications to be worked out before a really comprehensive approach to the problem could be devised.[75] As Lush explained in 1951:

> The information needed most for constructing selection indexes is, first of all, a description of the ideal and a means of measuring each character included. Second is the heritability of each character; that is, the amount of genetic and environmental variation in it. Third are the genetic and environmental correlations between the various characters ... It is still too soon to be sure how the use of selection indexes will actually affect animal breeding practice, but the indexes worked out have already reconciled several apparently conflicting ideas. With this development of selection indexes and related aspects of population genetics, we are becoming able to compare the efficiencies of two or more breeding plans which in part may be mutually exclusive ... But at any event the quantizing of breeding plans is beginning to replace fairy tales and wishful thinking.[76]

It was C.R. Henderson who made the breakthrough. Hazel's selection index methods were replaced by Henderson's mixed model methodology, under what would be known as BLUP, or Best Linear Unbiased Prediction. Henderson noted that "we do not yet have any suitable tool for evaluating breeding values for milk production except milk production records on the individual and its relatives," and that accurate evaluation of a dairy bull is possible only through a progeny test.[77] The advent of frozen semen, and the international trading of semen, however, had changed the guidelines that should have been used to structure progeny testing. He explained: "Varying selection practices among artificial insemination (AI) studs and different goals among dairymen, who have the opportunity to select any one of many hundreds of bulls, have destroyed the essentially random distribution across herds of bulls within the area of operation of certain studs that existed at one time. This past random distribution made sire evaluation, at least within regions, relatively accurate when simple methods were used. The situation now is different."[78]

He showed that the accuracy of AI progeny testing of bulls had to be assessed on the basis of certain variables: the number of tested daughters from a bull; the number of herds in which daughters were tested; contemporary comparisons in evaluating the milk production of the progeny of AI bulls; as well as corrections of the production records for age, year, season of calving, and the herd in which the record was made.[79] Henderson explained that "in order to account for genetic trend and for different selection policies of AI studs and of dairymen's choice of sires for natural service, bulls to be evaluated [had to] be divided into groups and the evaluation [had to] be the sum of the estimate of the group and the selection index type evaluation of the deviation of the individual sire from this mean."[80] Henderson's work, which began with his doctoral thesis completed in 1948 at Iowa State, was more fully developed at Cornell throughout the 1950s and 1960s.[81] His most elaborate explanations of BLUP emerged in the 1970s.[82] By this point in time, computers had advanced enough that BLUP could be more effectively applied.

Quantitative versus Population Genetics

It is clear that the development of quantitative genetics in relation to livestock breeding, through the work of scientists like Robertson and Henderson, was initiated by the work of J.L. Lush. But is it fair to describe Lush himself, the founder of theoretical livestock genetics, as a quantitative geneticist? This question might seem minor, but answering it helps elucidate the development of quantitative genetics, and distinct from that field, the development of livestock genetics. They should not be seen as one and the same thing. It seems clear that Lush theories fit into the approaches of quantitative genetics, and in later times, his work would be described as being early quantitative genetics. In the 1980s W. Hill, for example, stated that "the main pioneer in the application of quantitative genetics to animal breeding was Jay L. Lush of Iowa State University."[83] In the same period, the geneticist, J. Crow, also referred to the Lush school as being one of quantitative genetics.[84] In the late 1990s, Lynch and Walsh stated that quantitative genetics was founded independently by Wright and Fisher (both played a critical role in Lush's thinking), and served as the theoretical basis of most animal breeding programs for over half a century; that is, until advancements in molecular genetics changed the picture at the end of the twentieth century.[85]

Comments made after 1970 about the nature of Lush's work, and of animal genetics from 1930 to nearly 1950, seem to come from the perspective of hindsight. At the time that Lush was teaching, and Wright was formulating his theories that would become so important to Lush, theoretical genetics as applied to livestock was viewed as a form of population genetics. Certainly Lush saw it that way into the 1950s.[86] He looked for variation of characteristics, but he did so from the perspective of studying them across, as much as within, population groups.

Other geneticists working with livestock tended to share Lush's view that his field was an aspect of population genetics. I.M. Lerner, writing in the late 1940s, described the work of Wright, as well as the Lush school, as being population genetics.[87] In 1969, F. Pirchner, in his *Population Genetics in Animal Breeding*, called standard hybrid breeding a form of population genetics.[88] It is worth noting what Wright himself, as late as 1977, said that animal breeding and genetic shifts within groups were caused by certain selection practices. Wright saw, in his assessment of colour coats of Shorthorns, for example, not quantitative genetics, but rather a study in population. Populations could become altered qualitatively as to colour coat.[89] "The results are of interest from the standpoint of breed history in showing that a gene frequency may change enormously in a single generation ... as a by-product of the wholesale use of sires from a particular strain," Wright wrote, and added, "The chief interest in the inbreeding histories of various breeds of livestock is less in the inbreeding as such than in the light thrown on the process of improvement of domestic animals."[90] R.C. Lewontin, a geneticist trained at Harvard, offered a way to distinguish a population geneticist approach from that of a quantitative geneticist. For Lewontin the deciding factor that divided population from quantitative genetics was an emphasis on inbreeding, which he associated with the Wright/Lush school. He stated: "The science of inbreeding in animals and plants is clearly dominated by the Wrightian scheme, because inbreeding always deals with the probability of homozygosity [or uniformity] – that is, with gene frequency. But in dealing with progress under selection for continuous variation we depend on Fisher's scheme that talks about rates of change of means and of variance."[91] Perhaps one reason Hill believed that very slow advancements were being made in quantitative genetics from the 1920s until about 1950 was that advancements in theoretical animal breeding studies in the period fit into the Lewontin definition of population rather than quantitative genetics.[92] Most research on artificial selection strategies to improve livestock relied on some system of inbreeding. The shift to an emphasis on quantitative genetics as applied to experimental animal breeding would come after 1950, and would be heavily influenced by work done at Edinburgh University under Alan Robertson.

By the 1970s livestock geneticists went to considerable effort to define their field as quantitative genetics and to distinguish it from population genetics. At the first quantitative geneticist conference, held at Ames in 1977, O. Kempthorne, in his introductory address, argued that quantitative genetics was no longer simply biometric genetics or a branch of population genetics. He stated:

> We wished to organize a conference, then, in quantitative genetics and not in population genetics as it is conventionally understood. The distinction is in some ways very much one of emphasis rather than intrinsic basis ... Many of the ideas of conventional population genetics are important to quantitative genetics. And, contrariwise, it seems clear that many of the ideas of conventional quantitative genetics are relevant to population genetics ... Part of the distinction between the two areas is simply that quantitative genetics should be called experimental population genetics, with the emphasis on the word 'experimental,' connoting that we make genetic populations by controlled operations, while conventional population genetics is primarily observational population genetics, trying to understand populations that have arisen by natural and not humanly directed processes.[93]

Clashes: Quantitative Genetics versus Purebred Breeding in Holsteins

The USDA began, in the late 1960s, to advise AI organizations, known as AI studs, to use Henderson's BLUP methods in their assessment of young bulls coming into the stud. Indexing under BLUP did not take into account conformation type. In the eyes of some breeders, this distain for physical looks, coupled with an appreciation of the individual's phenotype, brought about an overreliance on numbers for selection purposes. Statistics might play a role but other factors did as well. It was the old mathematic/biology dilemma, as well as a revisitation of tension between the purebred and geneticist approach to breeding. For many practical breeders, numbers should not have overridden the quality they saw in living animals, or their decision to take into account that quality when it came to breeding. When the AI studs in the United States began to choose young bulls solely on the basis of indexing – a percentage worked out on the basis of combined production data collected on the animal's sire and dam – one important American Holstein breeder decided to move his operations to Canada, where conformation and type still counted with AI studs.

In 1973 two New York breeders, Peter Heffering and Kenneth Trevena (who worked for Heffering), moved to Port Perry, Ontario and started their new breeding operation under the name of Hanover Hill.

Heffering and Trevena stressed the importance of cow families, type, style, and longevity in their breeding programs, features that went unrecognized in the indexing system used by American AI studs when selecting young bulls. Over the next twenty-five years, more than one hundred Hanover Hill bulls passed successfully through progeny testing at Canadian AI studs. The most important of these was Hanoverhill Starbuck, a bull born in 1979. Starbuck was bought at the age of seven-and-a-half months by Centre d'Insemination Artificeille du Quebec (CIAQ), owned by three groups of producers in Quebec: Federation des producteurs de lait du Quebec, or federation of milk producers; the Conseil quebecois des races laitiere inc., or breeder associations; and the Conseil provincial des cercles d'amelioration du betail inc., or breeding clubs. Today, CIAQ owns 45 per cent of shares in Semex Canada, indicating that the cooperative bull-owning structure remains a strong part of the Canadian AI industry.

Starbuck's first progeny test results appeared in 1984, but it was his second set of daughters in 1986 that started the excitement. Many of these daughters dominated the show ring and commanded a lot of money. Starbuck was the leading sire of All-Canadian and All-American show cattle from 1987 to 1993. His semen would be exported to forty-five countries, and he was the top selling bull of Semex Canada for five years. By 1994, 25 per cent of Canada's AI sires were sired by Starbuck sons. He lived to be nineteen, dying in 1998. Starbuck produced daughters and sons of good quality, from conformation and production points of view, and on a worldwide scale.[94] He effectively excelled, via progeny testing under traditional BLUP conditions, as a sire producing high milk-yielding daughters. But he also progeny tested as a sire of good conformation, and with next-generation breeding bulls. Heffering noted that no American AI stud would have even considered Starbuck.[95]

The story of Heffering and Starbuck illustrates that tension between traditional purebred approaches and genetics continued to exist, despite the malleability of the breed associations and their receptiveness to innovative ideas from science. The AI industry might well adopt genetic principles when deciding what bulls to acquire for testing, and the breed associations might support that trend, but purebred breeding philosophy would continue to exist within the ranks of breeders. The story also shows that purebred emphasis on conformation and type did not necessarily lead to poor results, even from a production point of view. Starbuck was an exceptional bull on all accounts, and he proved that under genetic principles. The Starbuck story implies, too, that there had been considerable globalization of Holstein breeding in particular.

The AI Industry and the Holsteinization of the World's Dairy Cattle

The use of AI, quantitative genetics, and purebred breeding ultimately made the North American Holstein the most important dairy cow in the world. The history of the breed's rise to hegemony is interesting. The Holstein began as a derivation of the Dutch Black and White. Widely known for their good milking ability, the Friesian Black and Whites tended to be bred throughout the nineteenth and early twentieth centuries in the Netherlands and northern Europe to enhance potential beefing qualities as well as to maintain good dairy characteristics.[96] In North America, where beef cattle dominated, the breed was appreciated primarily for its unusually good milking ability. Interest in the breed grew quickly in the United States and Canada, and it would be this increasing North American demand for the Dutch cattle that triggered the establishment of herdbooks for the Black and Whites in the Netherlands.[97] The name appeared to become "Holstein" as the result of a mistake made by an American import official.

Black and White Friesian cattle were not new to North America by the late nineteenth century, but they had had virtually no impact before that time. The first Dutch cattle arrived in the new world in 1613, imported by the Dutch East India Company (known as the VOC). These intermingled with local cattle, as did an importation in 1810. Small nuclear herds began to appear in the mid-nineteenth century, owned by wealthy men from prominent families who were prepared to invest heavily in an agricultural hobby (it cost about $300 a head to import a cow at a time when the average man made a dollar a day). Winthrop Chenery became interested in Dutch cattle after acquiring one in 1852 from the master of a Dutch sailing ship. He established the world's first Holstein herdbook in 1871 (more would follow in the United States, multiple herd books being allowed in that country, a practice that was illegal in Canada after 1900) and it was at that time that the breed's name in the United States was officially changed from Dutch to Holstein Friesian cattle. (In 1978 Americans dropped the word Friesian from the breed's name, and in 1983 Canada did as well.) Gerritt S. Miller began importing in 1869, but the real impetus for the importation movement began in the late 1880s. The transatlantic heydays for the trade continued into the early 1920s.[98]

Selection for milking ability in North American Black and Whites resulted quite early in astonishing milk yields. Individual animals could yield as much as 20,000 pounds a year by 1916; this, at a time when the average dairy cow gave less than 4,400 pounds a year.[99] Over the years, the cattle would increasingly look different under the breeding programs in North America and Europe. Between 1900 and 1960 Black and Whites in the Netherlands and Germany, for example, became progressively smaller compared to their North American counterparts, which tended to be tall. (It has also been stated that Canadian Holsteins came to differ from American in type, a situation that lasted until at least the 1980s. The Canadian cattle were even taller than American animals and tended to be known for better conformation and udder structure.) Some wondered if the animals in North America and Europe should be described as belonging to two separate breeds.[100]

Because Holsteins were developed under purebred breeding, style and type were virtually as critical as milking ability in the cattle. Addressing what constituted good style and type was therefore imperative. In 1921 American Holstein breeders agreed that there was a need to standardize judging by creating models of the perfect male and female. A committee was set up to do so in 1922, and models, known as the True Type Model, were built the following year. Over the years, alternations came to the models. By the 1960s word descriptive classification of conformation was more important than models.[101] In the late 1970s purebred breed associations in North America began to coordinate their efforts to establish a uniform idea of what made ideal type in conformation. The Holstein association was central to this movement and, in 1979, joined NAAB in an effort to establish a single type evaluation for all American dairy cattle. Traits were to be more accurately measured, in an effort to get away from the idea of overall ideal show type. After a great deal of work classifying cows in different herds, the Holstein association presented a system known as the Linear Classification System, to become effective in 1983.[102] Most countries in the world also had type classification standards and have continued to use a form of linear trait measurement. A great deal of effort went into harmonizing the methods of linear type of classification used internationally, to make it easier to interpret foreign bulls. Linear measurement involved visual appraisal. By the late 1990s there was mounting evidence that some linear traits did correlate with economically important characteristics such as: milk production, longevity and disease resistance, live weight, and feed intake.[103]

In the 1970s a study under Maria Stolzman was conducted in Poland on the relative quality of Holstein and/or Black and White Friesian breeding bulls from different countries when used on Polish cows.[104] This was possible, of course, because of AI and due to the fact that bovine semen freezes well. Hybrid response for growth and yield traits was estimated for first generation crosses of bulls from Canada, Denmark, Israel, the Netherlands, New Zealand, the United Kingdom, the United States, and West Germany on Polish Black and White cows. Cooperating countries sent semen from 348 young unproven sires to Poland in 1974 and 1975 to be used on Polish cows. It was found that hybrid effects relative to the Polish mean was greatest for the Canadian strain (about 110 per cent for milk and fat yields) and the U.S. strain (106 per cent for milk yield and 110 per cent for fat yield). The Canadian strain was consistently superior for all traits – birth weight, growth rate, milk yield, fat in milk – while the United States was second and New Zealand third. The Netherlands scored poorly and Britain was last for hybrid effects on Polish cows.[105] The study would have an enormous effect on the North American AI industry.

Exports of North American Holstein genetics increased steadily from the 1970s into the 1990s, and the North American AI industry grew correspondingly. It also went through a period of contraction and mergers. In 1981 eleven companies processed 90 per cent of semen. By 2006, that same volume was handled by only six companies. The historic structure of cooperative bull owning that had developed late in the nineteenth century has not disappeared in the American AI industry. Four of the six were cooperatives and therefore owned by farmers. The most significant of the four were Select Sires, Genex Cooperative, and Accelerated Genetics. All six functioned on a global scale.[106]

The infiltration of North American Holstein genetics in Europe changed the structure of European herds. A good example is the situation in the Netherlands. After 1900 breeders of Black and Whites in the Netherlands attempted to increase the meatiness of their animals, believing that extreme dairy type was conducive to bovine tuberculosis. Milk recording was continued, and while Dutch breeders did not ignore the milking aspects of the cows, they did tend to favour a beefy type, which led to a dominance of stocky, short cows by 1950.[107] The Polish study indicated how much the milking ability of the Dutch Black and Whites had declined relative to that of their derivative, the North American Holstein. The result was extensive importation of Holstein genetics and the general "holsteinization" of Dutch cattle, a trend that

was more or less complete by the late 1990s when the proportion of Holstein genetics in the Dutch Black and Whites approached 100 per cent.[108]

The globalization of dairy genetics affected all the dairy breeds and
was encouraged by the foundation of an international research organization – the International Bull Evaluation Service (Interbull) – in Sweden
in 1983. The objective of Interbull was to standardize and publish information on methods various countries used to test bulls of various dairy
breeds, the Jersey and Holstein being the most prominent. Between 1995
and 2003, for example, geneticists compared characteristics of progeny
testing for Holstein bulls in Australia, Canada, Denmark, France, Germany, Italy, New Zealand, Sweden, the Netherlands, and the United
States against each other. Quantitative geneticists developed statistical procedures called multiple across-country evaluations (MACE) to
evaluate all bulls under one system for each breed. By 2006 Interbull
had over forty member countries.[109] The quantitative genetic approach
to the pure line breeding of dairy cattle, in conjunction with AI and
the freezing capacity of bovine semen, as well as the work of Interbull,
brought spectacular results to milk yields in dairy cattle worldwide, but
particularly in Holsteins. The American dairy cattle population in the
United States peaked in 1944 at 25.6 million animals, which produced
53.1 million kilograms of milk a year, for example, whereas the number
of dairy animals had fallen to 9.2 million head with total production
of 70.8 million kilograms a year by 1997 – representing an astonishing
increase of 369 per cent per cow in just over fifty years.[110] Better coordi-
nation of breeding programs around the world via a standardization of
progeny testing eased international trade in bull genetics, a factor that
also encouraged inbreeding.

The Spectre of Inbreeding in Dairy Cattle: An Outcome of AI

Full brothers entered AI units in Europe and North America and competed on a global scale for markets, and this trend became a critical
factor in promoting inbreeding. The massive export of North American genetics to Europe generally encouraged increased inbreeding in
both Europe and North America. Figures indicating alarming levels
of inbreeding abound in the literature on dairy cattle breeding.[111] For
example, it has been stated that the average genetic relationship of sires
of sons born in 1990 would be six times greater than that of sons born in
1970 – as a result of this cross pollination.[112] By 2005 only two American
Holstein bulls accounted for 30 per cent of the gene pool of Holsteins

in the United States, and by that time 20 per cent of Canadian Holsteins were related to Hanoverhill Starbuck.[113] By 2008 it was estimated that the effective size of the world's Holstein populations, genetically speaking, was between thirty-five and sixty animals in a total global population of close to thirty million animals.[114] (It should be noted that in the United States less that 5 per cent of dairy cows were not pure Holsteins or Holstein grades by 2002.)[115]

It was apparent by the early 1990s that reduced fertility in dairy cows was an increasingly common pattern. Conceptions rates, for example, had dropped alarmingly between 1950 and the mid-1990s. First-service, conception rates for cows was 65 per cent in 1951, and 40 per cent in 1996 in the United States.[116] The decline in fertility seemed to reach its lowest ebb in 2002 and then reversed itself.[117] Many continued to believe, however, that inbreeding had caused the problem and that it remained a critical issue. A study done in Canada in 2006 revealed something about the increasing levels of inbreeding. The inbreeding level of Canadian cows between 1980 and 2004 was lower than in the United States, but even so rates were rising at an alarming rate. The trend was particularly significant after 1990, and by the twenty-first century at least 96 per cent of Canadian cows were inbred enough to be of concern. Inbreeding in Britain showed similar trends: increasing dramatically after 1990. By 2002 some 98 per cent of bulls and 96 per cent of cows in Britain were inbred to some degree, compared to about 50 per cent overall in 1990.[118]

Crossbreeding has long been the accepted way to counteract the detrimental effects of inbreeding, just as inbreeding can correct problems in crossbreeding. It is always a question of balance. The intense inbreeding, which quantitative genetics encouraged through increasingly superior ways to progeny test bulls, and the ability to crossbreed on a global scale because of AI and organizations like Interbull, resulted in a growing interest in crossbreeding.[119] By the early twenty-first century, for the first time in over seventy-five years, crossbreeding attracted real attention internationally.[120] New Zealand served as a model because that nation had practiced crossbreeding to a considerable degree for some time, and much earlier at any viable level compared to other countries. In 2000 some 18 per cent of the national New Zealand dairy herd was composed of Holstein/Jersey cross cows. After twenty-five years of data collection on production and costs, it was apparent by 2000 that these crosses had a higher merit for farm income than the pure Holsteins, which represented 57 per cent of the national dairy herd.[121]

At the same time, less than 0.5 per cent of American dairy cows were crossbreds.[122] There was considerable interest in the idea in North America, however, and the AI studs offered information on crossbreeding.[123] Generally, they did not openly support crossbreeding. For example, Genex Cooperative, Inc. told its clientele: "Our member survey suggests that there is a perception among dairy producers that problem-breeder cows may settle more easily using a sire of a different breed as a service sire. Yet, recent university research does not confirm this and suggests it may be myth more than fact ... Crossbreeding has some short-term advantages and disadvantages. To date, it has not proven to be the magic pill to improve cow profitability for all dairymen ... Long-term generational effects of crossbreeding in dairy cattle are not well documented or understood."[124]

While AI organizations did not condone crossbreeding, they did offer their clients advice on to proceed with the method, if a dairyman wanted to try it. Select Sires, for example, laid out a three-way cross plan, in a document entitled "Crossbreeding 101: What Every Producer Should Know."[125] Using purebred sires and maintaining the rotation were the keys, though. "In a two-breed system scheme," the cooperative company, Select Sires explained, "if crossbred sires are used on crossbred cows, only 50 per cent of the hybrid vigour would be sustained as opposed to the 67 per cent sustained through rotation of purebred sires."[126] A three-way purebred rotational system resulted in 85 per cent levels of hybrid vigour, and a four-way breed rotation brought as high as 95 per cent. Select Sires did not recommend a four-breed crossing system because it was too cumbersome. But the proven benefits of crossbreeding do not come free, the AI organization warned. "To achieve these you will likely have more variety in the overall size and conformation of cows, sacrifice some overall conformation (especially udders) and, in the long run, will have slightly lower yield of milk components," Select Sires said, continuing, "It is also important to realize that the early results of crossbreeding will likely be better than what can be sustained in the long run ... Select Sires cannot recommend crossbreeding with confidence at this point because we don't have sound comparisons on currently available dairy breeds. Research is underway that will help provide these answers and Select Sires supports this work."[127] Generally speaking geneticists supported the view that crossbreeding could only be done successfully by using purebred stock for the crossing. Crossbred bulls on crossbred cows did not provide either hybrid vigour or any consistency.[128] Since accepted crossbreeding systems relied on the use of purebred bulls, it was not necessarily self-interest that drove the reluctance of AI studs to promote crossbreeding programs. After all, the semen they offered was virtually entirely from purebred bulls.

Dairy journals tried to explain the pros and cons of crossbreeding to ordinary dairymen. *Hoard's Dairyman*, for example, discussing the dichotomous results that come from crossbreeding, wrote:

> Crossbreeding in dairy cattle is generating renewed interest. Crossbreed-ing allows producers to reduce inbreeding, take advantage of heterosis (improved performance of offspring over the average performance of par-ents), and exploit superior fertility and health traits found in other dairy breeds. On the other hand, crossbred cattle usually have lower milk than Holsteins, and it can be difficult to maintain heterosis and uniformity in the resulting offspring from the crossbreds ... Despite the beneficial effects of crossbreeding, there is one major concern – the loss of milk yield com-pared to pure Holsteins ... No single answer can be given as to whether the improvements in fertility and health traits caused by crossbreeding compensate for the loss of milk yield.[129]

Beef Cattle Breeding, Quantitative Genetics, and the AI Industry

The situation in the North American beef-cattle world provides an inter-esting contrast to what happened in the dairy cattle breeding world as a result of quantitative genetics and the AI industry. Beef cattle produc-tion had always relied on hybrid vigour from crossbreeding, but the structure of the breeding industry had not been conducive to an accep-tance of organized hybrid breeding with biological locks, as the Shaver Blend's fortunes make clear. How did quantitative genetics interface with beef cattle breeding? Beef cattle semen, of course, freezes as well as dairy cattle semen, but the structure of the beef industry (large herds with minimal management) discouraged the extensive use of AI. That fact alone hindered quantitative genetic work in beef cattle: the data needed for statistics did not exist. To add to this situation the beef breed associations were not as receptive to the innovative thinking that arose from academic genetics as the dairy breed associations had been. The purebred beef cattle associations in North America, in complete control of any genetic advancement from their beginnings in the 1880s, contin-ued to evaluate breeding worth by subjective visual appraisal until into the 1960s. (As an aside, it might be noted that meat characteristics are observable in livestock in a way that milking capacity is not.) The first association effort to provide a more objective view of stock quality was the Red Angus Association, which in 1959 required weaning weights to be provided before pedigree registration was possible. Over the 1960s other beef breed associations developed performance-recording programs, although the Angus association remained the only one that required data reporting. It was largely the demands of feedlot operators in the 1940s and 1950s that led to improvements in performance record-ing of purebred beef cattle, under systems slowly being developed by quantitative genetics. A move to better orchestrate improvement pro-grams began in 1965 and resulted in the Beef Improvement Federation, formed in 1967. The breed associations maintained a strong voice in that organization.

In the 1970s performance-evaluation systems for beef cattle designed by geneticists began to reflect BLUP properties, and the breed asso-ciations tried to establish methods of testing animals under different farm conditions. Evaluation systems were breed specific and revolved around what was known as estimated progeny difference (EPD) for a variety of traits – the numbers of which expanded over the years and in relation to the capacity of computers to handle complicated statis-tics. An EPD is a prediction of an animal's likelihood of passing on a trait in relation to breed average for that trait. Over time EPDs within a breed changed as producers emphasized different traits. The most common EPDs calculated were for birth weight, weaning weight, and yearling weight as gain per day after birth. The beef breed associations did not adopt such innovative ideas quickly and it has been suggested that when they did so, the primary reason was competition with each other for markets that involved the commercial cow/calf operators. These farmers assessed the EPDs of the various breeds before deciding which bulls to use, a fact that encouraged each association to produce better and more comprehensive data.[130] It was the fractured nature of the beef breeding industry that forced the adoption of quantitative genetics, in sharp contrast to the cooperative nature of the dairy breed associations. Nevertheless, it has been argued that systems designed to evaluate the breeding worth of beef cattle do not work well. Pre-dictions tend to be presented without context to farmers functioning under a wide range of conditions. The one-size-fits-all scenario has not been a sensible approach, and over-reliance on statistics to do the job is also a mistake, according to R.M. Bourdon.[131] While the beef breed associations clearly played a role in how quantitative genetics would be applied to the beef-breeding industry, other factors were part of the story.

Artificial Insemination and Pigs

The situation with respect to AI and other livestock was different from that of cattle. The biological fact that cattle semen freezes better than the semen of other species affected how much AI interacted with breeding of these species. A brief look at AI and the pig industry in Britain illustrates fundamental differences. In the 1940s good pig breeding in Britain proceeded under the authority of the breeders who dominated the breed associations, and the authorities that licensed boars for public breeding used the breeders' standards.[132] Some effects of "genetic" breeding had infiltrated pig breeding by the 1960s. Although results had been discouraging, pig breeding did revolve around hybrid breeding by the late 1970s, a fact that led to the rise of breeding companies that could protect their interests by a biological lock.[133] A study of AI and pigs in Britain reveals how important the breeding companies, rather than AI technology or geneticists as such, were to the spread of AI use. Even though issues of low conception rates plagued pig AI until into the 1970s – therefore stalling extensive use of the technology – other trends played a part in the substantial change that took place in pig breeding and its connection to AI. As early as the 1950s scientists at Cambridge who had worked on cattle AI became interested in pig AI, but it was the involvement of pig breeding companies by the 1970s that played the most significant role in the growth of pig AI. The Cotswold Pig Development Company and the Pig Improvement Company believed that AI would enhance their ability to progeny test, or "performance" test boars used in their hybrid breeding programs.[134] The semen buyer took on faith that the quality of a boar had been proven "genetically."[135] By 1990 pig AI was firmly part of the British pig industry but it still was by no means dominant. Only 19 per cent of pig producers used it on more than half their sows. (In comparison, those figures were 80 per cent in East Germany, 23 per cent in West Germany, 71 per cent in Norway, 51 per cent in the Netherlands, 25 per cent in Denmark, 19 per cent in France, and only 7 per cent in the United States.)[136]

Pig breeding came to represent a curious blend between chicken and dairy cattle breeding. It relied on hybrid breeding and companies who worked with a biological lock, as did the chicken-breeding industry. The chicken-breeding industry, however, did not involve itself publicly with the selling of AI services. In the case of pigs, AI became attached to privatization via company/hybrid breeding. Dairy cattle breeding did not emanate out of companies using biological locks, and AI studs linked equally to geneticists, breeders, and producers.

Producer/Growers and Breeding in the Dairy and Chicken Industries

On the surface, many aspects of the dairy industry seemed similar to those of the chicken industry before its hybrid breeding revolution in the late 1940s and 1950s. Both proceeded under principles of "art" breeding and bred for true breeding lines. Both had breed organizations that regulated the structure of breeding. Both developed systems of collecting data on performance of females that emphasized individual worth and ancestry breeding. Both used the data for promotional reasons, and data collection in both industries aroused conflicting points of view. But one critical difference lay in the fact that the relationship of the producer/grower to the breeder was not the same across the two industries. In the dairy industry the two were not as clearly separated as they had been in the chicken industry. The breeding structure in many ways remained much as it had been before 1940 and the advent of AI; the ordinary producer stayed a part of the system in a way that the producer of chickens did not. This situation prevailed in both North America and Europe. The producer/grower used the product of the purebred breeders in their herds and supplied milk-yielding data back to breeders, information that proved vital for them in making breeding decisions. Dairymen were always able to make a choice in what bulls they used on their milking cows.[137] This factor would become increasingly important when sire indexing and progeny testing became more widely and comprehensively used. Progeny testing, for example, could not be done without the cooperation of dairymen because they provided the numbers needed to estimate daughter response.

Technology, in the form of incubation and the resulting hatchery industry, played a role in furthering the cleavage in the chicken industry between breeder and producer, but the corresponding advent of AI did not remove the dairy producer/grower from the breeding structure.[138] Farmer AI cooperatives in North America and northern Europe relied on the internal management of breeding by the purebred breeders when buying bulls for the units (as did the Milk Marketing Board in Britain). AI organizations (known as studs), from their inceptions, did not go into the business of breeding. They bought young bulls. They might contract out to a breeder to "rent" a breeding cow to produce a son, but they did not own breeding herds.[139] This would prove to be an important point, because increasingly after the advent of frozen semen, three major groups participated in the process of disseminating superior bulls to be used by ordinary dairymen: the studs, livestock geneticists, and purebred breeders. Although the studs dealt only with males, the entanglement with female production through males made producer/growers an integral part of the breeding structure. AI studs did not attempt to use a biological lock in order to regulate the use of semen. They did not, in other words, demand that semen be used only on commercially milk producing cows, which would restrict its use in elite breeding herds. In the twenty-first century, the geneticist for Semex Canada, J.P. Chesnais, explained the historical background of the studs' interface with breeding this way:

> Dairy cattle breeding organizations are in a very vulnerable position because they do not protect what they create, through patenting or trade secrets, or by restricting the distribution of seedstock. In the swine and poultry industries, parent stock from selected lines is never released to competitors or the public. In the plant breeding industry, parent stock is either not released or the strains are patented so they cannot be used for breeding. In dairy cattle, anyone can use a top bull an AI unit has selected as a breeding sire to produce a son.[140]

Dairy Cattle Breeding: Purebred Breeding combined with Quantitative Genetics

The cattle industry stayed with the pure line breeding philosophy that had traditionally been advocated by agricultural breeders, particularly in North America. It was geneticists, not breeders, who shifted in orientation on this fundamental point, and when they did so, livestock genetics became part of dairy cattle breeding. But breeders also readjusted their breeding approaches. Quantitative genetics managed to bring purebred breeding back to Bakewellian roots, and away from the prevailing Thoroughbred horse breeding culture, which emphasized purity, individual worth, and ancestry breeding to the exclusion of progeny testing and breeding via population groups. Purebred breeding, albeit altered in form to resemble its original Bakewellian roots, apparently triumphed in the case of dairy cattle.

What else can we say about the overall nature of livestock genetics' relationship, historically speaking, with the dairy breeding industry? What made it so different from the chicken situation? First, the cost of the individual animals and slow reproduction made breeding for hybrid vigour a difficult and lengthy process in dairy cattle, and therefore discouraged the involvement of corporate enterprise. Dairy cattle breeding remained, as a result, impervious to the early livestock genetics. Second, the breed associations in Canada and the United States were prepared to work with academic institutions from the beginning. The dairy breed associations seemed more malleable than the American Poultry Association had been to pressure rising from science. The dairy associations had already moved decisively, by 1930, to emphasizing progeny testing, for example, thereby attempting to overcome the historic concern of purebred breeding with ancestry breeding. They also organized extensive systems of data collection (the chicken industry did not do so to anywhere near the same extent) that complemented government efforts, even if some competition between the two did exist. They initiated data collection before the advent of genetics itself, and collected a huge amount of useful material over the years. Third, ordinary North American dairymen stayed part of the breeding structure in a way that was not true of ordinary chicken producers. They therefore did not abdicate a concern with how breeding proceeded.

Fourth, the advent of AI made data collection by the purebred breed associations extremely important for geneticists, encouraging them to cooperate with the breeder organizations. Before the advent of AI, there had been rumblings about the self-serving interests of the breed associations regarding their data collection methods, as the USDA's 1936 report on dairy cattle breeding made clear. There was pressure for reform, and for dairy cattle breeding to follow the path of chicken breeding when the purebred breed association, the American Poultry Association, lost its hegemony. AI gave the dairy cattle breed associations new life. Fifth, when geneticists developed more sophisticated statistics that introduced the capacity for effective sire indexing, the breed associations were prepared to modify their outlook to make it match that of geneticists. It is impossible to overstate the importance of AI and the ability of bovine semen to freeze to the successful joint work of purebred breeding and livestock genetics. There would be tension between purebred breeders and geneticists, in much the same way that had been the case between standardbred chicken breeders and scientists – and over many of the same beauty and utility issues. The scientists, however, failed to reduce purebred breeders and ordinary dairymen interest in conformation for breeding purposes. The combined work of the purebred breeders and geneticists would allow the dairy breeding industry to capitalize on advanced work in molecular genetics, something not possible in any other livestock industry. That is the story of the next chapter.

Chapter Five

Molecular Genetics, the Rise of Genomics, and Livestock Breeding

By 1960 agricultural genetics was characterized by two quite distinct theoretical approaches to artificial selection. These had affected practical breeding in different ways: hybridizing dominated the chicken-breeding industry, and selection within true breeding lines shaped the purebred dairy cattle world. Geneticists working in the field of animal breeding tended to divide into one section or the other – demarcated by the species they studied. A specialist in dairy cattle breeding worked within very different parameters compared to a specialist in chicken breeding. All livestock geneticists, however, continued to follow the 1918 infinitesimal model of Fisher; namely, the theoretical concept that quantitative traits are inherited via the interaction of multiple genes, all of which individually have a very small effect, and therefore trying to isolate major genes was a non-constructive exercise. Added to that approach was the problem of recombination of genetic material via mating for the next generation. Interest in inheritance at the gene level per individual – that is, the genetic architecture of trait inheritance – was not a priority for geneticists theorizing how to develop better artificial selection strategies.[1] Advances in molecular genetics, however, brought about different approaches by the 1980s. I begin this chapter with a few brief comments on advances in reproductive technology, transgenics, and cloning in the 1980s and 1990s, describing how they affected practical artificial selection. The main focus of the chapter, however, is on the historical development of animal genomics and its subsequent effects on artificial selection after 2000.

Reproductive Technology, Transgenics, and Cloning

Technology, in the form of AI, had allowed for the practical use of theoretical genetics in animal breeding. The focus had been on males, in the classical way that breeders starting from Bakewell had done. By the 1980s, however, a new reproductive technology allowed for an emphasis on females when it came to artificial selection. Technologies known as embryo transfer (ET), and more importantly multiple ovulation embryo transfer (MOET), allowed for extension of female influence on a breeding herd. Mammals that normally produced one offspring a year could now generate a number. Cows could be made to super-ovulate (a procedure that could be done as early as the 1950s) and then impregnated by either AI or natural service. Seven days later the resulting multiple embryos could be flushed from her uterus and either implanted in another cow, known as a recipient, or frozen for later use. Elite breeders of dairy cattle quickly adopted the technology for use on their best cows. MOET interested some AI companies. European AI organizations began to buy good females and run MOET programs. The resulting female progeny could subsequently be bred to their AI bulls. The hope was that genetic improvement in next generation bulls would be faster if the input of good females was added to the breeding program. In the United States, MOET was more likely to be used by AI companies in conjunction with elite breeders, who might well contract with an AI company to flush certain cows.[2] By the mid-1990s, however, it was apparent that MOET would not be the powerful tool for herd improvement that had originally been assumed. It clearly carried with it the danger of increased inbreeding as well.[3]

Other breeding techniques emerged over the 1980s and 1990s from work in molecular genetics. Transgenics, the trans-positioning of certain genetic material from one organism into another, advanced rapidly after 1980, and various techniques were developed for application to animals. With respect to livestock improvement, transgenics would be most successful in enhancing certain pig characteristics. The milking capacity of sows, for example, concerned geneticists because of its important impact on the early growth of piglets.[4] Transgenics, however, aroused fierce opposition from the public, especially in the United States and Britain. Its poor image seemed to explain partially why less than 1 per cent of research grants offered by the USDA between 1999 and 2012 provided for the study of transgenic food animals. As of 2012, the U.S. Food and Drug Administration had not issued a decision on any transgenically-altered animal submitted for approval.[5]

Cloning technology provided for the replication of special animals. The first cow was cloned in 1986. (Cloning is easier in cattle than in other species and is, therefore, most prevalent in cattle. Today it costs about $20,000 to clone a single calf.) The technique used today to clone animals was developed in the 1990s at the Roslin Institute of Edinburgh, Scotland. Dolly, the famous cloned sheep born in 1996 was produced via the Roslin technique, and arrived after 237 eggs were used to create twenty-nine embryos that produced three lambs of which only one of lived.[6] Cloning is an elite service, and clearly makes no sense on a large scale. The focus on a cloned individual also goes against all principles of livestock breeding that emphasized breeding by population or group worth. Far more important to the future of artificial selection strategies than reproductive technology like MOET, transgenics, or cloning would be different developments in molecular genetics/technology, namely genomics. Background information on the rise of molecular genetics, and critical technology's relation to it, is important for understanding the evolution of genomics.

Rise of Molecular Genetics and Recombinant DNA Technology

The genetic architecture responsible for a human genetic condition, sickle cell anemia, was discovered in 1949, and initiated the birth of molecular genetics. After American James Watson and Briton Francis Crick proposed the double helix structure of DNA in 1953, an explosive growth in the study of molecular genetics followed. Molecular geneticists, who were especially interested in human genetics and health, came from a fundamentally different standpoint when compared to scientists interested in artificial selection. They, unlike animal geneticists, were not focused on breeding for the next generation, and therefore segregation and recombination of genes was not central to their research. They hoped that molecular genetics could help solve many medical conditions by finding genes, often major genes, which singularly affected genetically inherited diseases. Quantitative traits did not command much of their attention and they therefore did not use quantitative genetic methodology. Classical genetics (in either its population or quantitative form) tended to be backgrounded when it came to funding research.

It was the ability to transfer short pieces of DNA from one set of DNA into another (a process known as recombinant DNA) that really opened the doors to studying and identifying genes at the DNA level in a way that had been impossible earlier. By removing DNA from one organism and putting it into simple fast-growing bacteria, one could isolate genes or short sequences of genes. This process could be accomplished because DNA molecules, which consist of two long chains of nucleotide subunits twisted around each other forming a right-handed helix, have the same chemical structure in all organisms. Each nucleotide subunit contains one of four bases: adenine (A), thymine (T), guanine (G), or cytosine (C).

In the two strands of DNA, where one strand has an A, the other has a T, and where there is a G on a strand, its partner has a C. Species differ only by the sequence of the A, G, T, and C nucleotides. Changes in genetic material could be seen in the sequence of the bases; that is, the substitution of one nucleotide for another, or the deletion of one or more nucleotides. Alternations, which imply difference from what was normal within a population, are said to be polymorphic, and the location of such changes in the DNA are called single nucleotide polymorphisms (SNPs). By 1977 F. Sanger had developed DNA-sequencing techniques that took advantage of the fact that DNA-cutting enzymes produced fragments of different lengths in different individuals, which were called restriction fragment length polymorphisms (RFLPs). The development of automated machines capable of working out sequences of A, T, G, and C in samples of DNA in 1986 made this work easier.[7]

Advances in molecular genetics and biotechnology opened up new possibilities in terms of genetic research. Since scientists no longer had to rely on statistics that might reveal gene expression, but not the structure of the DNA itself, quantitative genetics was viewed increasingly as having been superseded. Livestock genetics for many was a backwater.[8] Animal geneticists were disparaged for their avoidance of the genetic architecture underlying traits.[9]

Quantitative Genetics: A Thing of the Past?

Quantitative geneticists gathered at their first international conference in 1976 at Ames to discuss their work. In spite of the huge success they had brought to the dairy cattle breeding world, as had earlier agricultural geneticists focusing on hybrid breeding had brought to chicken breeding, there were deeply embedded seeds of concern within their ranks. Excitement over potential transgenics, the eventual possibility of cloning, and the possibility of patenting new forms of technology implied, not just to molecular geneticists but also to some quantitative geneticists, that future advancements in theoretical breeding lay elsewhere. The ability to locate and study individual genes might have been changing the research landscape for many scientists, but quantitative geneticists had not been able understand the genetic architecture of metric traits, O. Kempthorne noted.[10] R.E. Comstock enlarged on the shortcomings of quantitative genetics: "Quantitative genetics has made major contributions to the bases for the major design decisions involved in [breeding] … At the same time there are significant issues in the realm of quantitative genetics that have not been resolved. Some of these appear tractable, others relatively intractable … It appears from the perspective of the breeder that quantitative genetics still has a challenging future."[11] R.C. Lewontin addressed the problem of black box thinking in quantitative genetics where quantitative genetics was used to predict results of certain selection methods; at the same time, nothing was known about the genetic architecture that would dictate the effectiveness of any such method.[12] Lewontin elaborated as follows:

> We need to know the relations between gene and organism, how gene action … is translated into phenotype. The knowledge about these questions can come to us only by opening up the black box whose outer shape we have so far been describing, and seeing what the machinery inside really looks like. This is the task of molecular and developmental genetics and some general knowledge is already available to us from the recent activity in these fields. Our models of quantitative genetics must either take cognizance of these findings or else how show how they are, in fact, irrelevant because of the robustness of our theory.[13]

Other worries concerning the value and future of quantitative genetics, but unrelated directly to molecular genetics, were brought up at the meeting. D.L. Harris raised the issue of statistics and its overdominance in the study of livestock breeding: was animal genetics more about statistics than genetics? Emphasis on statistics had made many forget, he argued, that the roots of livestock genetics lay as much in methods of breeding developed by farmers as in Mendelism. "It is

well to remember," Harris stated, "that animal breeding was a serious
activity of many stockmen prior to the discovery of Mendelian basis of
inheritance."[14] Reviewers of the conference made equally gloomy comments. P.E. Smouse wrote:

> In recent years, the once thriving field of quantitative genetics (QG)
> has fallen on hard times. Both the discipline and its practitioners have
> been criticized for a total failure to deal with the genetic and biochemi-
> cal details underpinning the phenotype. Such attacks are more than a
> little unfair, considering that QG was designed to deal with the sort
> of continuous phenotypic variation which largely defies Mendelian
> analysis or exact biochemical characterization, the sort of variation
> which is so complex as to admit of little more than statistical summa-
> rization. Quantitative genetics is nothing more (and nothing less) than
> a convenient statistical construct whose prime function is to permit
> estimation and testing of a set of summary measures. The grist for this
> statistical mill comes in the form of phenotypic resemblances between
> biological relatives. Since nary a gene is seen, the fact that these sum-
> mary measures convey any genetic information at all must be viewed
> as a splendid accomplishment. It is precisely for the analysis of those
> phenotypes which are hopelessly complex that quantitative genetics
> was designed.[15]

Discouragement could be felt from the traditional hybridist geneticists as well. Poultry geneticists expressed various concerns about
the applicability of their work. Their hybridist approach did not even
seem to fit with quantitative genetics. One person at the conference
pointed out that there was no evidence that the application of quantitative genetic theory was important to the continuing success of a
chicken-breeding operation. While hybrid breeding did not dovetail
well with quantitative theory or the general thrust of livestock genetics after the 1960s, he wondered if the classic inbreeding/hybridizing
techniques of the past were really as effective as they should have
been.[16] Another geneticist pointed out at the 1976 meeting that the
competition between private breeding companies did not help the
situation.[17] An environment for sharing research information was not
conducive under a structure that was supported by biological patenting. Chicken-breeding geneticists seemed to sense that potential
future developments in molecular genetics would lead to their further
marginalization.

Quantitative Genetic Theory and the Rise of DNA Mapping

In spite of quantitative geneticists' attitudes to molecular genetics at the 1976 conference, the potential of molecular genetics for quantitative genetics had attracted at least some theoretical attention. Quantitative geneticists did not all ignore the possibilities of molecular genetics for animal breeding. Workability of any theorizing was, of course, another question. Long before the rise of recombinant DNA technology, Alan Robertson, at the Genetics Commission of the European Association of Animal Production meeting in 1969 at Helsinki, stated that quantitative geneticists should be paying more attention to information arising from molecular genetics.[18] He himself had tried to find a way to develop artificial selection strategies on the basis of DNA as early as 1961 when he experimented with ways of finding quantitative trait loci (QTL) – that is, the loci of genes or even just stretches of DNA known to hold material that related to productivity (and therefore contain the important genes in question) – by looking at blood groups found in highly producing dairy cows. Prior to 1980, the only suitable way to look for QTL molecularly speaking was to assess blood groups. Blood groups, however, had no visible effect on any traits of interest, and it soon became clear that the total number of polymorphic blood loci was quite limited. Evidently, using blood groups to find QTL in animal populations would not work. Furthermore, it was clear by the late 1970s that large populations were needed – at least 1000 individuals – to locate useful QTL in the advent that some form of technology made detection of them possible. A genome-wide scan for QTL on tomatoes using restriction fragment length polymorphisms (RFLP) in 1988 suggested that technology able to do so was on the horizon.[19]

Before RFLP technology it was impossible to study QTLs in a more constructive way from a DNA point of view, but theories on how it could be done and what was to be looked for, molecularly speaking, existed before technology made it possible. In order to locate QTL in DNA one had to find characteristics in the DNA that could be linked phenotypically – that is, via observable characteristics – with a known population. In order to "map" QTLs, then, families had to be measured for the trait of interest.[20] How did genes or stretches of DNA differ in families demonstrating high levels of productivity from those in lower producers? What, in other words, was polymorphic – that is, demonstrating various patterns – about DNA in the most valued animals? Could their DNA identify them in that way?

Here the theory of linkage disequilibrium (LD) was critical. LD is the non-random association of alleles (gene halves) at two or more loci, meaning that more variation (polymorphism) in genetic markers (at loci) occurs in a population than would be expected in random formation. Non-random associations between polymorphisms at different loci are measured by the degree of linkage disequilibrium. The stronger the LD the more meaningful it is when it can be associated with known high producing individuals. When the association of marker loci and QTL is in linkage equilibrium (that is, random association), nothing useful can be learned as to why an animal is genetically superior as to production. This type of theorizing was the beginning of a move away from black box thinking when it came to genetic architecture to one that supported an isolated gene/marker or DNA sequence model as the underlying conceptual way to look at inheritance of quantitative traits. Even so, this approach did not mean that Fisher's infinitesimal model was no longer useful. The question remained: how many genes/markers went into a QTL and therefore underlay any given quantitative trait?[21] The answer was many; locating a few genes or markers was not necessarily helpful.

Over the years Robertson (who died in 1989) became increasingly convinced that the number of loci responsible for valued traits was small, compared to the potentially large amount of DNA variation that might exist in an animal's genetic profile.[22] As well, some scientists began to question why the infinitesimal theory continued to have such sway when there was no real evidence that inheritance of quantitative traits did not proceed via a single or limited gene, rather than multiple gene process.[23] This tendency to move away from both the infinitesimal model and black box thinking bore strong similarities to perceptions of the early Mendelists, who believed that single genes could explain such traits as egg and milk production levels, and that such genes could be identified. The excitement surrounding breakthroughs in a new science seemed to provoke such a shift, evident after the advents of both Mendelism and recombinant DNA technology. Discoveries seemed to imply that at last simple answers to complicated questions might eventually become available.

In spite of the questioning that surrounded the wisdom of traditional ways of thinking and interest in QTLs, by the late 1980s livestock geneticists had not focused concertedly on how to use molecular genetics in prediction strategies for breeding or for developing new artificial selection theory. At the second quantitative conference held in 1987, it was apparent that black box thinking, statistical quantitative studies, and/ or hybrid outlooks – within the framework of infinitesimal model – still dominated livestock genetics.[24] (The hope that RFLP scans could locate QTL in animals ultimately proved to be elusive, even from an experimental point of view.)[25] M. Soller and J.S. Beckmann spoke differently, however, sensing that fundamental perceptions were about to change. They emphasized the value of specific DNA markers in understanding the way trait inheritance worked in livestock:

> The past few years have witnessed the introduction into genetics of methodologies that allow genetic polymorphisms to be examined at the DNA level ... These 'genomic' methods have the potential of uncovering a virtually unlimited number of genetic polymorphisms within the major agricultural species. This should allow the widespread extension to agricultural species of marker-based methods for the identification and manipulation of polygenic loci affecting quantitative traits ... When compared to classical biometrical approaches, marker-based methods can provide more general means of determining whether a genetic resource contains useful polygenic alleles not present in a commercial population and more effective means of introducing novel favourable alleles from the resource to commercial populations.[26]

Breeding Strategies and Specific Markers in DNA

By the 1990s the pessimism evident at the 1987 conference concerning the applicability of molecular genetics to animal breeding strategies had started to dissipate. DNA markers for QTL were located in the late 1980s with superior recombinant DNA technology, and by the 1990s, there was excitement over gene mapping for important polymorphisms by microsatellite technology.[27] Even single markers could be touted as important when it came to breeding selection, but it continued to be unclear how useful any form of marker technology was for breeding purposes and whether the expense of DNA testing animals made economic sense. Two markers for marbling and beef tenderness were located in cattle from experiments done on DNA of beef Shorthorns in Australia in the early 1990s.[28] By the early twenty-first century, three companies offered genetic testing for the two markers, and North American Shorthorn breeders in particular were targeted for sales. Favourable results could be useful to breeders, especially when it became evident that carriers of the markers were not particularly common. In 2002 Shadybrook

Farm in Canada, for example, advertised a cow that had been named "one of only three Shorthorn cows in North America who have, to date, tested for two stars in the GensSTAR Mapping Program for marbling genes."[29] A number of cows would be tested after that time, but finding both markers in an animal continued to be rare. In 2007 an independent scientific study was done to assess the viability of the companies' claims that genetic testing of this nature was worthwhile. Comments from the study were as follows:

> Although it is understandable that companies would want to protect their intellectual property from unauthorized use, such anonymity makes it difficult to compile some of the information that is important in using the marker in breeding programs (e.g. the frequency of the favourable allele in a range of breeds) … No single study can examine all of the breeds, allele frequencies, or environmental and management conditions that may factor into whether an association is found between marker and trait. Validation studies are therefore problematic because findings are dependent on the specific characteristics of the finite number of populations screened.[30]

In conclusion, the study announced, the GeneSTAR and Igenity tests *may* result in a higher percentage of choice carcasses, but "independent, third-party validation of commercial DNA tests [would] provide some assurance to producers that DNA-based tests perform in accordance with the claims of the marketing companies and [might] help to generate some of the data required to facilitate the integration of marker data into the national cattle evaluation."[31] It is clear from this example that DNA studies for markers could result in patenting of products that might not be of widespread practical use. One of the problems with developing DNA tests was/is the enormous cost, and companies undertaking to find answers were/are also seeking adequate remuneration.

Marker technology seemed no more useful to chicken breeders. Hybrid breeding did not lend itself to new ideas emerging from DNA studies because it did not promote the collection of adequate data. Privacy of breeding exacerbated the problem. As R.W. Fairhill, a chicken geneticist, wrote in the late 1990s: "Marker-assisted selection needs more development, especially empirically, before it can be considered for practical poultry improvement programs. Similarly, the use of markers to predict heterosis is an immature technology that awaits further verification and refinement before any practical implementation. In all considerations of marker technologies, real costs and relative benefit must be considered, not merely theoretical ones."[32]

The Rise of Livestock Genomics

As animal geneticists struggled with the usefulness for breeding programs of locating specific markers (that is, QTL in relation to certain genes), new directions were on the horizon. Genomics – specifically, the study of a species' DNA in its entirety – and improvements in genomic technology would change the situation with respect to breeding strategies based at the DNA level. In 1990 the Human Genome Project was initiated under James Watson of double helix fame. The effort, a joint international one, was completed in 2001. The Human Genome Project was important for the future of animal genomics for a variety of reasons. First, it proved that whole genome sequencing of large and complicated species was possible. Second, it indicated that international organizations could work together on complicated projects. Third, it set the precedent that sequencing information should be open to anyone wanting it and available over the Internet. Fourth, it established the use of a superior technology, not in existence at the beginning of the undertaking. Developed by Celera Corporation, shotgun sequencing allowed for completion of the project at an earlier date than had originally seemed possible.[33] DNA was sheared into random fragments, which were subsequently cloned into an appropriate vector. The clones were then sequenced at each end, and overlapping readings taken with other segments. A composite of sequences were then rejoined, a process that continued until the entire genome had been put together. The matching had to be done by computer. Put another way, segments of DNA were cloned in a bacteria and then sequenced. These sequences could be compared and assembled on the computer until the whole genome could be seen.[34] Fifth and finally, the Human Genome Project initiated a framework and rationale for undertaking the sequencing of animal genomes in a comprehensive way.

Livestock genetics would be catapulted into the world of genomics when it became apparent in 2002 that funding would be allocated for the sequencing of animal species' genomes. Technology that allowed gene transfer, the cloning of individual animals, and identification of genetic markers for better productivity had generated a lot of excitement in the livestock breeding world, but the impact would be minor with respect to strategies of selection for breeding when compared to that of genome scans. These would lead to genomic selection or breeding by an animal's total DNA profile. *Genome Biology* reported on the new genomic focus related to animals, stating that the US National Human Genome Research Institute (NHGRI) had released a high-priority genome list for sequencing. *Genome Research* urged research groups who were interested in particular organisms to work with one of these centres in putting forward a proposal and "making the case for sequencing their favourites."[35]

The U.S. National Institute of Health provided $13 million to sequence the chicken genome, a job undertaken by the Washington University Genome Sequencing Centre. In 2004 the International Chicken Genome Sequencing Consortium published the first complete draft of the chicken genome, and at least 2.8 million SNPs were identified.[36] In 2005 sixty bovine researchers met in Houston to discuss sequencing the cow. The Bovine Genome Project received public funding, with the Baylor College of Medicine in Texas orchestrating the sequencing, using whole genome shotgun assembly methods.[37] A first draft assembly was released in 2006 and another in 2007. The sequencing was completed in 2009 and published in *Science*. The Wellcome Trust Sanger Institute in Britain sequenced the pig genome and released results in 2009 (an earlier one done in China was released in 2005), and the sheep genome was completed in 2008. With the completion of whole-genome sequencing of these domestic animals, HapMap projects were developed.[38] The idea of HapMaps rested in the conviction that haplotypes were valuable for finding DNA markers. A haplotype is a section of a chromosome that is transmitted as a unit from one generation to another and contains a number of SNPs, or single nucleotide polymorphisms. With better technology to sequence them, studying SNPs became increasingly important. Highly significant, too, was the reality that the whole genome, not just sections of it, could be looked at as a unit.[39]

The sequencing of animal genomes and the establishment of animal HapMap consortia brought about a profound change in the outlook of livestock geneticists who dealt with quantitative genetic theory. The desire to understand the genetic architecture of quantitative traits, if at all possible, was forefront. As B. Walsh noted in 2009: "Perhaps the most important new direction over the last 20 years is a change in perception. Quantitative genetics has transformed from a field where the focus was on estimating and exploiting summary statistics (e.g. variance components, breeding values, inbreeding depression and heterosis values) to a new focus on finding and exploiting genes ... There is no question that, as a field, our focus has significantly shifted."[40] But important questions remained, not the least of which was: how were markers like SNPs linked to gene expression? Could DNA sequencing be linked to QTL effectively? Was QTL inheritance based on simple gene changes perceivable in a few markers, or did Fisher's infinitesimal model still apply when seeking to understand DNA architecture? Would black box thinking have to be applied to a DNA sequence, because only regions or markers in the sequence, not genes themselves, could be linked to a QTL? What about the inheritance of such traits as fertility and fitness? Were there QTLs for those features? The advent of genomics did not provide answers, but the stage was set for dramatic developments in livestock genetics.[41]

The AI Industry: Bull Testing via Progeny or DNA?

Advancements in animal genomics caught the interest of AI studs offering dairy bull semen. The ability to evaluate breeding bulls was critical to the whole the dairy industry breeding structure, and rested on progeny testing by the AI industry. Methods of progeny testing might vary between studs, but essentially the thinking behind the process was the same. Bulls were assessed on the basis of the milk production of their daughters. Therefore the number of cows artificially inseminated and the number of cows recorded for milk yields were both critical to comprehensive progeny testing. AI use in the dairy industry was high in North America. At least 80 per cent of American dairymen, for example, used AI by the early twenty-first century. (In contrast, only 7 per cent of beef cows in that country were bred by AI.)[42] Studs tended to have their own group of producers who supplied data on cows for the stud, which in turn measured that data for each bull. It has been estimated that data exists on over 16 million cows in Canada and the United States combined, collected between the 1960s and 2010.[43] Bull testing was accurate in these two countries, but it was an expensive and time-consuming business.

Studs had to acquire young bulls, raise them, and collect data on fifty to two hundred daughters before it was possible to estimate their breeding value.[44] The more daughters providing milk yield information the better. AI organizations in the United States expect to spend between US$25,000 and $30,000 to test a Holstein bull. The percentage of bulls that graduate as proven bulls is about 12 per cent. (Some estimates put the cost at $50,000, and argue one in ten bulls will be proven superior.[45])

An average investment of $30,000 per bull means that finding a proven bull costs about $250,000.[46] Obviously a $50,000 cost per bull – with a one in ten chance of finding a good one – makes the expense much higher. A 2006 assessment of the progeny testing costs by AI organizations in Canada painted a much more dramatic picture: approximately C$25 million a year on five hundred young bulls, of which twenty would be considered worthwhile.[47] Rates of success seemed to vary as well. Interbull ran a study on international success rates of international Holstein evaluations between 1995 and 2003 in various countries: Australia, Canada, Denmark, France, Germany, Italy, New Zealand, Sweden, the Netherlands, and the United States. The study found that the percentage of bulls graduating from progeny testing in those years ranged from 4.4 to 14.7 per cent, with Australia at 4.4, Canada at 4.5, U.S. at 7.5, and Germany at 14.7. Average across all bulls and all countries was 8 per cent.[48]

Even if cost estimates and success rates were not consistent, it was entirely clear that progeny testing was an enormous expense from a global perspective. If a DNA profile could be developed that accurately indicated the presence of desirable traits, testing incoming bulls in that fashion had the potential of hugely reducing costs related to progeny testing. The five to six years spent on progeny testing a young bull would end. Incoming potentially valuable bull calves would be genetically tested, and if they met certain standards, would go into service within months. This would aid both AI studs and producers who bought semen, which would be much reduced in price because no expensive progeny testing had been undertaken. The question remained, though: was any of this possible?

The Advent of SNP Testing

The theoretical groundwork for breeding via SNP testing across the genome was laid as early as 2001. In that year European T.H.E. Meuwissen and his colleagues did a simulation study that concluded that a dense marker map that covered all chromosomes could reveal the breeding value of an individual. While the group admitted that QTL might be located on shorter maps, this approach did not provide enough information to predict breeding value. They suspected that the development of DNA chip technology that could locate many SNPs across a genome would provide a way for selection via DNA.[49] They concluded:

1] By using a dense marker map covering all chromosomes, it is possible to accurately estimate the breeding value of animals that have no phenotypic record of the own and no progeny. 2] This requires the estimation of a large number of marker haplotype effects. Using least squares, all haplotype effects could not be estimated simultaneously. Even when only the largest effects were included, they were overestimated and the accuracy of predicting breeding value was low. 3] Methods that assumed a prior distribution for the variance associated with each chromosome segment gave more accurate predictions of breeding values even when the prior was not correct. 4] Selection on breeding values predicted from markers could substantially increase the rate of genetic gain in animals and plants especially if combined with reproductive techniques to shorten the generation interval.[50]

Meuwissen had worked with haplotypes in his simulation study, looking for multiple markers.[51] His theory would be critical to the development of genomic testing and a move away from the search for QTL in a single or limited marker way, a common approach in the 1990s and evidenced in the hunt for markers responsible for beef tenderness. (Geneticists would soon back up his assumption and begin arguing by 2009 that inheritance structure for QTL was in fact complex. As VanRaden noted, "The distribution of marker effects indicates polygenic rather than simple inheritance."[52] B.J. Hayes suggested the same thing.[53]) Even if the actual genes responsible for QTL might turn out to be limited, Meuwissen believed they were closely allied to many SNPs and that these were identifiable through whole genome sequencing and HapMaps. Fisher's infinitesimal model made sense within this theoretical environment. In a new sort of black box thinking it seemed that, by identifying many SNPs collectively and in the vicinity of genes affecting quantitative traits, information concerning the breeding worth of an individual could be known by its DNA. The idea that black box thinking should be re-entrenched seemed to be a curious reversal of theory for van der Werf, who wrote, "this [SNP] approach seems to revert back to the black box, where the emphasis is on predicting generic variability for observed traits, rather than on understanding the underlying biology."[54] Complexity, infinitesimal model theory, and black box thinking all reemerged with a genomic approach to breeding. Apparently genomics could not, in the foreseeable future, provide simple answers to the way genetic architecture worked.

Looking for individual genes in DNA that accounted for traits of interest was like looking for a needle in a haystack, which was the way *Holstein World* explained the situation to its readers in 2008.[55] It was here that SNPs were so valuable. While most SNPs are outside protein-coding gene regions in the DNA and are therefore phenotypically silent (meaning they could not be seen in the physical appearance of the animal and did not appear to even dictate what that appearance might be), the ability to track them proved to be invaluable. Because their linkage to the genes dictating quantitative traits was not clear, however, the only way to make use of them was to assess their presence in all the chromosomes, and therefore in all the DNA in a genome.[56] In this way, black box thinking returned. While inheritance was looked at the DNA level, actual genes were no longer centres of attention. As Meuwissen said in 2007:

> Genomic selection [GS] may be defined as the simultaneous selection for many (tens or hundreds of thousands of) markers, which cover the entire genome in a dense manner so that all genes are expected to be in linkage disequilibrium with at least some of the markers ... GS sounded like a crazy idea in 2001, when a major aim in marker assisted selection research was to reduce the costs of genotyping. However, the development in the technology of single nucleotide polymorphism (SNP) genotyping has been tremendous ... It is this incredible development of the genotyping technology that makes GS feasible.[57]

Advanced technology would be critical to this new approach as Meuwissen indicated, and instantiations of it used to sequence the human, the chicken, the cow, and other species did change dramatically after 2005. Shotgun sequencing was largely replaced by what was known as next-generation sequencing, a technology that allows for a high volume of data to be sequenced more rapidly. The stage was set for real advancements in understanding polymorphisms within a genome that might be useful in the search for QTL.

The USDA decided to fund the work of Illumina, a company that undertook an independent study to assess whether there was any validity to the idea that genomic testing was a reliable way to breed. Founded in 1998, Illumina had become a leader in biotechnology relating to human genomics before the USDA choose it to independently test a SNP chip. The company would be launched into a new international field with this livestock research. It collaborated with the University of

Missouri-Columbia and the University of Alberta, and had input from other academic institutions.[58] (By 2010 Illumina offered a test for pigs, the Porcine SNP60, or over 60,000 SNP located; for sheep, the Ovine SNP50; for horses, the Equine SNP50; two different tests for dogs; and for cattle, the 50K chip, but they were also working on a 3K chip and one with 700,000 SNPs. Chips for chickens were also available but with certain restrictions.)[59] AI studs across North America also provided money for the cattle SNP project, and perhaps even more importantly the genetic material used to test the validity of Illumina's SNP chip.[60] It was agreed that the studs would have exclusive rights to the genomic data on their sires for five years.[61] This tendency towards patenting worried some scientists. As M.E. Goddard stated, after the project was completed, at a conference for beef cattle:

> In U.S.A. only AI companies that participated in the U.S.D.A. experiment have the right to use the prediction equation on bulls. In other countries there are similar restrictions. In all cases these occur because companies believe they have an advantage over their competitors that they are unwilling to sacrifice. Since nearly everyone is investing in the same technology, in the near future nobody will have much advantage. The world's dairy farmers benefited greatly from the open system of exchange of information that has prevailed for traditional genetic evaluation based on pedigrees and phenotypes. They would also be best served by a similar approach to genetic evaluations that include genomic data.[62]

Some 3,500 Holsteins of known good quality phenotype, known as a training population, had been genotyped to provide the basis for mathematically formulating the SNP test. As of October 2008 some 14,720 Holsteins, 1,558 Jerseys, and 368 Brown Swiss dairy cattle had been tested with a high-density panel containing over 50,000 SNP markers, known as the Bovine SNP50 BeadChip.[63] The BeadChip proved to be highly accurate when assessed against the known phenotypic quality of the tested animals, but unfortunately that accuracy was breed-specific. Thousands of SNPs, polymorphic in Holsteins, were monomorphic – that is, they showed no deviation across individuals – in Jerseys or Brown Swiss.[64] There seemed to be no question either that the larger the training population, the higher the accuracy rate. The test, then, would have its most dramatic effects on the future breeding of Holstein cattle because the volume of data, used to verify what the DNA sequences of a known quality bull, allowed assessment of animals solely on the basis of their DNA sequences. No other dairy breed and no beef breed could amass data of observable and proven good value on that scale. Genomically enhanced sires were to be identified for the public in early 2009, meaning that producers could buy the semen of such bulls at that time.

In July 2008, *Hoard's Dairyman* explained the situation to its readers, telling them that AI bulls that had had no progeny testing might be available within a year. "If genome testing eventually becomes the gold standard prior to progeny testing, genome-tested bulls will replace the AI young sire group now familiar to producers," the journal stated, adding: "Some see [genomic predictions] as a replacement for progeny testing. Others don't quite believe any of this stuff yet."[65] Two things seemed clear, however. "First, genomic predictions will not approach the accuracy of large progeny tests for a long time. Second, even under very conservative expectations, genomic predictions will dramatically affect the dairy cattle breeding business ... Progeny information verifies – or refutes – relationships between gene sequences and progeny data in previous generations. Just how important this process will be is not yet known."[66] (Very shortly after the release of the test results, producers could also buy the 50K BeadChip and use it to test their cattle. By February 2009, the journal reported that genomic evaluation had been integrated into a number of Holstein and Jersey programs. It was clear that genomic testing would have a dramatic impact on how young animals were judged, the journal told its readers at that time.)[67]

SNP testing would bring about huge changes in the breeding of dairy cattle after 2009, but it did so in conjunction with the joint work of the purebred dairy breeders and producers over at least sixty years, and the research of quantitative geneticists using data collected by AI studs over the past fifty years. Purebred breeders supplied the male genetics that the AI studs tested. The AI studs relied on milk recording from producers to test bulls. Geneticists studied the results of both the producers and the testing of bulls, in order to improve the process of breeding selection for both purebred breeders and the producers who used their genetics. Hybrid breeding and the privatization that had occurred in the chicken (and pig) breeding industry did not generate public records that could be used by geneticists to make simulated models to study. Poverty of data in species such as chickens and pigs also resulted from the fact that there could not be an AI industry corresponding to that of dairying: their semen did not freeze – either at all or not well. In the case of chicken, cock semen (which could not freeze) was used within companies, which never sold genetic material for breeding purposes. In the case of pigs, breeding companies sold semen in order to generate data for performance testing their home-bred boars, which then could be marketed more successfully. Semen-selling, then, was allied with their private breeding operations.

Effects of BeadChip SNP Testing on Dairy Cattle Breeding

It was the potential to reduce the cost of progeny testing that commanded the real interest of the dairy-breeding industry in the results of the test. Clearly, simply screening young bulls that went into progeny testing would reduce costs, a feature that was attractive enough to the AI studs to make them join in the Illumina project. While the studs looked to reducing their costs of progeny testing, via cutting down on the intake of young bulls for progeny testing, there were other changes in the entire breeding industry that resulted from DNA testing. The experiences of one AI stud serve as an example of the larger impact that genomic testing had on the whole breeding structure. In 2006 the American AI company, Genex Corporation, progeny tested about 300 young bulls, but with the advent of Illumina's BeadChip, bulls were DNA tested before entering progeny testing, with only 180 to 200 being judged as being worth that greater expense. The need for milk recording data was therefore reduced. In early 2008 Genex Corporation was testing bulls in 2,000 herds, but by the end of the year that number has been reduced to 160. Under these new conditions, one in five bulls passed the progeny test, instead of one in ten.[68]

But what if producers sidestepped all progeny testing and wanted to use bulls tested only by DNA? A whole set of changes would come into play and raised many questions. What would the rate of accuracy look like with reduced progeny-tested animals, a situation that would arise if using genotyped sons of genotyped but unproven bulls? How many bulls would be needed for progeny testing to restore accuracy rates? Can genotyped bulls be accurately tested through the results of commercial sales and not through organized progeny testing programs? How much data would be collected from such commercial sales when progeny testing programs are reduced or eliminated? Organized progeny testing encouraged the keeping of milk production recording and type classification programs, but if it seemed that this data was no longer of any use, less of it would be collected, thereby reducing the amount of data needed to re-estimate SNP effects on either traits valued at present or traits that might be valued in the future.[69] In 2008 managers for Genex Corporation had a meeting to discuss the issue of producer purchase of non-progeny-tested bulls. The group came to the conclusion that because of intrinsic farmer conservatism, only about 15 per cent of the company's business would move from progeny-tested older bulls to genomically promising young bulls. To the company's astonishment, by mid-December of 2009, 40 to 45 per cent of their sales were from sires with no milking daughters.[70]

The move to using non-progeny-tested, DNA tested bulls by producers was rapid. Figures on the Canadian situation were available by early 2011. Although there were no figures on the American situation, *Hoard's Dairyman* saw little reason to think it was different in that country. In an article entitled, "Genomics being adopted at a swift pace," the journal reported:

> In less than two years since genomic-tested bulls were released for sale in Canada, young bulls have overtaken their progeny-proven barn mates by garnering 58 per cent of the country's AI services. While parallel observations on the US market aren't available, we believe there is a similar pattern. In their research, the Canadian Dairy Network evaluated insemination data from major AI and D.H.I. organizations over the past decade. A major finding was that bulls formerly known as 'sires in waiting' in the pre-genomic era are now in great demand thanks to genomic testing. Semen sales for these 2-to-4-year-old bulls hovered from 10 to 12 per cent for much of the past decade but jumped to 29 per cent last year. Their new-found fame came at the expense of newly progeny-proven 5-to-8-year-old bulls whose market share dropped from 49 to 32 per cent in just a few short years. These new trends certainly change the AI industry's structure. Bulls now have a much shorter shelf life ... The top 10 bulls for semen sales garnered 32.6 per cent of all AI services at the beginning of the decade. Now they account for just 17.6 per cent.[71]

With the completion of the Illumina study, the studs began to change their arrangements with the breeders: they asked for more planned matings, genotyped all resulting male calves, and increased collection of embryos from top females.[72] DNA testing young bulls was a more accurate way of deciding whether they should enter a progeny test program: 50 per cent for genomically tested bulls versus 27 per cent under the indexing system used by many studs – the same system that drove Heffering out of the United States and into Canada.[73] Another way to assess a young bull genomically was to DNA test both its parents and take a parent average of from that set of tests.

Interbull, International Standards, and Dairy Breeding by DNA

The international dairy breeding industry quickly adopted genomic testing. In late 2008 Interbull put out a survey to all its member organizations designed to assess the international effects of both genomic testing and the 50K BeadChip on the breeding of the world's dairy cattle. Answering rate was 100 per cent. The report stated that eleven countries planned to include genomic information in pedigrees by 2011. None had done so by 2008, and most of the eleven intended to start in 2009. Six members stated that they had no plans to include genomic information, while fourteen remained undecided. In virtually all countries, Holsteins were getting the most attention. Jerseys were second, but at a much reduced rate: ratios of Holstein to Jersey were ten to three, or nine to two, or eight to one. While Holsteins were the main focus of genomic testing, the report indicated that more attention was going to be paid to Jerseys.[74]

The early indication that increasing numbers of accurate phenotypic data was hugely significant made blocks of breeding organizations join forces to increase the volume of information pertaining to breeding animals. The desire to pool data led to amalgamations, first of data collection in the United States and Canada, and then that of various European countries. The United States and Canada had worked jointly for some time before the advent of genomic breeding. In 1992, for example, the North American Collaborative Dairy DNA Repository (CDDR) was formed, when major AI organizations in North America began banking semen on young bulls. Both countries started a national genomic evaluation program in 2009 using the Illumina 50K BeadChip, and had previously shared information of genotype research. B. Muir elaborated on the U.S./Canada coordinated efforts at an Interbull conference held in Paris in 2010. He explained that all laboratory work on genotypes was submitted to the USDA, which then forwarded all North American genotypes and pedigrees that were associated with those genotypes to the Canadian Dairy Network (CDN). "In addition," he added, "the U.S.D.A. and C.D.N. exchange pre-genomic genetic evaluations for genotyped females and their ancestors."[75] He also pointed out that procedures for genomic testing in the two countries were not identical. In the United States proven bulls and cows were used in the estimation of

SNP effects, whereas in Canada females were not part of the equation. In order to overcome this divided way of assessing SNP effects, "every genotyped animal in North America receiv[ed] a genomic evaluation in both the U.S.A. and Canada, regardless of where the animal was originally genotyped or evaluated."[76] Muir concluded that: "Increased ability to access information on animals due to the incorporation of genomics promotes greater opportunities for across border merchandizing of genetics ... Several studies have shown that ability to accurately predict future proofs for young animals using genomics improves as the number of proven genotyped bulls included in the reference population for SNP estimation increases ... The reliabilities for young animals nearly doubled with the inclusion of genotypic information in both countries."[77]

Muir noted that public access to obtain male genotypes on bulls used by AI organizations would, as per the five year agreement, be granted in April 2013, and that would profoundly change the situation. At the same conference, X. David explained the amalgamation efforts occurring in Europe. He started by noting that the 2008 North American merger of reference, or training populations, allowed for a pool of 9,300 bulls. In 2009 the situation triggered a response in Europe when the French initiated the pooling of data from various European countries – namely Germany, Netherlands/Belgium, France, and Finland/Denmark/Sweden – and the formation of EuroGenomics. The consortium added some scientific partners: INRA, University of Liege, University of Arrhus, and Nordic Cattle Genetic Evaluation. The Illumina 50K BeadChip was central to data collection. David continued: "EuroGenomics['s] common reference population is the largest reference population in the world. Furthermore, it is unique in its quality. All EuroGenomic partners run breeding programmes based on the best available Holstein genetics in Europe and North America. Therefore, the EuroGenomics reference population mirrors the full variation of international Holstein genetics perfectly. This will be certified by the next Interbull validation for national genomic evaluation systems once it is offered for official use."[78] MACE evaluations would soon be modified in attempts to include genomic data, and were labelled as GMACE. Creating a workable GMACE was not without difficulties.[79] Increasingly, countries that were not members of Interbull found it hard to function genetically in this international world. If non-member countries did import semen, they had no way of judging the effects that might have on national herds, because no statistics existed that could account for environmental differences under which data had been collected. This was perceived to be a serious problem for countries trying to develop better dairy herds via the use of international genetics.[80]

For better or for worse, by 2009 it appeared that the old progeny test, a selection system in place since the eighteen century, and finely tuned over the twentieth century by quantitative genetics with the aid of AI, might be on the way out in dairy cattle breeding. The swiftness with which producers adopted genomic testing was matched by the explosion in experimentation using the 50K BeadChip; that is, evaluation that aimed to further assess its reliability and see what other uses it could be put to. For example, could the 50K Beadchip be applied to crossbreeding? Or, could it ever be applicable to cattle generally, instead of being breed-specific?[81]

DNA Testing and Cows in Breeding Programs

The potential impact of genetic improvement through the DNA testing of females attracted attention almost as early as males. Studies indicated that when using female data, the rate of genetic improvement over generations would indeed increase, and do so more rapidly than if only males underwent the test. In 2011 N. McHugh and colleagues concluded from a simulation study that: "The potential use of females, with their own performance records, to estimate marker effects becomes increasingly important, especially in countries with small populations. This is a particularly pertinent question, as the cost of genotyping is likely to decrease considerably, either through advancements in genotyping technology or the development of smaller, less dense marker panels coupled with imputation algorithms, or both."[82] The rise of the 3K Illumina chip seemed to suggest a way to utilize female genomic data on a much broader basis; namely, for use on commercial dairy farms. The USDA collaborated with Illumina on a project to produce a 3K BeadChip, which would be substantially cheaper than the 50K one and which would offer some genetic information on breeding quality.[83] In September 2010, Illumina put a 3K chip on the market at the cost of US$30 to $50. Some thought this chip would be a better diagnostic breeding tool than the 50K chip.[84] Its accuracy was said to be nearly as high as that of the 50K chip.[85] But that was true only under certain circumstances; namely, in conjunction with 50K chip-use in stock similar to the individual. Animals tested with the 3K chip could have their missing genotypes imputed from relatives or other members of the population tested with the 50K chip. Imputing increased the reliability of the 3K chip almost 20 per cent. Average gains in reliability with the 3K chip for young animals were 79–88 per cent of those with 50K chip if imputing was used, but only 61–63 per cent without imputation.[86]

It soon appeared that its chief value would be in assessing commercially producing females in dairy herds – namely, helping farmers decide which animals to keep for milk production and which to cull. The 50K chip had been used primarily on AI bulls, bull dams, and other elite cows. Before the availability of the 3K chip, only 39 per cent of genotyped cattle in the United States were female, but by the end of 2011, 59 per cent of cattle tested by any chip were female. The majority of animals that had been tested with the 3K chip were female Holsteins, making any results of the test breed-specific.[87] Geneticists became interested in learning how this test could be used across general American Holstein herds. A simulation study was undertaken to evaluate the expected gains in lifetime net merit that would result from genotyping females on a commercial dairy farm with the 3K chip. This was the first study to look at genomic testing as a way to help the ordinary farmer make decisions about his milking cows on the farm, and it brought interesting results. Genomic testing proved to be most valuable when little other information on an animal's potential was available, meaning heifers and yearlings, not lactating cows. The 3K chip served best in selecting for replacement cows and culling poor producers before they entered the milking herd. Farmers should start by collecting females with pedigrees showing good performance data in relatives, and then genomic test the lot. It was then possible to know which animals within that group would be better than others, and farmers could screen out the poorer ones.[88]

Breed Variation and DNA Testing

Chips of larger sizes were also made available. It was apparent quite quickly that even denser coverage of SNPs – from 500,000 to 700,000 – might give more accurate prediction.[89] Larger SNP coverage seemed desirable if one was to construct an accurate test that was not breed specific; namely, a test that predicted the value of purebreds in a crossbreeding program. It was also hoped that a chip could be developed that would be useful for dairy breeds that did not have sufficient data to establish a DNA profile for quality. The breed-specific nature of the 50K chip seemed to bottleneck research on breeding strategies for breeds other than Holsteins and, to a more limited degree, Jerseys. The assumption was that an even more infinitesimal model was needed – the net had to be cast further and more widely to capture the multitude of genes acting on QTLs across the genome. Black box thinking and the Fisher model seemed even more relevant with advanced research simulations using the 50K BeadChip.

With the advent of a 778K BeadChip from Illumina and a 649K, a different type of array from the company Affymetrix, one study looked at how using the two combined might work to overcome the breed specific nature of SNP testing. It had been established that the interval between markers in the 50K chip were five times greater than what was needed to show consistent LD phase across breeds. This simulation study showed that each of the larger arrays alone could not reduce the marker interval enough to provide a suitable threshold for crossbreed testing. But the study did conclude that "The combined use of both platforms significantly improved coverage over either platform alone and this decreased the gap size between SNP, providing a valuable tool for fine mapping QTL and multibreed animal evaluation."[90] Then there was the problem of assessing crossbreds outside any reference to purebreds.[91] Another enormous issue with large and larger arrays and the amassing of huge amounts of data was how to sort and store the information.[92] By 2011 there was some evidence that estimating haplotypes across the genome (instead of SNPs) might work better, especially as arrays became denser and denser and required stronger software.[93]

The increased availability of data did not come without problems. The ability to gather data far outpaced the ability to analyse or organize it by 2012. The amount of information created globally grew by a factor of nine between 2005 and 2010. At the same time the cost of processing the data – that is, creating, managing, and storing it – has been drastically reduced: in 2012 just one-sixth the cost in 2005.[94] Some data sets were so large they could not be accommodated all at once by available software. The sets must be divided up as subsets, and extracted as such for analysis, via a process known as data mining. Computing power was also a problem in dealing with enormous amounts of raw data.[95] Time to process data and come up with results for questions addressed was another issue that needed consideration. Still, it seemed to some that hope lay in the ability to use ever more complex amounts of data, rather than in restraining its collection. At the heart of the matter was the hope that the functioning of biology itself would be better understood. As J.B. Cole stated in 2012:

> History indicates that we never have enough information; the more data we have, the more data we want. The ultimate value in big data lies in their ability to answer questions, which may range from routine calculation of genomic estimated breeding value to predictions of regulatory networks, and there are always new questions. Relationship among traits are of increasing interest as we try to understand the biology that underlies complex phenotypes in livestock species, which will require ever-increasing computational resources. Customers demand rapid turnaround, often in real time, and smart engineering will be required to deliver such services cost effectively. Ultimately, we hope that more really will be better.[96]

Cole recognized that the situation was ever more complex for animal breeders. He advised them "to seek out more formal training in programming, rather than depending primarily on self-learning."[97] Such thinking suggests the return of an old trend: genetics becoming less rather than more available to the general livestock breeding industry.

The BeadChip and Beef Breeding

Purebred beef cattle breeders watched the drama as well, and quickly assessed the value of the 50K BeadChip for their purposes. An article in *Ontario Beef Farmer* told beef breeders and producers that:

> The beef and dairy industries are going in opposite directions on genomics … There's been a highly developed genetic evaluation system for dairy farming for generations, now being fully integrated with genomics. But the beef genetic evaluation system has always been haphazard, underfunded, and if anything, has less momentum now than it had a decade ago … But there are a couple of other structural impediments to the development of genomics in beef cattle. The first is that [there are] so many breeds and crosses used that establishing good genomic benchmarks for what it needed for each breed will take a lot of work. The Angus association in the US is working in it and that makes sense. The other question is that there still isn't any consensus on what make great beef … In the dairy industry, the breeding leaders for North America got together to fund the expensive research to set genomic guidelines for Holsteins, with some help from the US government. There's not the same push in the beef side.[98]

Shorthorn Country warned breeders in the summer of 2009 that the 50K BeadChip had been designed for specific breeds, and therefore did not apply to all purebred animals. As Patrick Wall, director of the American Shorthorn Association, explained, "Since the original tests were not developed on a population of Shorthorns, it's likely any testing done previously on your purebred herd has been an utter waste of time and money."[99] Beef purebred breeders and producers do not rely on technology, he added. "All the marketing and technology available to producers over the past 30 years has just 7-8 [per cent] of them using AI. That's from an unbiased nationwide poll from the USDA, including purebred and commercial breeders of all sizes."[100]

Geneticists began to wonder if genomic tests for the beef breeds could be made to work by collecting data on the general crossbred beef herds. This approach would overcome the problem of insufficient data from training populations. L.D. Leachman, reporting to the Beef Improvement Federation in 2009, noted that performance data was most frequently collected for purebred herds and was important for crossbreeding. He added, though, "Since the backbone of beef cattle production is the crossbred herd, defining the aim of beef breeding programs in terms of crossbred performance would seem sensible. At present, only limited amounts of crossbred information are being utilized in formal genetic evaluation programs."[101] Simulated studies were undertaken the next year to test the hypothesis that genomic testing could locate traits of interest in crossbred herds. Toosi, for example, argued that it made more sense to focus on the traits of interest in crossbred populations of uncertain breed composition, which then could be extrapolated back to the purebreds that went into the crossbred herds. It seemed to him that one, the general herds would be improved under these conditions, and two, the purebred breeds that went into that composite would as well. Simulated results suggested that the strategy might work.[102]

Chicken Breeding, Molecular Genetics, and Genomics

As with dairy cattle, work on blood groups and chicken diversity had commenced as soon as studying genes at the DNA level was possible. In the 1980s scientists investigated the effects that blood groups could have on artificial selection for the breeding company DeKalb. Work with RFLP technology indicated that resistance to the worldwide Marek's disease had a genetic base.[103] The advent of DNA markers, the chicken genome, and HapMap intensified research on inheritance patterns in chickens. Characteristics such as quality of egg shells and resistance to disease commanded the most attention. One breeding company, Hy-Line International, claimed in a science journal for the poultry industry, that it had in-house research programs since roughly 2000 that studied molecular genetics, looking for marker effects on such issues of disease. What is perhaps more interesting, though, was the reality that any information coming out of that research was not public,[104] and therefore added nothing to the scientific community studying chicken genetics. By this time, however, far more sophisticated efforts at maintaining open dialogue between the breeding industry and academia were in place than had been the case in the 1950s and 1960s, when the Poultry Breeders Roundtables in the United States occurred.[105]

With the advent of the genome, the chicken HapMap, and Illumina technology new information concerning chicken genetics became evident – information that could not have been available with microsatellite technology. Assessments of LD in layer and broiler lines of chickens indicated that linkage disequilibrium was highest in layers and lowest in broiler lines. It was apparent from studies of LD as well, according to one study, that: "individual commercial breeding lines [had] lost 70 [per cent] or more genetic diversity of which only 25 [per cent] of this loss [could] be recovered by combining all stocks of commercial poultry. However, and interestingly, this did not mean that modern agricultural practices were the primary source of this allele loss as, in fact, the majority of alleles were lost prior to the formation of the current industry. These results emphasize the need for concerted national and international efforts to preserve chicken biodiversity."[106] Since short distance LD reflects population size many generations ago, while long distance LD reveals more recent population history, LD patterns in chickens made it clear that most genetic diversity loss occurred before hybrid corporate breeding.[107] An interesting point, perhaps, given that the modern chicken-breeding industry has been blamed for greatly reducing the genetic diversity of the world's domestic chickens. Scientists, as evidenced in the above quote, argue that the situation made it all the more important that existing diversity be preserved by the breeding companies. It seemed clear too that existing LD covered shorter distances than was true in other livestock, particularly cattle. Genetic diversity loss in cattle has apparently been more recent than in chickens.

Early attempts to measure LD involved assessment of only several of the thirty-nine chicken chromosomes, using a 3K or 6K chip designed by Illumina, on both layer and broiler lines. Geneticists argued that SNP testing of this nature could be useful in assessing relationships of lines to each other in breeding programs.[108] SNP testing across the genome, however, was more difficult in chickens than it was in cattle because the chicken genome was far more complex than that of other livestock species. It had important micro-chromosomes in addition to macro ones, a fact that made it difficult to obtain good coverage across the bird's genome. For reasons not clearly understood as late as 2011, genome assemblies seemed to leave the smallest micro-chromosomes highly under-presented; meaning coverage of the entire genome was not uniform.[109] This situation made DNA testing for QTL by SNPs along the theoretical lines applied to cattle – namely, that uniform coverage would capture QTL – difficult. The biological and technological problem meant that the advances in genetics, which had worked in dairy cattle breeding, could not be applied to chickens; somewhat ironic, perhaps, in view of the fact that sequencing the chicken genome had been completed earlier than that of cattle. H-J. Megens explained the difficulties that had to be overcome:

> While it is known that recombination frequencies are much higher for micro- as compared to macrochromosomes, there is limited information on differences in linkage disequilibrium (LD) and haplotype diversity between these two classes of chromosomes ... The limited haplotype structure and LD suggests that future whole-genome marker assays will need 100+K SNPs to exploit haplotype information. Interpretation and transferability of genetic parameters will need to take into account the size of the chromosome of the chicken, and, since most birds have microchromosomes, in other avian species as well ... All SNP based studies so far have used SNP densities that were insufficient to ascertain LD and haplotype structure in chickens ... The higher recombination rate on microchromosomes is expected to reduce LD compared to macrochromosomes ... A systematic survey of differences in haplotype structure between micro- and macrochromosomes in birds has not been done.[110]

By 2014 a different approach to the enhancement of chicken breeding through genomics was undertaken through a project launched by the UK government. This project hoped to elucidate how evolutionary biology, the process of domestication, and the effects of chicken-breeding strategies could be explained by a comparison of modern DNA with ancient DNA. While understanding the process of domestication via chickens (not dogs, pigs, or cows as is more normal in such studies) is a top priority for researchers, they also expect to help the chicken-breeding industry. Certain difficulties (for example health issues involving avian influenza, leg weakness, and fertility) might be tackled more successfully with better knowledge of how the bird's modern genetic structure relates to its historical structure.[111]

Quantitative Genetics in the Genomic Era

While hybrid breeding as applied to chickens had become the "old" genetic approach to the breeding of livestock by 1960, and quantitative genetics had taken the lead, the situation changed again with genomic breeding. At the forefront now was selection by DNA. Men trained in the old Lush school of what they called "population genetics" felt the shift particularly keenly. The whole foundation of livestock genetics, from its hybrid/quantitative base within the framework of classical genetics, had changed. Livestock genetics had become part of molecular genetics. A move away from quantitative genetics to genomics in animal breeding research resulted in the aging of a research population that studied animal breeding. A decrease in public funding in the area accompanied this trend. I. Misztal claimed that students found looking for specific genes more alluring than assessing the levels of their expression, "Especially because the latter [approach] require[d] extensive training in quantitative genetics, statistics, and programmes."[112] Black box thinking, however, remained part of any study of livestock genetics, he pointed out, and said: "The new trend in animal breeding is genomic selection using SNP chips. In this methodology, one estimates effects of individual haplotypes, and genomic [values] is estimated as a sum of those effects. No effort is made to identify QTLs. The genomic selection is based on the assumption opposite from previous efforts in markers but the same as in 'black box' genetics: that a large number of genes are responsible for a trait."[113] These trends worried the USDA and led the department to develop a *Blueprint for USDA Efforts in Agricultural Animal Genomics* in 2007. "The plan," the document stated, "was built on the foundation that quantitative genetics has been used for many years in selecting animals for improved production ... and has achieved remarkable results, while pointing to the fact that addition of animal genomic technology to quantitative genetics programs has the potential to lead to more accurate and rapid improvement."[114] The blueprint elaborated as follows:

Although the opportunities of this new Renaissance period of animal breeding and genetics are exciting, several daunting questions exist that must be addressed to ensure success. The reduction in number of academic animal breeding programs around the world, and particularly in North America, has resulted in a deficit of human talent and resources ready to take on these challenges. Not only are quantitative geneticists lacking, those few who are being produced in the remaining programs are entering a highly competitive job market where few are remaining in academia, with many of the trained animal breeding scientists being pulled into the plant and biomedical arenas.[115]

By 2012 it appeared to some that the collection of ever more data, in conjunction with the developments in genomics, offered real hope for overcoming the old black box problem and the tarnishing of quantitative genetics by it. It seemed that in the near future, breeding could proceed via an understanding of the genetic architecture that explained why markers were in linkage disequilibrium. What one might even call "black box genomics" would end, thereby removing the stigma inherited from classical quantitative genetics. As J.B. Cole and associates wrote in early 2012, "Genotypes for each individual require a large amount of storage, increasing as SNP chips increase in density, and will continue to grow as individual sequences become routinely affordable. Full sequence data are of great potential value because the accuracy of genomic evaluations will increase once causal mutations can be tracked, rather than markers in linkage disequilibrium with those mutations."[116]

Regardless of the decline of quantitative genetics, its relationship to black box thinking, and developments in genomic technology, the most advanced form of genetic breeding (breeding by DNA tests) still relied heavily, somewhat ironically, on purebred breeding. The dramatic changes that the dairy industry had undergone from the 1940s until the 1960s resulted from the work of purebred breeders in conjunction with that of quantitative geneticists such as Robertson and Henderson. With the advent of genomic breeding, its applicability was breed-specific. The old "art" of breeding had become once again the "science" of breeding, or, at least, part of the science of breeding.

Chapter Six

Biology, Industry Needs, and Morality in Livestock Breeding

Patterns described in this book indicate that technology cannot be separated from the application of principles arising from genetics. While technology has often been described as the poor sister of science, when it comes to approaches to animal breeding in many ways it seems to be the other way around. In sequencing SNPs for the 50K BeadChip, for example, it could be argued that science is the poor sister of technology. Genetics has been increasingly closely linked with technology since the advent of AI, when that technology became a critical partner of quantitative genetics. With the rise of molecular genetics, and more particularly recombinant DNA as well as genomics, technology can no longer be separated from the functioning of genetics applicable to livestock. Animal breeding as science/technology has been metamorphosed. It is more accurate to describe it today as biotechnology.

But the technology/science complex, or later what might be described as biotechnology, interfaces with agricultural concerns in complicated ways, and therefore presents many faces. One aspect might be described as an industry-needs/ethics one. Problems arising from the combined effects of agricultural industry-needs and biology of the animals, in conjunction with contemporary levels of technology, provoke ethical responses.[1] In this chapter I outline how issues arising from the complex – specific to certain livestock industries – were dealt with over time: namely, handling unwanted male offspring in chickens and dairy cattle, and horns on cattle. I concentrate on farmer reactions to the entangled problems from the nineteenth to the twenty-first century, not so much on the involvement of humane societies and animal rights groups within that framework. My intent is to present the information in an illustrative rather than analytical way. It might seem that I have moved away from the general topic of historical artificial selection. Not so. The phenotypic response of livestock to the breeding strategies applied to them plays a large part in this story, and has done so increasingly since the late twentieth century. Biology and strategies of breeding cannot be separated from each other, because artificial selection's ability to alter an animal can create industry problems as often as it solves others.

Dealing with Male Baby Chicks and Dairy Calves

A particularly important question that plagued both the dairy and table egg industries was how to deal with the male young born every generation. Dairy products and eggs come from only one sex, the female. Males could be useful (outside of reproductive purposes) only in a subsidiary way – that is, as meat producers. That ancillary use became increasingly untenable with the advent of ever more effective artificial selection strategies. With the extreme specialization that had been bred into dairy cattle and egg-laying Leghorn hens, their male counterparts came to provide so little meat when reaching adulthood that the cost of raising them no longer warranted the activity. About half of each new generation would be male. The problem, increasingly evident over time as a result of breeding strategies, was what to do with them. The situation was different in livestock used specifically for meat – that is, beef cattle, broiler chickens, and pigs. Breeding strategies had led to better meat production in both sexes and both, therefore, continued to be useful economically. Although males might be preferred over females, the females could justify the cost of raising them.

The commercialization of the dairy industry in the latter half of the nineteenth century did not immediately lead to breeding strategies that overly specialized dairy cows as milk producers. But the rise of better markets for dairy products made farmers more interested in the milk their cows could produce than in the calves that had to be born in order for the cows to give milk. In limited numbers, heifer calves were valued as replacements. Virtually all bull calves, under these conditions, were completely useless to a farmer who collected the cow's milk for sale in the form of butter or cheese. But there was no way to prevent bull calves from being born. What to do with them? The nineteenth century situation in Ontario serves as an example of how early dairymen handled this industry problem.

In the late nineteenth and early twentieth centuries, the only way to deal with unwanted calves was to kill them at birth. By the early

1880s calf killing had become extensive in Ontario, with at least 200,000 killed each year.[2] The dead calves were skinned and the carcasses left out to feed the hens.[3] Not all of these were male either, to the disgust of the *Farmer's Advocate*, which saw heifer killing as an abomination.[4] Sometimes very young calves were sent instead to market, in the hopes that some income could be realized from them. Feeding the calves separately, even for a short period, made no economic sense because veal generated poor returns. "The sale of this class of meat is brought about by the high price of butter at this season, and also the low price of veal," the *Farmer's Advocate* explained. "The consequence is, in plenty of cases the calves are sold at a few days old, or kept, at the most, a couple of weeks."[5]

The increasing emphasis on dairying and its separation from beef farming made many people anxious about a potential shortage of meat.[6] The general move towards dairying meant that fewer farmers bred and fed cattle for meat. While dairy cattle could of course also supply meat, feeding them to do so made no economic sense. The dairy and beef industries were, increasingly, separate industries and were perceived to be supplied by cattle of different types. Dairying, in effect, decreased the beef supply because it promoted the birth of unusable calves. "We think that some of our correspondents hardly take fully into account the continued strength of the dairy business, the demand for cows of the dairy type, and the annual slaughter at birth of thousands of calves, steadily going on in the [cheese] factory sections, all of which tends to diminish the ranks of beef cattle," explained the *Farmer's Advocate*.[7] While the *Advocate* did not condone extensive calf slaughter, the journal believed that many of these calves were simply not worth keeping – for even a limited period – because the poor beefing quality of their sires made them useless from a meat point of view.[8] The practice of calf killing began to attract attention in urban areas of Ontario, and in the process an industry problem became for many irrevocably attached to morality. Slaughtering newborn babies was ethically repulsive. The *Farmer's Advocate* tried to keep the focus on practical matters surrounding the problem, and refused to be drawn into a moral discussion, saying, "We have no desire to kindle anew the ludicrous discussion that emanated last season from urban centres concerning the slaughter of calves."[9]

Another way to handle the males was to funnel the calves into a viable veal industry. Before the twentieth century no such industry existed in Canada because there was little demand for such meat.

A veal industry did not develop quickly after that time either. A more viable veal industry arose in Britain, but it was clear from the British experience that an industry of this nature did not completely diminish moral concerns over the fate of male calves. By the late twentieth century, the handling of these babies had attracted the attention of British animal welfare supporters who pressured for some form of regulation concerning how the calves were housed. Taken at birth from their mothers, calves were kept individually in what was known as calf hutches, or crates. Generally this procedure was followed to prevent excessive illness in the youngsters, which were susceptible to many contagious diseases, and to keep the meat they provided white rather than pink. In 1990 government regulation prohibited the use of small crates to house the animals. Calves had to have enough room to turn around and many would now be penned in groups. (Similar legislation was proposed in the United States but failed to pass into law.)

One of the veal industry's problems was that the meat was not attractive to people in some countries were dairying was widespread. Demand for veal meat was never particularly robust in either North America or Britain. It was strong, however, in northern Europe. As a result, by late in the twentieth century, most of Britain's veal calves were shipped to the veal-consuming nations of France and the Netherlands. The way the animals were treated – no feed or water during the trip – also worried animal welfare people. The British government passed legislation in 1995 that regulated travelling conditions for the livestock. The ban on British livestock due to foot and mouth disease halted the export of calves, which reduced the protesting. In 2006, pressure to stop the whole export industry of veal calves gathered momentum again. British dairy farmers argued that if the export of the calves was discontinued, they would have no other option but to kill the calves at birth. No adequate veal market in Britain existed to sustain the rearing of specialized dairy calves. (By this time there was also pressure in the United States to stop the use of small veal crates.)[10] The veal approach to the problem of male calves triggered as much moral outrage as had calf killing, and by the late twentieth century, the only tools available to solve the problem were the two nineteenth-century methods.

Industry needs and specialization began to affect the lives and breeding methods practiced on dairy heifer calves by the twenty-first century in the United States. Shifts in fundamental strategies towards how to practice artificial selection on these dairy heifers could alter the framework of the dairy breeding industry. Outsourcing heifers became increasingly common as American dairy operations increased in size, while farmers known as heifer growers took over the raising of young calves when dairymen boarded the stock off the farm. In 2003 a study of the industry indicated that heifer grower operations housed from thirty to twenty thousand heifers, with the average number being roughly twelve hundred. Farmers who were heifer growers derived most of their income from that occupation. The heifers were returned to their owners when bred. Methods of breeding them changed in this environment. Heifer growers used a fair amount of natural service for breeding, although AI was used to some degree. The use of pasture bulls was surprisingly prevalent when compared with the national average.[11] It is not clear why heifer growers relied on walking bulls to such a degree, but apparently increased segmentation of dairy operations could affect how breeding would proceed. The rearing of heifer calves in this fashion brought about no moral outrage, but it did influence breeding methods and potentially could change selection forces. AI bulls were progeny tested far more accurately and for designated traits than walking bulls.

By the early twenty-first century excitement mounted over the development of a new technology that had the potential to solve the sex problem: sexed semen. A dream for a number of years (Lush discussed its possibility as early as 1945), the technology of sexing semen had its real beginnings in the early 1980s with the rise of flow cytometry/cell sorting.[12] By 2003 developments in fluorescence-activated cell sorting, a technology known as Beltsville Sperm Sexing, indicated that at last it was possible to sort female from male sperm. Two issues had to be overcome: price and level of conception rates from the sperm. Conception rates of virgin heifers (the most fertile females) ranged from 35 to 40 per cent with female sexed semen, compared with 55 to 60 per cent with unsexed semen.[13] Sexed semen began to be marketed by AI companies in spite of these difficulties. In 2005 Select Sires Inc. would become the first major American organization to offer sexed semen. In order to keep the price down on sexed semen, straws were loaded with only 10 to 15 per cent of the number of sperm that went into unsexed semen straws, a factor that limited its conception success rate. A straw of unsexed semen contained 15 to 20 million sperm. Sexed semen contained about 2 million sperm. If pregnancy resulted, 90 per cent of the calves would be female. Even though sexed semen had been available as early as 2005 it was not until 2008 that cattle breeders began to use it with any frequency.[14] (The sexing of semen did not work as well in horses or pigs as it did in cattle.)

In 2008 there was a considerable growth in numbers of replacement heifers, but prices for them remained high, thereby increasing even more production of them. "While you would think that the large supply of heifers relative to the milking herd would push replacement prices down, that has not been the case as record milk prices more than off-set the heifer supply growth and pushed the prices to record levels as well," *Hoard's Dairyman* told its readers.[15] The journal provided testimonials from dairy producers who had used sexed semen. One person reported using sexed semen since 2006, particularly on virgin heifers. Sexed semen cost US$50, but at a conception rate of 1.8 services, or $90, this farmer could get a $400 day-old heifer calf. Conventional semen might cost only $35 but he was just as likely to end up with a $25 bull as a $400 heifer.[16] Others reported similar results.[17]

In the winter of 2007–8 seventeen Wisconsin Extension Agricultural Agents decided to conduct a study on the use of sexed semen by dairy farmers and heifer growers. The results showed that almost twice as many heifer growers used sexed semen as dairymen. Nearly half of dairy farmers surveyed had never used it, versus 30 per cent of heifer growers. A few from each group had tried it and discontinued its use. About half of those surveyed claimed that growth of the dairy herd – and many emphasized the desire to increase the genetics of their best cows – was the primary reason they used sexed semen. Cost concerned most dairy farmers, but the greater value of heifer calves often counter-balanced that effect. The study indicated that if conception rates reached 80 per cent of that of conventional semen, it did pay off.[18] The sexing of semen is a technology, and a technology that could solve the sex problem from a moral and an industry point of view. Desirability for more heifers and for fewer bull calves drove interest in sexed semen in spite of its greater cost and lower conception rates. Other clear advantages to using it were the ending of calf killing and the veal industry – both of which had attracted such unfavourable commentary over the years. Farmers hoped to expand the best part of their female genetic base with sexed semen and to increase the number of cows in their dairy herd from that base.

It is difficult to separate the implications of artificial selection strategies from a technology as powerful as sexed semen. It might have changed how breeding proceeded, although no one was sure exactly what shifts would actually take place. Some thoughtful comments on the genetic/technology intersection were made in *Hoard's Dairyman* early in 2008, under the title "Where Does Sexed Semen Fit?"[19] Bulls in high demand, and perhaps the best bulls, would not have their semen sexed, because the owners could sell all the conventional semen they could produce. Furthermore, sexing semen wasted at least 80 per cent of it, so the expense alone of doing it would restrict bulls with sexed semen. The result would be that there might not have been as many bulls to choose from for sexed semen, and certainly not the best bulls. On top of this, using sexed semen for first service on virgin heifers did not increase the number of heifers as much as one would think. With conception rates at 48 per cent producing 90 per cent heifers, conception rates with conventional semen being 52 per cent and producing 48 per cent heifers, of 100 heifers bred once to sexed semen and then to conventional semen, 68 heifer calves would be expected. But conventional semen alone would produce 48 heifers. The return on the sexed semen for 100 heifers would be 20 heifers.[20] By breeding from cow to cow and ignoring the potential input of males, generational changes might look different. It was unknown whether that breeding strategy, that is, ignoring males, would reduce the rate of genetic gain over generations.

The problem of male sexed young played itself out differently in the chicken world, and for one very good biological reason. It is impossible to sex cock semen: the males do not determine the sex of the young as is true in mammals. In birds it is the female that dictates whether the chicks are either male or female. In mammals the sex chromosome is XX for females and XY for males. In birds the sex chromosome is ZW for females and ZZ for males. In the period before intense specialization of egg-laying hens, spent hens and excess cockerels supplied meat. Effectively the meat industry was a subsidiary of the egg industry, and even forty years ago there was a market for day-old cockerels of egg-laying strains. Ever more effective hybrid breeding of egg-laying chickens has changed that picture. Male egg-laying chicks have become completely worthless from a meat-producing perspective.

Interest in developing a sexing technology that could divide the newborn by sex (it is impossible to separate the sexes of newborn chicks by sight alone) developed rapidly with the rise of a commercial egg industry and before truly specialized breeding existed. It was clearly advantageous to know that a person was rearing only females rather than the less economically valued males. Newly hatched chicks could be sexed as early as 1919 by feather colour, when breeds such as the Plymouth Rock dominated the poultry table egg world, and in the 1930s by vent sexing (detection of sex organs). Today sexing is done on the basis of feather growth, colour not being useful in the all-white Leghorn.[21] Separating the sexes at birth, possible as early as 1919, was useful because male chicks could be identified earlier. Still, by the time breeding strategies had rendered these males useless for any meat purpose, merely identifying them did not solve the problem of what to do with them.

Millions of baby male chicks around the world are killed at birth. Half of the chickens of egg-laying strain born are useless to the industry. In 2001 it was estimated that 93 million chicks were killed in Germany alone, 283 million in the entire European Union, and some 226 million in the United States. While much of this chicken meat could be used to feed to zoo animals,[22] it struck many that the wastage was still tremendous. The chicken industry attracted increasing public concern over this issue: its wastage and cruelty to the baby chicks. The poultry industry would benefit from the ability to eliminate the male half of its chick crop before the babies were born, and efforts have been made to sex chickens while they are still within the egg. So far there has been no success in finding a way to make this work. One scientist, E.F. Kaleta, pointed out in 2008 that male chicks could be allowed to live if consumer attitudes changed, but not under market conditions as they now exist. He put it this way:

> The methods currently available to circumvent hatching of numerous undesired male chicks of the layer type are rather academic, expensive and time consuming ... Thus, further thinking, reviews of the previously described alternatives and subsequent experimentation is required to contribute to a solution of the persisting issue of killing male chicks ... A more promising approach to solving the dilemma between ethics and economy may be to develop a niche market for a premium product based on slow growing cockerel of egg-type breeds. If enough consumers are willing to pay a higher price for less meat to satisfy their ethical demands, the industry would soon find a way to grow and process these cockerels.[23]

By using the only method available, though – that is, killing the chicks – the chicken industry triggered the same sort of moral indignation in the public that calf killing and the veal industry did in dairying. To this day, the problem of sexing in the table egg industry remains unsolved.

The Dehorning Issue

The story of cattle dehorning serves as another interesting example of an industry needs/morality complex involving artificial selection practices. Growth of both dairying and feeding of beef stock for market led to increased crowding of animals over the years, making horns on cattle a problem. The rise of larger and larger feedlots, for example, where cattle were closely penned and fed, brought with it the increased danger of the animals hurting each other with their horns. This situation, of course, threatened to decrease the value of their meat. Crowded animals with horns were also dangerous to people. Clearly industry needs had made horns a problem. What to do about them? One way to solve the problem was to cut off the horns. The growing tendency to do so aroused intense furor near the end of the nineteenth century in Ireland, Britain, the United States, and Canada.

The practice of dehorning cattle in the Anglo world was initiated in Ireland about 1870, and from there was taken to first England and then the United States by 1883. It was introduced to Ontario from Illinois in 1888. Dehorning was a subject of litigation as early as 1874. The first important case was brought up in Ireland in 1884. The main question surrounding dehorning was one of morality: was it cruel and did it inflict pain? Clearly the available technology to perform the operation caused pain. But did the advantages offered by dehorning – limited damage to other animals, ease of handling, quieter nature – justify the infliction of suffering? Could it be said that was morally right to cause pain if it was for the animal's good? And could the animal's good be separated from what was perceived to be the good (or at least in the interests) of people? The courts failed to find any answers, and dismissed most cases against dehorning over the mid-1880s. Even so, humane societies ultimately managed to secure success. Dehorning was deemed a violation of the act for prevention of cruelty to animals, and by 1889 the practice had been made illegal in England, Ireland, and Scotland.[24]

A review of the dehorning controversy in Ontario between the late 1880s and 1920s illustrates how the interplay of conflicting interests surrounding the practice played itself out. Canadians were well aware of dehorning before it arrived in Ontario. There was little approval for the procedure. In 1885 *The Canadian Breeder* commented that not long ago no one wanted a cow without horns. Times have changed, the journal noted. It asked readers, what about simply crossing a Galloway or an Angus – both of which were breeds without horns – with a horned cow?[25] The journal's comment is interesting: it indicates that it was fully understood before the advent of genetics that polling (no horns) was a dominantly inherited characteristic, and that by breeding a polled bull to a horned cow the resulting calf would almost certainly be polled. In other words, genetics could be used to dehorn most cattle. It was not a common way to remove horns, and would not be for some time. Inherited horning, but having no horns, seemed to be what was wanted, although some beef cattle producers in both Britain and Canada did use Angus (polled) bulls on Shorthorn (horned) cows in order to get non-horned calves to feed for the market.[26] Even if breeding for polling was not common, and horns were not wanted, the idea of dehorning procedures horrified the Canadian farming community. The *Canadian Live Stock and Farm Journal* stated in April of 1886: "We had been congratulating ourselves with the thought that three thousand miles of the Atlantic lay between us and the abominable practice of dishorning [*sic*] cattle, but it seems that men as cruel are found not nearly so far away. One Mr. Haaff, a farmer of Atkinson, Illinois, was prosecuted by the Illinois Humane Society not very long ago, and we regret to have to chronicle that they failed to secure a conviction. We trust that the cruel and needless practice will never disgrace our Dominion."[27] One reader responded enthusiastically:

I am glad to see you take the stand you did in the April number of the journal regarding this cruel and barbarous practice – a practice much better adapted to the dark ages than the age of progress and enlightenment. It appears that some men are so thirsty for notoriety that they care little how it is attained, so long as they can get their names before the public – whether it be by knocking off the horns of innocent calves, as in the case of Mr. Haaff, or knocking the heads of their fellow men ... It is very surprising to me that a paper, a stock journal, like the *Western Rural*, should sanction such a nefarious and cruel practice.[28]

Two men, Kinney and Johnson, farmers from Oxford County, introduced dehorning to Ontario in 1888. The practice was not common until 1890 when Chauncey Smith, a farmer's son, returned to Oxford County from Illinois and dehorned his father's herd. Soon his neighbours began to dehorn. The practice expanded quickly and by 1891 at least 700 head had been dehorned. Controversy arose over the amount of pain inflicted on the animals, and in 1891 W.V. Nigh, a farmer of Avon, Middlesex County, was prosecuted before two Justices of the Peace at London on the charge of cruelty. The case was dismissed after ten witnesses argued that the operation was beneficial and the pain short-lived.

Opponents of the operation, on the grounds of cruelty, brought about new court proceedings in January 1892. The defendants were unhappy with the way the second trial evolved; thus, they requested that the government set up a commission to look into the entire matter. In March the government complied. The commissioners, a group comprised of members of the legislature, a veterinarian, J.J. Kelso (who spoke for the Humane Society), a dairyman, and a farmer, planned to approach various U.S. agricultural college experiments stations over the matter, and review the precedent set in Britain and elsewhere. They:

> sought to ascertain what commercial advantages accrued from the operation; what were the humane considerations as shown by the conduct of the animals towards each other before and after the operation; the amount of pain inflicted by the operation as judged by the anatomy of the part, the actions of the animal during and following the operation, and the probable duration of the suffering; the effect upon the general condition as evidenced in the flow of milk, loss of appetite or weight or undue rise of temperature; the possibility of fraud as to age when the horns were removed; the extent to which knobbing or tipping the horns was serviceable as a preventative of goring; the best age, the proper season and the most suitable instruments for the operation if it should be permitted; the relative advantages of taking the horns off when developed, preventing their growth by means of caustic at two weeks old, or cutting out the embryo horn at the age of a month or six weeks – in fact the Commissioners endeavored to elicit information on every phase of the question.[29]

The commissioners watched dehorning operations and noted the short duration of pain inflicted. They also observed the calmness of the stock after the dehorning procedure. The commissioners saw horned cattle at the Toronto Stock Yard harming each other. They had meetings with the London Humane Society and the Toronto Humane Society – all animal welfare people in the organizations opposed dehorning. "A great deal of opposition to the practice was met with members of Humane Societies and others who believed that the operation was purely for commercial considerations," the commissioners noted, "and therefore unjustifiable, and that the pain inflicted was excessive."[30] The commissioners spoke to numerous farmers and veterinarians – all who had seen the operation approved. Those who had not were opposed.

"With regard to the amount of suffering involved in the operations," the commissioners reported, "farmers and others accustomed to the care of cattle, who had either seen the operation of dehorning or had performed it themselves, testified that the animals did not apparently suffer much pain at the time or afterwards, that they manifested no symptoms of shock, but partook of water and food immediately, and that the secretion of milk was not diminished or changed for more than a day." Veterinarians and doctors generally agreed with these statements, but not all were in accord. The commissioners noted that some prominent vets stated that there must be excruciating pain "and their evidence, given in various legal cases affecting the practice, has been a strong factor in arousing opposition." The president of the veterinary college in Edinburgh, for instance, argued that there was great pain. This type of testimony influenced those who had no experience with the practice but who opposed when speaking to the committee.[31] The extreme differences of opinion on the matter interested the commissioners who stated: "While many of the Humane Societies have expressed strong opposition to the practice as being cruel and unnecessary, it is a remarkable fact that all who favour it claim that it is a humane operation and in the best interests of the animals themselves." For example, J.W. Robertson, Dairy Commissioner of Canada, stated that the animals were more hurt by horns than by having them removed. I.P. Roberts, Director of Cornell University Agricultural Station said, "If by hurting one animal for a few minutes we can prevent it from inflicting many sever and painful wounds and hurts on many other animals for many years I feel that that is an act of humanity to take the horns from the offending animals."[32]

The commissioners revealed the conflicting points of view surrounding the practice of dehorning, but could not reach the conclusion that it should be prohibited. And in fact dehorning continued in Ontario, but not without controversy. Numerous reports in the farm press supported the practice throughout 1895.[33] In 1896 one person wrote to the *Farmer's Advocate* expressing his conversion to favouring dehorning, indicating how much opinions had shifted in only a few years. "When the idea of dehorning was first brought before me I was very much opposed to it, and, living in a large city (Toronto), was influenced to some extent by the views of humane societies. But after engaging in the veterinary profession, and coming in contact with the subjects of the operation (which most members of humane societies never do), I was convinced that dehorning was commendable on the score of humaneness. The operation itself is not a very painful one. I consider it a humane act to dehorn a cow," he continued. "One cow will often keep all others away from water. Horned cattle gore each other, and this hurts far more than the operation." Horned cattle are also dangerous to attendants – the writer knew a number of people who barely escaped death from horns.[34]

Dehorning continued, and in fact became more widespread, but it remained a contentious practice. In 1901, for example, the *Farming World for Farmers and Stockmen*, in an article entitled "The Dehorning of Cattle, the Advisability of this Practice Discussed by Breeders and Others," revealed that while many farmers and stockmen supported dehorning, many did not for a variety of reasons.[35] In 1905 Thomas Crawford, a member of the Ontario Legislature, introduced a bill to the House which called for the compulsory dehorning of cattle under the age of one year. A frenzy of letters appeared in the *Farmer's Advocate* over this topic. Most did not oppose dehorning, but plenty rejected the idea that the government had the right to force dehorning. The issues of pain and morality were again played out in the letters.[36] A few members of the legislature spoke out strongly against the bill, and Crawford, a large cattle exporter of beef animals to Britain, was forced to withdraw it.[37] Dehorning continued to attract attention, but increasingly support for it revolved around the question of the age of the animals. Dehorning adults was no longer favoured. Instead, the early budding of horns on calves should be dehorned so that adults would never experience them. Even better, caustic potash should be applied to calves within three days of birth to prevent horns growing.[38] From 1913 until 1920 there was a lot of discussion around this subject and the technology used in the practice.[39] The move to breeding for non-horns had gained little ground in the beef cattle world, and none at all in the dairy cattle world. Genetics, which could have solved the problem, was not used to any great extent.

The issue of pain and dehorning seemed to go underground for a number of years, but with the resurgence of animal welfare support in the 1990s, it again came to the fore. Here it pertained particularly to dairy cattle, because by that time the vast majority of beef cattle had been bred to be hornless. The continued use of a technology to dehorn, even if there had been improvement in the way that technology worked (rather than a genetic solution) could still trigger negative reactions in the public. M.A.G. Von Keyserlingk reported in 2009 in the *Journal of Dairy Science*: "Considerable research has shown that all methods of dehorning and debudding cause pain to calves, and this can be shown with a variety of physiological and behavioral measures ... It is now also becoming clear that use of a local anesthetic alone does not fully mitigate this pain, and does not provide adequate postoperative pain relief ... A second consideration is that animals respond to both the pain of the procedure and to the physical restraint."[40] Dairy farmers in the United States became increasingly conscious of the public concerns with pain inflicted on calves during the dehorning process. *Hoard's Dairyman* reported that one farmer felt that dehorning without using painkillers was an abomination. He explained, "If we are to be good stewards of this earth and its creatures, it behooves us to take advantage of every available improvement over 'the old way.' Even for us 'old dogs,' progress and change can be good things."[41] In another article, *Hoard's Dairyman* asked its readers, "Would your dehorning pass the '60 minute' test? Imagine a film crew watching this painful, but necessary, procedure." The journal continued, "If you would not want your farm to be featured in a film that was an embarrassment to you and the dairy industry, then it would seem like a good time to address how necessary, yet painful, situations, such as dehorning, are addressed on your dairy."[42] Get up to date on how things can be done, the journal added. Calves should be dehorned under two months old, there should be good facilities to do the job, a nerve block and a painkiller should be used, and employee behaviour should be strictly monitored.[43]

Breeding and Wider Ethical Issues

The male sex/dehorning issues were only a few of the problems that attracted ethical concerns when it came to animals. Both were interconnected with breeding practices, but breeding practices themselves became entangled with a vast array of other issues that worried people when it came to how animals were/are treated. In a general way, it was quality of life that was at the root of the problem, and there can be no question that quality of life was hugely affected by the way artificial selection shaped life. The whole field of animal rights in relation to livestock industry-needs, inside and outside breeding issues, has mushroomed in the last number of years, and a profusion of material has been generated as a result. Reviewing the literature on the subject is beyond the scope of this book, but some overview commentary is helpful to show how wide ranging questions over the quality of animal life have become, and also how farmers fit into the movement. A brief discussion, designed primarily to outline historical trends and to situate farmer-reactions within the story, therefore follows.

Formal organizations set on the ethical treatment of animals by preventing cruelty (in the form of pain, starvation, etc.) to them arose over the nineteenth century: the Royal Society for the Prevention of Cruelty to Animals (RSPCA) being an important example. As early as 1838, a number of American states had legislated against cruelty to animals. Protection was aimed at dogs and horses.[44] The idea of cruelty and what it meant would be modified over the years. Changes in agriculture – which became more commercialized and industrialized – and the corresponding expansion of urbanization enlarged the focus of humane society supporters in both North America and Britain to include the welfare of livestock. Humane treatment of farm animals was promoted in a general way by the farm press by the latter half of the nineteenth century, even if breeding practices as such did not come under scrutiny.

Canadian attitudes in this early period illustrated what sorts of issues attracted attention when it came to the welfare of livestock. Treat a bull kindly, the *Canadian Breeder* told its readers in 1885, for example. Bulls were dangerous, the journal admitted, but treat him badly and you have a really dangerous animal on your hands.[45] "At about one year old bulls frequently become playful, and this is often mistaken for vice," the paper added some time later. It is just the feelings of youth. But be mean to him and he will never forget it.[46] Industry needs, particularly in the beef cattle industry, also provoked commentary in the farm press. The beef industry called for considerable movement of cattle from breeding to feeding grounds and from there to terminal slaughter points. Conditions for cattle travelling on trains, stockpiled in railway yards, and on board transatlantic ships (when the beef cattle trade became international and transatlantic) concerned the Canadian press. One article, called "Abuse of Stock in Transportation," written in 1885 called upon humane societies to look into the problem. Cattle in transit on trains and in yards went for days without food and water – an outrage to decency.[47] By 1902 the *Farmer's Advocate* thought much had improved with respect to treatment of animals. "In this age, happily, cruelty to dumb animals is vastly less common than was the case half a century ago," the journal told its readers, "though there is yet room for much improvement in some lines in this connection."[48] In 1918 the journal pointed out that real progress had been made:

> Humane societies have done a good work, but at first they were looked upon as a fanatically sentimental group of 'busybodies' with nothing to do but interfere in the affairs of people who had their tasks to perform and life's battle to win. No doubt there were those in these organizations who showed too much zeal and too little practical knowledge of what was right and what was wrong, but on the whole we must admit that a good work has been done and expression has been given to a growing sentiment that dumb animals should not be submitted to cruelty and torture at the hand of civilized man.[49]

Transatlantic cattle shipping provoked outrage in Britain as well. The farm press in Canada kept close track of British reactions to the cattle ships. Any threat to the lucrative trade emerging from outside the country was serious, as far as the *Advocate* was concerned. The journal might want to better conditions for cattle in transit, but never in a way that would threaten the transatlantic industry. In 1890 a British member of parliament, Samuel Plimsoll, began a campaign to improve outbound North American cattle ships, which he saw as dangerous for men and inadequate for animals. Plimsoll argued that animals were inhumanely treated in overloaded vessels and men were at risk for their lives. His work would lead to what is known as the Plimsoll line on a ship's hull, a load line that indicated the amount of weight a particular ship could hold. While it was the treatment of cattle that triggered Plimsoll's campaign to stop the overloading of ships, the eventual adoption of load line rules (by France, Germany, and the Netherlands) did not result from efforts to regulate conditions for cattle on ships. In 1930 at an international convention held in London, thirty nations signed an agreement providing for a fixed Plimsoll line.[50]

The twentieth century brought about ever more intensive farming practices that affected the lives of agricultural animals – a pattern that accelerated over the second half of the century. Animal welfare concerns correspondingly became much more complicated after the mid-twentieth century, particularly in Britain. While the ongoing effects livestock genetics has had on breeding arguably played a subtle role in the rise of modern concerns with animal welfare, it continues to be difficult to separate genetics from technology/industry needs when looking at the problem, as Ruth Harrison's *Animal Machines – The New Factory Farming Industry* makes clear.[51] Published in 1964, it initiated animal rights in its modern shape. The book aroused widespread concerns in Europe over what life for farm animals had become, as a result of breeding, housing/confinement, and even feeding (viz. genetics/technology/industry needs). Natural behaviour was no longer possible for them.

The British government responded by appointing a formal commission, the Brambell Committee, to examine how animals fared on farms. The committee set forth what it described as 'Five Freedoms,' which generally restricted attention to issues that caused physical pain. All animals required: freedom from hunger and thirst; freedom from discomfort caused by extreme temperatures; freedom from pain, injury, and disease; freedom to behave normally; and freedom from fear and stress. While no legislative action was taken as a result of these ideas, they caused a number of people in U.S. and Europe to think more about animal treatment, and about what science had done to shape animal lives. The 'freedoms' would also act as a guideline for future animal welfare programs.[52] Clearly industry needs and technology used in livestock farming made it difficult to ensure the freedoms would be preserved to an adequate degree. So did breeding via livestock genetics/biotechnology. How could MOET, for example, be considered natural? Newer genetic work with livestock (particularly molecular genetics), as a result of recombinant DNA technology, worried many people interested in animal welfare. Cloning received a great deal of attention from animal behaviour scientists. How did cloned animals act? Were they normal? If not, was it morally wrong to create them?[53] How would transgenics affect the health of animals? Evidence existed to show that using the technology had been detrimental to pigs. In fact, had not all breeding procedures injured the health of animals?[54] It was clear to animal welfare supporters that artificial selection had entangled animal breeding with larger issues concerning the right to a normal life.

The scope of the freedoms seemed to become larger and increasingly vague over the years. Understanding how to enforce them, and within the framework of an ever-widening definition of what a freedom meant, required more knowledge of normal behaviour in animals. What could be defined as stress? What was normal breeding behaviour? The growth of research concerning animal behaviour in conjunction with animal welfare was rapid over the 1970s and 1980s, and focused on attempts to scientifically understand life from an animal's perspective. Frustration in hens was studied in 1971, for example, and fear in hens received attention in 1974. The publication in the late 1970s of Marian Dawkin's *Animal Suffering: The Science of Animal Welfare*, which promoted the study of the animal perspective, was an important milestone in animal behaviour research.[55]

As animal welfare specialists moved more decisively towards broadly based research on animal behaviour, the public's general interest in that research continued to expand, most particularly after the early 1990s. It is not clearly understood why the movement, now worldwide, has been so strong and continues to grow, although a great deal of material has been written about the subject. A.B. Lawrence argued that one reason seemed to be an increasing conviction that animals are sentient; that is, they have the capacity to "feel" in a way humans can relate to. Pain and reasoning appeared to be irrevocably linked, historically speaking, in moral/ethic thinking around the meaning of cruelty to animals.

In the seventeenth century, René Descartes had argued that animals were not capable of any mental activity or feeling as well. His attitude derived from his mechanistic theories concerning the physical world, and his widely accepted philosophy did not encourage any animal welfare movement. The general mood by the eighteenth century, certainly with respect to animals, was decidedly away from the Cartesian conviction. David Hume rejected Descartes' dogmatism and argued that animals had both minds and capacities to reason. They could, in Hume's view, feel pain and do so in a human way. Modern acceptance of animal sentience seems to have resulted from scientific evidence that behavioural, physiological, and neurological similarities between humans and animals exist. Lawrence concluded: "Scientific evidence relating to animal sentience may help further advance the moral status of animals by heightening the 'moral intensity' we associate with animal welfare issues. Recent work has shown that belief in animal sentience is important in determining attitudes to animals. This suggests that further scientific support for animal sentience should increase the moral importance of animal welfare through widening societal acceptance and belief in animal mind."[56]

By the early twenty-first century, the mushrooming animal welfare movement attracted renewed attention from the farm press. *Hoard's Dairyman*, for example, published a great deal of information about animal welfare activists and warned dairy farmers that they should be aware of the consuming public's views because that public was increasingly influenced by activist views.[57] "The spotlight on animal welfare in the United States has intensified over the past two decades for several reasons," the research of various philosophers indicated, the journal told its readers.[58] They cited four reasons: urbanization of the U.S., proliferation of philosophers writing about animal rights, extensive media coverage of animal issues making the consuming public more aware, and the widespread anthropomorphic representation of animals skewing consumer views. *Hoard's Dairyman* pointed out that legal protection of farm animals in the United States was limited compared to the situation in Europe; the public was increasingly critical of that state of affairs. Farmers should know too that animal rights groups had grown, raised considerable amounts of money, and worked actively for change – change, that is, of many practices normally deemed acceptable by farmers. Whether acceptable or not to farmers, animal rights activists adamantly opposed any practice that inflicted pain or led to undue confinement, the journal emphasized. "Keeping animals confined and performing painful procedures (such as dehorning) without pain control are high on the agenda of these groups. A well-publicized example is the recent partnership of the Humane Society of the United States with Farm Sanctuary to get measures aimed at regulating aspects of farm animal production placed on state ballots. Many of these measures have targeted continuous physical confinement of farm animals, such as crates for sows and veal calves, and most have passed by wide margins."[59]

Animal welfare and morality became interwoven with technology, genetics, and industry needs over the years, as is evident from a review of how male calves and chicks were/are handled in the dairy and chicken industries, and in the way dehorning has been dealt with. The idea that animals should be treated properly is not new, and evidence suggests that the rising strength of animal rights activists, the study of animal behaviour within that framework, and advanced technology and genetics did not initiate farmer or public concern with pain and morality as much as would first meet the eye. The right to inflict pain and the meaning of pain provoked huge anxiety in the farming community in the nineteenth century. Calf killing, the functioning of the veal industry, and even dehorning aroused strong reactions concerning animal welfare. Artificial selection can be used to remove horns on cattle and has done so for the majority of beef cattle. No such breeding strategy has been applied to dairy cows: all the dairy breeds are naturally horned. Milking capacity seems to be related, at least to some degree, to horning. Better technology can alleviate moral problems as the sexing of semen shows. But the use of sexed semen may result in genetic changes that will not work well with industry needs, and may also provoke new outrages on different moral grounds.

The modern and enduring belief in animal sentience had focused the minds of many people in an increasing fashion on such overarching questions as: how do animals feel about the conditions we have made them live with? Have we tried to make them into machines through breeding? Do we or did we have the moral right to change them in that fashion? And perhaps, above all else, what does all of this say about us? Farmers have asked and continue to ask these questions. Those outside the livestock industries are not the only ones to question practices and think about effects and consequences.

Conclusions

Animal breeding is as old as domestication, but the rise of what might be described as structured, artificial selection methodology is relatively recent. British Thoroughbred horse breeding strategies and culture in the late seventeenth century laid the groundwork for an organized approach to breeding. Developments in the eighteenth century during the Enlightenment expanded on that groundwork. Livestock breeders like Robert Bakewell set out breeding strategies that relied on both inbreeding and progeny testing (particularly in males), in order to create livestock that better matched agricultural needs. The breeders showed no desire to understand why they could change animals in this fashion; the question for them was which artificial selection strategy worked with natural laws to achieve the desired ends. They wanted to create, effectively, new breeds that would reproduce traits consistently generation after generation. At the same time naturalists, seeing the outcome of breeding strategies that brought about variation in domestic animals, questioned what hereditary laws made those changes happen. They began to experiment with breeding methods that might show how heredity worked, and arrived at a certain strategy. They inbred separate lines and then crossed them. The progeny deviated from each parent line in telling ways, but would not breed truly in the next generation – a critical disadvantage for livestock breeders.

In the nineteenth century the Colling brothers used Bakewell's principles to create the Shorthorn, and it was in the breeding of Shorthorns that by the mid-nineteenth century purebred breeding was born. The late nineteenth century saw the solidification and extension of purebred breeding and also the rise of standardbred breeding. These systems, used by livestock breeders of various species, varied in fundamental ways from Bakewellian principles. Inbreeding took on new meanings and progeny testing no longer played such an important role in artificial selection theory. Naturalists continued to follow their breeding methods and the two groups continued to be divided by that primarily different outlook: breeding for true producing lines and breeding of hybrid progeny that served as an end result but achieved no breeding purpose itself. Naturalists became increasingly involved in the process of breeding via Darwinism. Could artificial selection be related to natural selection? What made species change? Or did species not change? Biometry, an approach to heredity based on complicated mathematics, supported the idea behind Darwinism; namely, continuous inheritance. Biometry would compete with Mendelian theory's apparent support of discontinuous inheritance. The dominance of Mendelism and the concurrent acceptance of discontinuous inheritance over biometry's approach of continuous inheritance meant that the potential growth of science in relation to animal breeding took a back seat. While Mendelism could (and did) explain the inheritance of quantitative characters, as well as support the theory of continuous inheritance, that reality was not evident for some time. Early Mendelists focused on the idea that all traits showing variation were qualitative and therefore discontinuous. Mendelists also suggested that traits of interest to farm animal breeders were inherited in the simplistic way, a pattern that breeders would have rejected from experience. The inapplicability of Mendelism to agricultural animal breeding, particularly its stance that only mutation could change animals, made the work of practical breeders of little interest to new geneticists. They did not believe they learned much for their own research from the practical breeders. A cleavage that demarcated breeding as either "science" or "art" became clearly evident in a way that had not been true in the eighteenth century.

The science of livestock genetics was not possible before the synthesis of Darwinism and Mendelism and the subsequent rise of population genetics. Of the three geneticists who were important to the development of population genetics – namely, J.B.S. Haldane, Sewall Wright, and R.A. Fisher – Wright and Fisher were particularly significant to livestock genetics. Wright was interested in the way inbreeding tended to pool the way inheritance worked, by bottlenecking genes. Fisher focused on quantifying the process of variation in inheritance within a population. It was J.L. Lush who synthesized these ideas into a formal structure that would be known as livestock genetics. With these concepts in place agricultural geneticists focused on using them within the framework of hybrid breeding – that is, inbreeding and quantification of its levels and results within a population, and crossing those lines within the same conceptual framework (that is, quantification of results on a population basis). The approach matched the standard eighteenth century breeding methodology of the naturalists. Livestock geneticists believed that inbreeding both purified the lines for certain characters and also provided an extra kick to hybrid vigour in the cross. Quantification, as such, underlay much of this thinking, but the particular emphasis of such breeding approaches was on the process of hybrid breeding. All work in animal genetics up to this stage was more theoretical in nature than practical, and livestock breeding was virtually unaffected by it. Breeding continued to proceed under the structure of purebred/standardbred systems.

Developments within the chicken-breeding industry made it susceptible to geneticist theorizing along the lines of hybrid breeding, and over the 1940s and 1950s the methodology came to dominate chicken breeding. The terminal cross that was part of hybrid breeding acted as a biological lock, thereby forcing the producer back to the breeder at every generation for new stock. The structure of the chicken-breeding industry was conducive to this separation of the breeder function from that of the producer. The producer/grower had, by the early twentieth century, become detached from the practice of breeding. The division resulted from a gender divide (men were breeders and women were producers), complicated breeding methods that required time and skill, the rise of the hatchery industry and the selling of baby chicks, and the obsession of the breeders with beauty points. The declining influence of the main poultry breeding organization, the American Poultry Association, tended to reduce the effects of any hegemonic voice within the breeding industry, making it easier for new developments to take place. Hybrid breeding provided a natural patent because of the biological lock, making it attractive to corporate enterprise. The dominance of breeding companies in the industry and the privacy of breeding methods that accompanied hybrid breeding shaped chicken breeding from a world perspective. Secrecy was paramount to breeding success.

The advent of artificial insemination technology (AI) changed the framework of livestock genetics by making it possible to progeny test males on a vastly more comprehensive scale than had been possible earlier. The data generated by AI organizations attracted geneticists, particularly in Britain where the centralization of the new industry provided an especially comprehensive set of data generated under varying environmental conditions. The Wright/Lush concern with inbreeding and hybrid vigour would be replaced by the Fisherian approach. The rise of AI would change patterns in livestock genetics by shifting a primary focus away from hybrid breeding to the breeding of lines that reproduce truly and progeny testing to find the best breeding males. Quantification of certain specific characteristics became the sole objective behind male progeny testing. The capacity of bull semen to freeze, and the existence of collected data with which to work, encouraged agricultural geneticists to focus on the dairy industry and bulls. The work of geneticists would never have been possible without the input of purebred dairy cattle breeding or the input of ordinary dairymen. The structure of the dairy breeding industry was different from that of the chicken-breeding industry. The breeder arm could not be separated from the producer arm. Producers used the bulls of the breeders, but the breeders followed a breeding program based on data collected from producers on milk yields of daughters. The AI industry functioned between the two and acted as a unifying force between them. Increasingly geneticists worked with the output of all three groups, thereby becoming part of the breeding structure.

Livestock geneticists began to describe themselves as quantitative rather than population geneticists, as had been true in the Lush school. This was true worldwide by the early 1970s – where Lush's work at Ames had evolved from what was called population genetics to what was now called quantitative genetics. Men alive today, and trained at Ames by either Lush or his school as late as the 1960s, still call themselves population geneticists. Livestock geneticists at Wageningen in the Netherlands also describe themselves as population geneticists today. It is interesting, then, that many present-day livestock geneticists who are interested in the past of their profession tend to see historical agricultural genetics only as forms of population genetics.

Central to quantitative genetics was statistics, and developments in this area (specifically through the work of Americans L.N. Hazel and C.R. Henderson) worked with that of geneticists like Alan Robertson at Edinburgh. By the 1970s many scientists began to wonder if quantitative genetics was simply a matter of statistics. The importance of statistics to quantitative genetics and the application of quantitative genetics to dairy cattle breeding also bothered breeders within the purebred Holstein world. This phenomenon reminds us that no matter how much livestock genetics had penetrated dairy cattle breeding, purebred breeding still played a central role in that endeavour. When a bull was to be assessed by an AI unit for acceptance in a progeny-testing program only on the basis of an index figure calculated from values placed on its parents, many purebred breeders reacted negatively. American geneticists and the USDA pressured the AI industry to adopt indexing and abandon type standards, and by the late 1960s that had largely become the case. One important Holstein breeder, P. Heffering, opposed the idea of using indexing as the sole tool to judge a young bull so strongly that he decided to move his operations to Canada in 1973, where the AI studs took into consideration standards upheld by the purebred breeders regarding physical looks and general family background. The result was that Hanoverhill Starbuck, a world-class bull from the point of view of indexing and type as well as the quality of his male and female progeny, was born in Canada. Heffering believed that Starbuck would never have even made it into U.S. studs for testing. The Heffering/Starbuck story indicates that tension between geneticists and purebred breeders still exists within the dairy cattle breeding world, which remains a fusion of the traditional and the modern when it comes to breeding.

The advent of indexing did not reduce the importance of the breed associations within the breeding industry. Interest in physical type, for example, stayed in place with the establishment of linear classification. While straight show room style was not necessarily bred for, aspects of phenotype were part of the standards under which cattle were classified, along with production, longevity, and characteristics such as fertility. How much scientists were in agreement with the continued emphasis on looks is not clear. Today there seems to be evidence that looks do dictate levels of productivity, especially when productivity is related to overall cost issues (feed, calving difficulty etc.). One is reminded of the Standard of Perfection in poultry breeding, and the failure of the American Poultry Association to convince the general public that standards based on looks mattered in chickens. In the case of dairy cattle, dairymen were influenced by phenotypic (appearance) characteristics of bulls used on their cows. The physical looks of resulting females were important to them: they believed that good udders and legs promoted profit for them. There is much, therefore, in modern dairy breeding that dates back to the late nineteenth century purebred breeding standards, in spite of the huge advancements that genetics has brought about within that framework.

With the advent of molecular genetics, quantitative geneticists working on livestock issues seemed to be falling behind, at least that appeared to be the case in the eyes of molecular geneticists. Central to the division between the two geneticist groups was attitudes to research approach: molecular geneticists focused on locating genes while quantitative geneticists used statistics to study the physical outcome of gene action. They did not look at the genetic architecture behind the perceivable appearance. It was a problem of whether black box thinking was applicable. Before genetic structure could be studied at the DNA level, effectively before the rise of recombinant DNA technology, a black box approach to genetic architecture did not seem unreasonable. The illusive search for the gene that would explain quantitative inheritance of production traits was, in fact, as old as genetics itself, and played a role in how theoretical and experimental genetics proceeded for over a hundred years. Mendelists had focused on identifying genes and livestock breeders were dismissed as uneducated because of their reluctance to think in terms of trait inheritance on the basis of unit character, or a particular gene. Biometry argued for continuous inheritance and used statistics to show how that would work. Repression of the fundamental idea of continuous inheritance by the Mendelians had only thwarted any incipient development of livestock genetics. The roots of livestock genetics as a science in fact owed as much to the theorizing of biometry as to Mendelian theory, which explained how segregation and transmission of hereditary material worked, but did not offer any ideas about how to utilize that information from a breeding perspective.

Molecular genetics did lead to the identification of certain markers on specific genes responsible for quantitative traits – marbling in beef cattle, for example. But how that gene could be used in conjunction with breeding remained anything but clear. Livestock geneticists were pressed to look for such markers, but the effort seemed fruitless to many. The human genome project would change the landscape. With its completion, the international consortia that had worked on the human genome and HapMap received public funding to sequence livestock species. With the development of the chicken and cattle genome and corresponding HapMaps, the world of livestock breeding was launched in new ways into molecular genetics. It was AI, the purebred dairy breeding industry, and quantitative genetics that capitalized on advances in genomics. With the rise of dense arrays that covered the entire genome and located thousands of single nucleotide polymorphisms (SNPs) it became possible to profile what a superior bull looked like by his DNA. This could be done only because bulls proven through the system and subsequently genotyped and served as a test population. A non-progeny tested bull with a similar DNA profile to that of a known quality bull (due to his progeny testing) could be assumed to be good. Markers in the form of SNPs and their particular positioning in DNA, not genes, were used to assess the breeding potential of an animal. It is perhaps ironic that breeding by DNA was only possible within the framework of purebred breeding and quantitative genetics.

It is interesting to note the rapidity with which dairymen have moved to non-progeny tested bulls that have a SNP profile. This situation made many initially think that the progeny test was on the way out. Until the advent of SNP testing, all advances in genetic improvement in dairy cattle had resulted from progeny testing males, and its elimination struck these people as being dangerous. If all efforts at maintaining quantitative data were abandoned it would be difficult to back up SNP tested bulls after several generations. It was argued as well that, because females could be so satisfactorily tested with SNPs, progeny testing could be used differently. Simulations that would predict how utilizing female and male input could be done together without extensive (but at least modest) progeny testing might make sense. Initial fears that progeny testing would be bypassed showed no evidence of coming to pass by at least 2014. Attempts to unite traditional breeding values generated by quantitative genetics with genomic testing via SNP chips through statistics commanded the attention of animal breeding scientists almost as soon as breeding by DNA became possible. Considerable research was undertaken to find the simplest way to incorporate quantitative genetic data with the new genomic data, in order to produce one standard for selection purposes based on all forms of information available. Interest in pedigree and phenotype were not discarded in many of these studies. Simulations showed that one system, known as single-step evaluation, was the least cumbersome.[1]

The complexity of the chicken genome did not lend itself to the immediate application of breeding by DNA. But neither did hybrid breeding, or the privacy of breeding records that corporate enterprise preserved for biological patenting reasons. The hybrid approach of the chicken breeders had not generated the public data needed for DNA breeding, and the privatization of poultry breeding did not encourage the use of public funding for research tailored to meet the needs of the breeding companies. But other ways of assessing genetic variation within the genome are possible without reference to SNPs, and at least one molecular biologist working within a company believed this approach to research might better serve the chicken-breeding industry.[2] The pig breeding industry, which relies on hybrid breeding and privatization, remained tied to purebred breeding even after it adopted that genetic methodology. However, it has been able to incorporate quantitative genetics in a way that the chicken-breeding industry has not. Pig breeding is based on hybrid breeding, purebred breeding, and quantitative genetics, a curious combination of outlooks not found in other livestock industries. How pig breeding will be able to utilize genomic evaluations for selection purposes has not yet been fully worked out. But the move to single-step evaluations – which utilize quantitative genetic data, pedigree information, and SNP profiles – suggest that pig breeders will be able to utilize new scientific approaches for selection strategies.[3]

The privatization of breeding that resulted primarily in the chicken-breeding industry, but increasingly in the pig breeding industry as well, change the dynamics of structures relating to research and education.[4] Academic chicken geneticists were concerned about the trend as early as the 1960s. Company-dominated experimenting programs seemed incompatible with good science to some university geneticists. "What breeder," the chicken geneticist, F.B. Hutt, asked in the 1960s, "having discovered something that gives him an advantage over his competitors, will be altruistic enough to tell [others] about it?" "Furthermore," Hutt added, "there are many kinds of research in which breeders have no interest whatever, and [other kinds that] might almost put them out of business."[5] Many others working at agricultural colleges and experiment stations noted the growing trend in the 1970s with dismay.[6] By 2000 the privatization of research regarding reproductive technology and genetics had become much stronger because of increasing funding from the private sector. The growing strength of the pig breeding companies provides a good example of this trend. The dominance of purebred breeding in the pig breeding system, however, has curbed private investment in the industry, because control over intellectual property in the form of genetics is not complete. The independent structure of the beef breeding industry has made it impervious to large amounts of private capital. The only arm of the dairy breeding industry that has attracted private enterprise to any great extent is the AI industry, but the success of any AI company is built on the pyramid of dairymen in conjunction with purebred breeders and geneticists. The AI industry is only one arm of an intertwined breeding structure that exists for the breeding of dairy cows. Private investment in a breeding industry is clearly made more attractive if breeding systems provide intellectual property control over genetics. The biological lock of hybrid breeding is one of its hallmarks and therefore has been central to major trends in breeding industries that have gone beyond the mere methodology of breeding, or the incorporation of genetics within a breeding program.[7]

The infiltration of genetics to livestock breeding is based on many factors – none of which have much to do with the failure of farmers to adopt innovation on a timely basis. To begin with, the ideas concerning selection were not in any way new in their own right. Chief among factors dictating the incursion of genetics into traditional breeding programs were two issues: the ability of the science to interface properly with the historic structure of any specific livestock industry, and the critical development of forms of technology upon which genetic theories and practices could be based. The general structure of the chicken industry supported the emphasis on hybrid breeding that so attracted the attention of geneticists. This was not true of the dairy industry. The separation of breeder from producer was not strong in the dairy industry, for example. The failure of the American Poultry Association to adopt utility standards was in sharp contrast to the cattle breed associations' data collection based on scientific methods. The Record of Performance structures functioned entirely differently in the two industries. Timing as well as culture explains why changes occurred in certain livestock industries but not in others. Artificial insemination technology was critical to new developments in livestock genetics; the fact that bull semen froze easily went a long way in dictating how the AI industry would function and which species of livestock would consequently attract the attention of geneticists.

The complex of biology/industry-needs triggers interesting dynamics in the livestock world, some of which relate to ethics. Breeding was at the heart of the matter because it could change the phenotype of the animal profoundly, and in doing so often introduced industry problems where none had existed before. The entanglement of breeding with ethics and the five freedoms received considerable attention in 2012 in the new on-line journal, *Animal Frontiers*.[8] While there is a profusion of literature on the general subject of industry and ethics, this book concentrates on the reaction of farmers and agricultural people to the entangled problems. Animal rights people and humane societies were not the only groups aware of these issues. A look at farm reaction to how male young in the dairy and table egg industries have been handled since the nineteenth century shows that agricultural people were concerned with moral questions, which rose and fell with the methodology available to solve problems. The dehorning of cattle is another example of the workings of the complex over the years. The activity troubled farmers. The dehorning story also indicates that when genetics alone can solve a problem it is not always used for industry-needs reasons, thereby providing for the continuation of exacerbated moral contentions. It is entirely clear that farmers and agricultural people were deeply worried about the moral implications of killing male young, raising veal calves, and dehorning cattle in the nineteenth century. That situation has not changed. The difficulty of dealing with male young in the chicken and dairy industries, and of controlling pain in dehorning operations, remain contentious for many people from both an ethical and a breeding point of view.

It is perhaps sensible to explain briefly why sheep breeding did not receive attention in this book. The primary reason is that it dovetailed quite closely with characteristics of the beef cattle breeding industry in most countries of the western world, and therefore by looking at the beef cattle situation, it would be evident what the sheep situation was. Sheep breeding continues to be dominated by purebred breeding in nuclear herds, and rams from these herds are used on producer flocks to generate crossbred market stock. The interest here is in hybrid vigour. In North America sheep tend to fill a dual-purpose role: they supply wool, but also meat in the form of lambs. Mutton is less likely to be consumed in North America compared to parts of Europe and Britain. The breeder structure, then, is dominated by purebred breeding, which is used by the multipliers/producers who have some choice in how breeding for hybrid vigour will be done. Generally speaking, quantitative genetics looks at sheep breeding the same way that it addresses beef cattle breeding: namely, by assessing EPDs in the purebred sector and attempting to generate EPDs for crossbred performance. The feeder structure is less defined in the sheep industry, where the multiplier/producer might breed for lambs and raise them for market consumption as well. Sheep breeding is often dual-purpose in nature; raising sheep for wool and meat. That had not always been the case, as Merino sheep breeding in eighteenth and nineteenth century Spain makes clear, and as specialization for meat breeds occurred in Britain over the same period.[9] Regardless of levels of specialization, breeding prediction equations do not work any better in the sheep breeding industry than they do in the beef cattle breeding industry, as has been argued recently.[10] The sheep industry is more important in countries like Britain and Australia/New Zealand than it is in North America.

A few comments might be appropriate as well on horses and genetics. The horse-breeding situation received considerable attention from the state before the birth of genetics. Stallion enrolment plans and stallion licensing dominated state intervention with horse breeding over the late nineteenth and early twentieth centuries, and dovetailed with the interests of purebred breeders. But horse breeding had a deeply embedded culture that predated purebred breeding and the rise of breed associations. Issues of purity and type were as old as the mid-eighteenth century when the Thoroughbred was created. The culture of horse breeding was so deeply engrained in the minds of the breeders that they remained unreceptive to ideas emanating out of population genetics; or the idea that breeding predictions had to rely on a group or population basis.[11] Horses, however, were not generated in the numbers that cattle, chickens, pigs, or even sheep were, making it difficult to collect sufficient data on them to quantify results. Horse breeders, like all livestock breeders, were aware of hybrid vigour, but breeding within a complicated hybrid program made no sense. It is perhaps significant too, that horses were declining in use as agricultural animals over the period that livestock genetics took shape: the 1940s with hybrid breeding and the 1950s and 1960s with quantitative genetic breeding. Horses increasingly were pleasure animals and not farm livestock. An inbuilt prejudice against the use of AI did not encourage the incorporation of genetic principles into the breeding of horses. Arabian horse breeders did not accept AI for registration purposes, for example, until the 1980s. The Thoroughbred horse industry still does not accept the use of AI. When stallions can only breed by natural service, they do not produce as many progeny as would be possible with the use of AI. Restriction of progeny numbers encourages higher stud fees. Molecular genetics was used to look at the historical breeding of horses. Interesting information came to light. Mitochondrial studies of the genetics comprising Thoroughbreds and Arabians, for example, revealed that many more Arabian mares went into the genetic make-up of the Thoroughbred than tradition claimed.[12] It was evident from another research project that pedigrees for some of the ancient lines of Polish Arabians were inaccurate.[13]

This study of livestock genetics history raises interesting questions about the future of breeding practices in several livestock industries. For example, will the dominance of breeding companies in the chicken-breeding industry, and the privacy of breeding methods that they uphold, become a thing of the past? Could we see a return of lines that breed truly in the chicken-breeding world? If we do, that would be the end of the patent via a biological lock. Or will the companies develop genomic selection via other avenues that better suit both the chicken genome itself and the privatized hybrid breeding that they follow?

How will breeding by DNA solve issues, which chiefly revolve around problems of increasing levels of inbreeding as a result of favouring certain males, in breeding industries (such as the dairy industry) that rely on the production of true breeding lines? How will SNP testing affect the AI industry? Even though all bulls acquired by the studs for progeny testing now have genomic profiles, and the companies have therefore incorporated genomic selection into their programs, that does not mean the studs will be able to sell as much semen as in the past. If a dairy farmer can DNA test all the bull calves born on his farm, he could locate a very good one without any of the progeny testing expense that AI organizations incur. He could even sell the semen of this bull to other dairymen, completely bypassing the AI companies. Will it be possible to DNA test any animal, and not just purebred members of breeds for which there is a great deal of data on productivity within breeding lines?

Purebred breeding in dairy cattle has promoted many problems, and quantitative genetics has played a role in this phenomenon. Although not necessarily condoned or encouraged by geneticists, inbreeding goes hand in hand with quantitative genetic breeding that works with lines that breed truly. Good bulls will tend to come from good families, and their sons are more likely to be kept and to be used on related females. The overwhelming selection emphasis on volume of milk produced by a cow, and not on the many aspects that dictate her fitness – that is, livability, structural soundness, and resistance to disease – has created problems in the general dairy herds. Declining fertility has become a problem, as has calving difficulty. The cure for many of these difficulties is, ironically, crossbreeding. Can DNA testing ever be done without the supporting data that purebred breeding supplies? Will such technology then bring about the end of purebred breeding?

Changes in livestock genetics have been rapid and overarching over the past ten years, but many features of it have ancient roots: the importance of black box thinking, for example. Identification of genes responsible for quantitative traits has not affected livestock breeding as much as molecular geneticists hoped, or indeed what the early Mendelists had hoped before the thinking behind biometry reoriented livestock genetics' approach to come more in line with the thinking of the eighteenth century livestock breeders. In fact, black box thinking seems to be more entrenched than ever, as geneticists cast wider and wider nets to assess the genome for more and more markers. Fisher's 1918 infinitesimal model, which argued that quantitative traits were inherited on the basis of many genes all acting in a minor way if looked at individually, seems more in place than ever. Much of livestock genetics still looks, then, as a process that might be described as back to the future.

All of this could change very rapidly as a result of bioinformatics and functional genomics, which studies the dynamics of so-called "junk DNA" – that is, parts of DNA like SNPs that do not code for protein and therefore do not (in any way we can understand now) transmit traits. At present, the Encyclopedia of DNA Elements (Encode) project studies the human genome with this end in mind. So far the project has found that "many non-coding variants in individual genome sequences lie in ENCODE-annotated functional regions; this number is at least as large as those that lie in protein-coding genes" and that "single nucleotide polymorphisms (SNPs) associated with disease by genome-wide association studies are enriched in or near ENCODE-defined regions that are outside protein-coding genes."[14] The consortium elaborates as follows:

> 80 [per cent] of the genome contains elements linked to biochemical functions, dispatching the widely held view that the human genome is mostly 'junk DNA' ... The vast majority of the human genome does not code for proteins and, until now, did not seem to contain defined gene-regulatory elements. Why evolution would maintain large amounts of 'useless' DNA had remained a mystery, and seemed wasteful. It turns out, however, that there are good reasons to keep this DNA. Results from the ENCODE project show that most of these stretches of DNA harbour regions that bind proteins and RNA molecules, bring these into positions from which they cooperate with each other to regulate function and level of expression of protein-coding genes.[15]

There is little reason to think the situation will be found different with livestock genomes. The mystery seems only to be deepening with further research. With the sixtieth anniversary of the discovery of the molecular structure of DNA in 2013, *Nature* commented on how little we understand the functioning of DNA in spite of advances in genomics, certainly in relation to appreciating how evolution works, and offered the following thoughts: "When the structure of DNA was first deduced, it seemed to supply the final part of a beautiful puzzle, the solution for which began with Charles Darwin and Gregor Mendel. The simplicity of that picture has proved too alluring."[16] Genomics has brought answers to many questions, but it has also brought new queries into the picture.

The problem of what is practice and non-educated, versus what is science and educated in breeding needs to be separated from concepts involving hybrid breeding and true line breeding. There has been a tendency to see hybrid breeding as science breeding, and purebred or pure line breeding as art breeding. A history of chicken and dairy cattle breeding in relation to genetics proves conclusively that such an approach is false. Part of the difficulty in disentangling art and science from hybrid and true line breeding is that inbreeding is central to both concepts and is not simply the opposite of crossbreeding. Inbreeding can be done in crossing for hybrid vigour or in conjunction with breeding lines that breed truly. The reason that inbreeding is seen as the opposite of crossbreeding is that it can correct the problems of inbreeding, just as inbreeding can correct the problems of crossbreeding. The art/science division forced on livestock breeding has always been artificial, and one might say, the result of a cultural logic that was established in the eighteenth century. That scientists needed to distinguish their efforts from what farmer breeders practiced does not make the division less artificial. What is perhaps more astonishing is how long that sense of division has stayed in place. A study of the history of artificial selection theory and practice reveals that science has been slow to develop workable theories that differed from practical approaches, and that technology and statistics played a critical role in any move to breeding by geneticist theory.

Notes

Introduction

1 "Charles Darwin to Asa Gray, 20 July 1857," Darwin Correspondence Project, accessed 14 October 2012, www.darwinproject.ac.uk/entry-2125.
2 See, for example, Ann Millán Gasca, "The Biology of Numbers: the Correspondence of Vito Volterra on Mathematical Biology," in *The Biology of Numbers: The Correspondence of Vito Volterra on Mathematical Biology*, edited by G. Israel, and Ann Millán Gasca, vol. 26 of *Science Networks. Historical Studies* (Berlin: Birkauser Verlag, 2002), 1–54;.and S.E. Kingsland, *Modeling Nature: Episodes in the History of Population Ecology* (Chicago: University of Chicago Press, 1995); and her "Mathematical Figments, Biological Facts: Population Ecology in the Thirties," *Journal of the History of Biology* 19 (1986): 235–56.
3 See, for example, J. Grey, "History of Mathematics and History of Science Reunited?," *Isis* 102 (2011): 511–17.
4 R. Cassidy, *The Sport of Kings: Kinship, Class and Thoroughbred Breeding in Newmarket* (Cambridge: Cambridge University Press, 2002); M.E. Derry, *Bred for Perfection: Shorthorn Cattle, Collies, and Arabian Horses since 1800* (Baltimore: The Johns Hopkins University Press, 2003); and her *Horses in Society: A Story of Breeding and Marketing Culture, 1800–1920* (Toronto: University of Toronto Press, 2006); and her *Ontario's Cattle Kingdom: Purebred Breeders and Their World, 1870–1920* (Toronto: University of Toronto Press, 2001); H. Ritvo, *The Animal Estate* (Cambridge: Harvard University Press, 1987); and her *The English and Other Creatures in the Victorian Age* (Cambridge: Harvard University Press, 1987); and her *The Platypus and the Mermaid and Other Figments of the Classifying Imagination* (Cambridge: Harvard University Press, 1997); N. Russell, *Like Engend'ring*

Like: Heredity and Animal Breeding in Early Modern England (Cambridge: Cambridge University Press, 1986); M.L. Ryder, *Sheep and Man* (London: Duckworth, 1983); J. Serpell, *In the Company of Animals: A Study of Human-Animal Relations* (Cambridge: Cambridge University Press, 1996); K. Thomas, *Man and the Natural World: Changing Attitudes in England 1500–1800* (London: Allen Lane, 1983); C. Grasseni, "Designer Cows: The Practice of Cattle Breeding Between Skill and Standardization," *Society and Animals* 13 (2005): 33–49; S. Swart, *Riding High: Horses, Humans and History in South Africa* (South Africa: WITS University Press, 2010); S. Swart and G. Bankoff, eds., *Breeds of Empire: the 'Invention' of the Horse in Southeast Asia and Southern Africa 1500–1950* (Denmark: NIAS Press, 2007); J. Marie, "For Science, Love and Money: The Social Worlds of Poultry and Rabbit Breeding in Britain, 1900–1940," *Social Studies of Science* 38 (2008): 919–36.

5 See, for example, S.D Jones, *Valuing Animals: Veterinarians and Their Patients in Modern America* (Baltimore: The Johns Hopkins University Press, 2003); and her *Death in a Small Package: A Short History of Anthrax* (Baltimore: The Johns Hopkins University Press, 2010). See also A. Woods, *Manufacturing Plague: The History of Foot-And-Mouth Disease in Britain* (London: Earthscan, 2004).

6 Examples of science and practice in plant breeding outside corn are P. Dreyer, *A Gardener Touched with Genius: the Life of Luther Burbank* (Berkeley: University of California Press, 1985), and J. Harwood, *Technology's Dilemma: Agricultural Colleges between Science and Practice in Germany, 1860–1933* (New York: Peter Lang Publishing Group, 2005).

7 For corn, see D. Fitzgerald, *The Business of Breeding: Hybrid Corn in Illinois, 1890–1940*, (Ithaca: Cornell University Press, 1990); and her "Farmers Deskilled: Hybrid Corn and Farmers' Work," *Technology and Culture* 34 (1993): 324–43. For emphasis on corn and for the general use of hybridizing as a biological lock, see J.K. Kloppenburg, Jr., *First the Seed: The Political Economy of Plant Technology, 1492–2000* (Cambridge: Cambridge University Press, 1988). See also D.B. Paul and B Kimmelman, "Mendel in America: Theory and Practice, 1900–1919," in *The American Development of Biology*, ed. R. Rainger, K Benson and J. Maienchein (Philadelphia: University of Pennsylvania Press, 1988).

8 For a discussion of hybrid corn in relation to the study of innovation diffusion see E.M. Rogers, *Diffusion of Innovations*, 4th ed. (New York: The Free Press, 1995), 31–6, 53–5.

9 See J. Harwood, *Technology's Dilemma: Agricultural Colleges between Science and Practice in Germany, 1860–1933* (New York: Peter Lang Publishing

Group, 2005); and J. Harwood, *Styles of Scientific Thought: The German Genetics Community, 1900–1933* (Chicago: University of Chicago Press, 1992). See also S. Castonguay, "The Transformation of Agricultural Research in France: The Introduction of the American System," *Minerva* 43 (2005): 265–287.

10 A few examples are B. Kimmelman, "Mr. Blakeslee Builds his Dream House: Agricultural Institutions, Genetics, and Careers 1900–1945," *Journal of the History of Biology* 39 (2006): 241–280; C. Bonneuil, "Mendelism, Plant Breeding and Experimental Cultures: Agriculture and the Development of Genetics in France," *Journal of the History of Biology* 39 (2006): 281–308; T. Wieland, "Scientific Theory and Agricultural Practice: Plant Breeding in Germany from the late 19th to the Early 20th Century," *Journal of the History of Biology* 39 (2006): 309–43.

11 Edited by Sharon Kingsland and Denise Phillips, *New Perspectives on the History of the Life Sciences and Agriculture*. The book will published in the Archimedes series by Springer in 2015.

12 See K.J. Cooke, "From Science to Practice, or Practice to Science? Chickens and Eggs in Raymond Pearl's Agricultural Breeding Research, 1907–1916," *Isis* 88 (1997): 62–86.

13 G.E. Bugos, "Intellectual Property Protection in the American Chicken-Breeding Industry," *Business History Review* 66 (1992): 127–68. For the patenting of agricultural plants and animals more generally, see D. Kevles, "The Advent of Animal Patents: Innovation and Controversy in the Engineering and Ownership of Life," in *Intellectual Property Rights and Patenting in Animal Breeding and Genetics*, ed. S. Newman and M. Rothschild, (New York: CABI Publishing, 2002), 18–30; and D. Kevles "Patents, Protections, and Privileges," *Isis* 98 (2007): 323–331, and D. Kevles, "Protections, Privileges, and Patents: Intellectual Property in Animals and Plants since the Late Eighteenth Century," in *Contexts of Invention*, ed. M. Biagioli, P. Jaszi, and M. Woodmansee (Chicago: University of Chicago Press, 2010); and H. Ritvo, "Possessing Mother Nature: Genetic Capital in Eighteenth-Century Britain," in *Early Modern Conceptions of Capital*, ed. J. Brewer and S. Staves (London: Routledge, 1995), 413–26.

14 D.L. Harris et al., "Breeding for Profit: Synergism between Genetic Improvement and Livestock Production (a Review)," *Journal of Animal Science* 72 (1994): 2178–200.

15 See, for example, A.Gobin, "Mendelism in Animal Breeding as Developed by Porfessor Leopold Frateur, Louvain (1877–1946)," *Argos* 23 (2000): 111–8.

16 See R.J. Wood and V. Orel, *Genetic Prehistory in Selective Breeding: A Prelude to Mendel* (Oxford: Oxford University Press, 2001). Their earlier work is also valuable. See also Wood and Orel, "Early Development in Artificial Selection as a Background to Mendel's Research," *History and Philosophy of the Life Sciences* 3 (1981): 145–70; "Scientific Animal Breeding in Moravia Before and After the Discovery of Mendel's Theory," *Quarterly Review of Biology* 75 (2000): 149–57. See also Orel, "Selection Practice and Theory of Heredity in Moravia Before Mendel," *Folia Mendelianna* 12 (1977): 179–99.

17 R.J. Wood and V. Orel, "Scientific Breeding in Central Europe during the Early Nineteenth Century: Background to Mendel's Later Work," *Journal of the History of Biology* 38 (2005): 239–72.

18 B. Theunissen, "Breeding Without Mendelism: Theory and Practice of Dairy Cattle Breeding in the Netherlands, 1900–1950," *Journal of the History of Biology* 41 (2008): 637–76; and "Breeding for Nobility or Production? Cultures of Dairy Cattle Breeding in the Netherlands, 1945–1995," *Isis* 103 (2012): 278–309.

19 See S. Wilmot, "From 'public service' to Artificial Insemination: Animal Breeding Science and Reproductive Research in Early 20th Century Britain," *Studies in History and Philosophy of Biological and Biomedical Sciences* 38 (2007): 411–41; and "Between the Farm and the Clinic: Agricultural and Reproductive Technology in the Twentieth Century," *Studies in History and Philosophy of Biological and Biomedical Sciences* 38 (2007): 303–315; C. Grasseni, "Managing Cows: an Ethnography of Breeding Practices and Uses of Reproductive Technology in Contemporary Dairy Farming in Lombardy (Italy)," *Studies in History and Philosophy of Biological and Biomedical Sciences* 38 (2007): 488–510; and P. Brassley, "Cutting across Nature? The History of Artificial Insemination of Pigs in the United Kingdom," *Studies in History and Philosophy of Biological and Biomedical Sciences* 38 (2007): 442–461.

20 H.A. Herman, *Improving Cattle by the Millions: NAAB and the Development and Worldwide Application of Artificial Insemination* (Columbia: University of Missouri Press, 1981).

21 B. Matz, "Crafting Heredity: The Art and Science of Livestock Breeding in the United States and Germany, 1860–1914" (PhD diss., Yale University, 2011).

22 S. McCook, *States of Nature: Science, Agriculture, and the Environment in the Spanish Caribbean, 1760–1940* (Austin: University of Texas Press, 2002).

23 B. Charnley, "Experiments in Empire-building: Mendelian genetics as a national, Imperial, and global enterprise," *Studies in History and Philosophy of Science* 44 (2013): 292–300.

24 See S. Swart, *Riding High: Horses, Humans and History in South Africa* (Johannesburg: Witswatersrand University Press, 2010); and S. Swart and G. Bankoff, eds., *Breeds of Empire: The 'Invention' of the Horse in southeast Asia and Southern Africa* (Copenhagen: Nordic Institute of Asian Studies Press, 2007).

1 Artificial Selection Theory

1 J. Clutton-Brock, *A Natural History of Domesticated Animals*. 2nd ed. (Cambridge: Cambridge University Press, 1999), 40.
2 T.S. Kuhn, *The Structure of Scientific Revolutions*. 4th ed. (Chicago: University of Chicago Press, 2012), 10–22.
3 B. Tozer, *The Horse in History* (London: Methuen & Co., 1908), 203–4.
4 J. Clutton-Brock, *Horse Power: Horse and Donkey in Human Societies* (London: Natural History Museum Publications, 1992), 61.
5 M.E. Derry, *Bred for Perfection: Shorthorn Cattle, Collies, and Arabian Horses Since 1800* (Baltimore: The Johns Hopkins University Press, 2003), 4–5.
6 R.J. Wood and V. Orel, *Genetic Prehistory in Selective Breeding: a Prelude to Mendel* (Oxford: Oxford University Press, 2001), 72.
7 V. Orel and R.J. Wood, "Early Development in Artificial Selection as a Background to Mendel's Research," *History and Philosophy of the Life Sciences* 3 (1981): 151.
8 N. Russell, *Like Engend'ring Like: Heredity and Animal Breeding in Early Modern England* (Cambridge: Cambridge University Press, 1986), 104.
9 Wood and Orel, *Genetic Prehistory in Selective Breeding*, 72.
10 V. Orel and R.J. Wood, "Scientific Animal Breeding in Moravia Before and After the Discovery of Mendel's Theory," *Quarterly Review of Biology* 75 (2000): 151; V. Orel, "The Spectre of Inbreeding in the Early Investigation of Heredity," *History and Philosophy of the Life Sciences* 19 (1997): 321.
11 For attitudes to breeding that existed before the Enlightenment period, see N. Russell, *Like Engend'ring Like: Heredity and Animal Breeding in Early Modern England* (Cambridge: Cambridge University Press, 1986).
12 P. Kitcher. *In Mendel's Mirror* (Oxford: Oxford University Press, 2003), 63–4.
13 For information on Bakewell see R. Trow-Smith, *A History of British Livestock Husbandry, 1700–1900* (London: Routledge & Kegan Paul, 1959); H.C. Pawson, *Robert Bakewell: Pioneer Livestock Breeder* (London: Crosby Lockwood, 1957), and "Some Agricultural History Salvaged," *Agricultural History Review* 7 (1959), 6–13; Wood and Orel, *Genetic Prehistory in Selective Breeding*; R.J. Wood, "Robert Bakewell (1725–1795) Pioneer

Animal Breeder and His Influence on Charles Darwin," *Folia Mendelianna* 8 (1973): 231–42; D. Wykes, "Robert Bakewell (1725–1795) of Dishley: Farmer and Livestock Improver," *Agricultural History Review* 52 (2004): 38–55; M.E. Derry, *Horses in Society: A Story of Breeding and Marketing Culture, 1800 –1920* (Toronto: University of Toronto Press, 2006), and *Art and Science in Breeding: Creating Better Chickens* (Toronto: University of Toronto Press, 2012).

14 Bakewell to Culley, 8 February 1787, "The Bakewell Letters," in *Robert Bakewell: Pioneer Livestock Breeder*, ed. H.C. Pawson (London: Crosby Lockwood, 1957), 107.

15 Quoted in George Mingay, ed., *Arthur Young and His Times* (London: The Macmillan Press Ltd., 1975), 77–8.

16 Wood and Orel, *Genetic Prehistory in Selective Breeding*, 89.

17 Pawson, *Robert Bakewell*, 60, 62, 69.

18 Sir J.S. Sebright, *The Art of Improving the Breeds of Domestic Animals* (London: John Harding, 1809), 30.

19 Ibid., 5.

20 Ibid., 10.

21 Ibid., 11.

22 Ibid., 17.

23 Ibid., 17–18

24 Ibid., 7.

25 C.J. Bajema, ed., *Artificial Selection and the Development of Evolutionary Theory* (Stroudsburg, Pa: Hutchinson Ross Publishing, 1982), 3, 4, 11–12.

26 S. Müller-Wille and V. Orel, "From Linnaean Species to Mendelian Factors: Elements of Hybridism, 1751–1870," *Annals of Science* 64 (2007): 177.

27 L.C. Dunn, *A Short History of Genetics, The Development of Some of the Main Lines of Thought, 1864–1939* (New York: McGraw-Hill, 1965), 27.

28 S. Müller-Wille and V. Orel, "From Linnaean Species to Mendelian Factors," 182.

29 R. Olby, "Mendel no Mendelian?," *History of Science* 12 (1979): 57. For a different point of view, see S. Müller-Wille and V. Orel, "From Linnaean Species to Mendelian Factors," 171–215.

30 P. Kitcher, P. *In Mendel's Mirror*, 64.

31 Wood and Orel, *Genetic Prehistory in Selective Breeding*, 276.

32 C. Trudge, *In Mendel's Footnotes: An Introduction to the Science and Technologies of Genes and Genetics from the Nineteenth Century to the Twenty-Second* (London: Jonathan Cape, 2000), 64–66; Wood and Orel, *Genetic Prehistory in Selective Breeding*, 7; V. Orel, "Selection Practice and Theory of Heredity in Moravia Before Mendel," *Folia Mendelianna* 12 (1977): 179.

33 Wood and Orel, *Genetic Prehistory in Selective Breeding*, 215, 237.

34 Bajema, *Artificial Selection*, 7, 50–9.

35 Wood and Orel, *Genetic Prehistory in Selective Breeding*, 215, 237

36 Ibid., 5, 6, 7.

37 Orel, "Selection Practice and Theory of Heredity," 179, 180, 182, 183.

38 Ibid., 187, 188.

39 Ibid., 180, 185, 187, 191–2, 194, 195.

40 Wood and Orel, *Genetic Prehistory in Selective Breeding*, 215, 222, 247, 237–8.

41 A. Sanders, *Short-Horn Cattle: A Series of Historical Sketches, Memoirs and Records of the Breed and Its Development in the United States and Canada* (Chicago: Sanders, 1900), 31, 34–5, 37–9, 44; S. Wright, "Mendelian Analysis of the Pure Bred Breeds of Livestock. Part 11: The Duchess Family of Shorthorns as Bred by Thomas Bates," *Journal of Heredity* 14 (1923): 405.

42 Sanders, *Short-Horn Cattle: A Series of Historical Sketches*, 75.

43 See Wright, "Mendelian Analysis of the Pure Bred Breeds of Livestock. Part 11," 405–22; Sanders, *Short-Horn Cattle: A Series of Historical Sketches*, 81–5.

44 J. Lush, *Notes on Animal Breeding* (unpublished manuscript, 1933), Chapter 111, 61.

45 H. Ritvo, *The Animal Estate* (Cambridge: Harvard University Press, 1987), 61.

46 *Proceedings of the Fourth Meeting of the American Shorthorn Association*, 1876, 22.

47 Derry, *Horses in Society*, 10–12; *Bred for Perfection*, 7, 17–28; *Ontario's Cattle Kingdom: Purebred Breeders and Their World, 1870–1920* (Toronto: University of Toronto Press, 2001), 39–44.

48 Lady Wentworth, *Thoroughbred Racing Stock* (New York: Charles Scribner's Sons, 1938), 60.

49 Wood and Orel, *Genetic Prehistory in Selective Breeding*, 264.

50 See, for example, *Farmer's Advocate*, January 1876, 13; February 1876, 27; March 1876, 46; 8 December 1910, 1927–8; *Farming World and Canadian Farm and Home*, 1 January 1906, 161.

51 In the 1870s the examination of agriculture at the Ontario Agricultural College was on the history of Shorthorns, and revolved around the methods of the Collings and also Bakewell. See Sessional Paper (referred to as SP) of the Legislature of Ontario, 13, 1875–6, 31–2; SP 12, Ontario, 1877, 48.

52 *Farmer's Advocate*, January 1876, 13.

53 *Farming*, February 1896, 337.

54 Derry, *Horses in Society*, 241–4.

55 See Harper, *Breeding of Farm Animals*, 169, 174.

56 *Breeder's Gazette*, 20 April 1882, 537–8; 12 January 1882, 162; 16 March 1882, 395; 8 April 1896, 272; 2 November 1898, 422–3; M.W. Harper, *Breeding of Farm Animals* (New York: Orange Judd Company, 1920), 138–9; M. Derry, *Horses in Society*, 35–8; P. Thurtle, "Harnessing Heredity in Guilded Age America: Middle Class Mores and Industrial Breeding in a Cultural Context," *Journal of the History of Biology*, 35 (2002): 43–78.

57 MacEwan, *Heavy Horses*, 42–3, 50–2; *Farmer's Advocate*, 7 November 1918, 1788; *Breeder's Gazette*, 5 March 1885, 357.

58 *Breeder's Gazette*, 27 April 1882, 565.

59 Ibid., 8 November 1883, 640.

60 Ibid., 18 October 1883, 530.

61 D. Goodall, *A History of Horse Breeding* (London: Robert Hall, 1977), 179.

62 *Breeder's Gazette*, 25 June 1885, 971.

63 Ibid., 10 March 1887, 381; 8 December 1887, 909–10; 28 March 1888, 317; 21 August 1889; 180.

64 Ibid., 28 March 1888, 317.

65 Ibid., 11 January 1888, 38.

66 Ibid., 13 June 1888, 572.

67 Ibid., 17 November 1887, 795.

68 Charles Darwin, *The Variation of Animals and Plants under Domestication* (1883; repr. Baltimore: The Johns Hopkins University Press, 1998), 447.

69 Darwin, *On the Origin of Species* (London: John Murray, 1859), 13.

70 Ibid.; The literature on Darwin and his knowledge of how heredity worked is extensive. See, for example, S.E. Kingsland, "The Battling Botanist: Daniel Trembly MacDougal, Mutation theory, and the Rise of Experimental Evolutionary Biology in America, 1900–1912," *Isis* 82 (1991): 484–5; and R. Olby, *Origins of Mendelism*, 2nd ed. (Chicago: University of Chicago Press, 1985), 40–47.

71 D.B. Paul and B.A. Kimmelman, "Mendel in America: Theory and Practice, 1900–1919" in *The Development of American Biology*, ed. R. Rainger et al. (Philadelphia: University of Pennsylvania Press, 1988), 289.

72 J.A. Secord, "Nature's Fancy: Charles Darwin and the Breeding of Pigeons," *Isis* 72 (1981): 166–171, 174, 177. For more on Darwin's interest in artificial selection see S.G. Alter, "The Advantages of Obscurity: Charles Darwin's Negative Inference from the Histories of Domestic Breeds," *Annals of Science* 64 (2007): 235–250; R.A. Richards, "Darwin and the Inefficiacy of Artificial Selection," *Studies in History and Philosophy of Science* 28 (1997): 75–97; S.G. Sterrett, "Darwin's Analogy between Artificial and Natural Selection: How does it go?," *Studies in History and Philosophy of Biological and Biomedical Sciences* 33 (2002): 151–168; L.T. Evans, "Darwin's Use of the Analogy

between Artificial and Natural Selection," *Journal of the History of Biology* 17 (1984): 113–140; J.F. Cornell, "Analogy and Technology in Darwin's Vision of Nature," *Journal of the History of Biology* 17 (1984): 303–344; H-J. Rheinberger and P McLaughlin, "Darwin's Experimental Natural History," *Journal of the History of Biology* 17 (1984): 345–368; and B. Theunissen, "Darwin and his Pigeons: The Analogy Between Artificial and Natural Selection Revisited," *Journal of the History of* Biology 45 (2012): 179–212.

73 Theunissen, "Darwin and His Pigeons."

74 B.A. Matz, "Crafting Heredity: The Art and Science of Livestock Breeding in the United States and Germany, 1860–1914" (PhD diss., Yale University, December 2011), 169–75, 207.

75 Paul and Kimmelman, "Mendel in America," 293.

76 Theunissen, "Knowledge is Power: Hugo de Vries on Science, Heredity and Social Progress," *British Journal for the History of Science* 27 (1994): 293, 301, 304.

77 S.E. Kingsland, "The Battling Botanist: Daniel Trembly MacDougal, Mutation theory, and the Rise of Experimental Evolutionary Biology in America, 1900–1912," *Isis* 82 (1991): 489–93.

78 Paul and Kimmelman, "Mendel in America," 289, 293; USDA *Yearbook of Agriculture*, "Better Plants and Animals," Part 1, 1936, 469; L. Carlson, "Forging his own Path: William Jasper Spillman and Progressive Era Breeding and Genetics," *Agricultural History* 79 (2005): 50, 52, 53, 58.

79 Matz, "Crafting Heredity," 212–16.

80 Ibid., 178–181.

81 The *Third Report of the Royal Commission on Horse Breeding*, 1890. See, for example, page 21 of that report.

82 Matz, "Crafting Heredity," 182, 191.

83 Ibid., 169–75, 207.

84 J.L. Lush, "Genetics and Animal Breeding," in *A Short History of Genetics, The Development of Some of the Main Lines of Thought, 1864–1939*, ed. L.C. Dunn (New York: McGraw-Hill, 1965), 494.

85 Quoted in M. Bulmer, *Francis Galton: Pioneer of Heredity and Biometry* (Baltimore: The Johns Hopkins University Press, 2003), 299.

86 See, for example, *OAC Review*, December 1913, 138–42.

87 "Eugenics on the Farm," *Journal of Heredity*, 7 (1916): 47.

88 A. Oleson and J. Voss, *The Organization of Knowledge in America, 1860–1920* (Baltimore: The Johns Hopkins University Press), 219–28.

89 N.R. Gillham, *A Life of Sir Francis Galton: From African Exploration to the Birth of Eugenics* (Oxford: Oxford University Press, 2001), 9.

90 Ibid.

91 W.C. Hill, ed., *Quantitative Genetics: Part I, Explanation and Analysis of Continuous Variation* (New York: Van Nostrand Reinhold Company, 1984), 10. For examples of Pearson's work on the subject see "On the Systematic Fitting of Curves to Measurements and Observations: Part I," *Biometrika* 1 (1902): 265–303; and "On the Systematic Fitting of Curves to Measurements and Observations: Part II," *Biometrika* 2 (1902): 1–23.

92 Pearson, "On the Systematic Fitting of Curves to Measurements and Observations: Part I," 266.

93 Hill, *Quantitative Genetics: Part I*, 9.

94 Gillham, *A Life of Sir Francis Galton*, 9.

95 K. Pearson, "On the Fundamental Conceptions of Biology," *Biometrika* 1 (1902): 320.

96 Ibid., 322.

97 Gillham, *A Life of Sir Francis Galton*, 306, 322.

98 A. Barrington and K. Pearson, "On the Inheritance of Coat-Colour in Cattle: Part 1. Shorthorn Crosses and Pure Shorthorns," *Biometricka* 4 (1906): 427–64.

99 See S.E. Kingsland, *Modeling Nature: Episodes in the History of Population Ecology* (Chicago: University of Chicago Press, 1995); and her "Mathematical Figments, Biological Facts: Population Ecology in the Thirties," *Journal of the History of Biology* 19 (1986): 235–56, for a discussion concerning the attitudes of ecologists to the role of mathematics in biological research.

100 Quoted in R.B. Churchill, "William Johannsen and the Genotype Concept," *Journal of the History of Biology* 7 (1974): 8.

101 Bulmer, *Francis Galton*, 315. See as well J. Sapp, "The Struggle for Authority in the Field of Heredity, 1900–1932: New Perspectives on the Rise of Genetics," *Journal of the History of Biology* 16 (1983): 315.

102 See E.P. Cunningham, *Quantitative Genetic Theory and Livestock Improvement* (University of New England, 1979), 8–9; and W.C. Hill, ed., *Quantitative Genetics: Part II, Selection* (New York: Van Nostrand Reinhold Company, 1984), 6.

103 See B.S. Weir, ed., *Proceedings of the Second International Conference on Quantitative Genetics* (Sunderland, MA: Sinauer Associates, Inc., 1988), 536.

104 E.J. Pollak et al., eds., *Proceedings of the International Conference on Quantitative Genetics* (Ames: Iowa State University Press, 1977), 56.

2 Early Developments in Genetics

1 For a review of the historical debate over Mendel's role in the development of genetics, see Jan Sapp, "The Nine Lives of Gregor

Mendel," in *Experimental Inquiries: Historical, Philosophical and Social Studies of Experimentation in Science*, ed. H.E. Le Grand (Dordrecht: Kluwer Academic Publishers, 1990), 137–166. See also R. Olby, "Mendel no Mendelian?," *History of Science* 12 (1979): 53–72.

2 P.K.Gupta, "Quantitative Genetics on the Rise," *Current Science* 93 (2007): 1051.

3 For a first hand discussion of the conflict, see R.C. Punnett, "Early Days in Genetics," *Heredity* 4 (1950): 1–10.

4 W.F.R. Weldon, "Mendel's Laws of Alternative Inheritance in Peas," *Biometrika* 1 (1902): 252.

5 W.B. Provine, *The Origins of Theoretical Population Genetics* (Chicago: University of Chicago Press, 1971), 25, 34, 40–44.

6 L.C. Dunn, *A Short History of Genetics, The Development of Some of the Main Lines of Thought, 1864–1939* (New York: McGraw-Hill, 1965), 120.

7 S. Wright, "The Relationship of Livestock Breeding to Theories of Evolution," *Journal of Animal Science* 46 (1978): 1194; Dunn, *A Short History of Genetics*, 120.

8 W.D. Terhmohlen, "The History of Development of Poultry Departments in the State Colleges or Universities of the United States," *Poultry Science* 46 (1967): 294.

9 Marcus, *Agricultural Science and the Quest for Legitimacy*, 61, 63.

10 W.H. Heape, *The Breeding Industry: Its Value to the Country, and Its Needs* (Cambridge: Cambridge University Press, 1906), v, vi, 21, 22, 24.

11 Ibid., 25, 26.

12 F.B. Hutt, "Seventy-Five Years of Poultry Genetics," Roundtable of Poultry Breeders Association, 1975, 152, accessed 17 December 2007, www.poultryscience.org/pba/1952–2003/1975/1975%20Hutt.pdf.

13 K. Cooke, "From Science to Practice, or Practice to Science? Chickens and Eggs in Raymond Pearl's Agricultural Breeding Research, 1907–1916," *Isis* 88 (1997): 67, 70–4, 76, 77, 82. See as well, R. Pearl, "Breeding Poultry for Egg Production," in *Annual Report of the Maine Agricultural Experiment Station*, Bulletin 192, 1911; and "Inheritance of Hatching Quality of Eggs in Poultry," *American Breeders Magazine* 1 (1913): 129–33; R.R. Slocum, "Poultry Breeding," *Journal of Heredity* 6 (1915): 484–6; Research Committee on Animal Breeding, "Live-Stock Genetics," *Journal of Heredity* 6 (1915): 21–2.

14 Slocum, "Poultry Breeding," 486.

15 Maine Experiment Station, Bulletin 305, 1913, 388. This quote appeared in the *American Poultry Journal*, May 1913, 847.

16 Ibid.

17 Ibid.
18 Ibid.
19 Ibid.
20 Ibid., April 1913, 672.
21 Ibid., May 1913, 847.
22 Ibid.
23 See J. Marie, "For Science, Love and Money: The Social Worlds of Poultry and Rabbit Breeding in Britain, 1900–1940," *Social Studies of Science* 38 (2008): 925 for more on this phenomenon.
24 Provine, *Population Genetics*, 89.
25 In 1906 William Bateson coined the word "genetics." This situation allied Mendelism primarily with genetics, and encouraged the further marginalization of biometry from mainline hereditary science.
26 *American Poultry Journal*, October 1913, 1278.
27 Ibid.
28 Ibid., December 1913, 1517.
29 E.B. Babcock and R.E. Clausen, *Genetics in Relation to Agriculture* (New York: McGraw-Hill Book Company, Inc., 1918), 449.
30 W. Johannsen, "The Genotype Conception of Heredity," *American Naturalist* 45 (1911): 142–3.
31 J. Sapp, "The Struggle for Authority in the Field of Heredity, 1900–1932: New Perspectives on the Rise of Genetics," *Journal of the History of Biology* 16 (1983): 336–337, 324.
32 Dunn, *A Short History of Genetics*, 125.
33 S. Wright, *Evolution and the Genetics of Populations*, vol. 3, *Experimental Results and Evolutionary Deductions* (Chicago: University of Chicago Press, 1977), 11, 29; and Shull, "What is 'Heterosis'?," 440.
34 D.F. Jones, "Dominance of Linked Factors as a Means of Accounting for Heterosis," *Genetics* 2 (1917): 471.
35 Ibid., 477.
36 D. Fitzgerald, *The Business of Breeding: Hybrid Corn in Illinois, 1890–1940* (Ithaca: Cornell University Press, 1990), 55; and J.R. Kloppenburg, *First the Seed: The Political Economy of Plant Biotechnology* (Cambridge, U.K: Cambridge University Press, 1988), 99.
37 Fitzgerald, *The Business of Breeding*, 64.
38 W.E. Castle, "The Mutation Theory of Organic Evolution, from the Standpoint of Animal Breeding," *Science* 21 (1905): 522.
39 Ibid., 524.
40 Ibid.,

41 See K. Pearson, "A Mendelian's View of the Law of Ancestral Inheritance," *Biometrika* 3 (1904): 109–112.

42 See Provine, *Population Genetics*, 111, 112.

43 W.E. Castle, *Heredity in Relation to Evolution and Animal Breeding* (New York: D. Appleton and Company, 1911), 2, 4.

44 Ibid., 151–2.

45 Ibid., 2, 4.

46 K. Rader, *Making Mice: Standardizing Animals for American Biomedical Research, 1900–1955* (Princeton: Princeton University Press, 2004), 31.

47 See W.E. Castle, "Pure Lines and Selection," *Journal of Heredity* 5 (1914): 93–7.

48 D.C. Warren, "A Half Century of Advances in the Genetics and Breeding Improvement of Poultry," *Poultry Science* 37 (1958): 3.

49 Ibid., 4–5.

50 It has recently been argued that that remained the case, even in the world of plant breeding. See N. Roll-Hansen, "Theory and Practice: the Impact of Mendelism on Agriculture," *C.R. Acad. Sci. Paris, Sciences de la Vie/Life Sciences* 323 (2000): 1107–16.

51 I.M. Lerner, "L.C. Dunn (1893–1974) and Poultry Genetics: A Brief Memoir," *Journal of Heredity* 65 (1974): 185–6.

52 Quoted in J. Marie, "The Situation in Genetics II: Dunn's 1927 European Tour," *Mendel Newsletter* 13 (2004): 3.

53 D. Falconer, "Quantitative Genetics in Edinburgh, 1947–1980," *Genetics* 133 (1993): 137.

54 Quoted in Marie, "The Situation in Genetics II," 1.

55 Ibid., 6.

56 For his study of Russia see L.C. Dunn, "Genetics at the Anikowo Station: A Russian Animal Breeding Center that has been Developed During the Reconstruction Period," *Journal of Heredity* 19 (1928): 281–4.

57 Quoted in J. Marie, "The Situation in Genetics II," 7.

58 Ibid.

59 See Dunn, "Genetics at the Anikowo Station," 281–4.

60 Quoted in J. Marie, "The Situation in Genetics II," 8.

61 J. Harwood, *Styles of Scientific Thought: The German Genetics Community, 1900–1933* (Chicago: University of Chicago Press, 1992), 159–62. See also B. Matz, "Crafting Heredity: The Art and Science of Livestock Breeding in the United States and Germany, 1860–1914" (PhD diss, Yale University, December 2011).

62 L. Ollivier, "Jay Lush: Reflections on the Past," *Lohmann Information* 43 (2008): 8.

63 See A. Gobin, "Mendelism in Animal Breeding as Developed by Professor Leopold Frateur, Louvain (1877–1946)," *Argos* 23 (2000): 111–8.

64 T. Theunissen, "Connecting Genetics, Evolutionary Theory and Practical Animal Breeding: Arend L. Hagadoorn (1885–1953)" (unpublished manuscript, 2012), 7–9.

65 Provine, *Population Genetics*, 81–9.

66 See A. Stoltzus and K. Cable, "Mendelism-Mutationsim: the Forgotten Evolutionary Synthesis." *Journal of the History of Biology* (May 9, 2014). http://link.springer.com/article/10.1007%2Fs10739–014–9383–2, for a revisionist approach to the classic view regarding developments in genetics leading to the rise of population genetics.

67 R.A. Fisher, "The Correlation between Relatives on the Supposition of Mendelian Inheritance," *Transactions of the Royal Society of Edinburgh* 52 (1918): 399–433.

68 S.E. Kingsland, "Mathematical Figments, Biological Facts: Population Ecology in the Thirties," *Journal of the History of Biology* 19 (1986): 252.

69 O. Kempthorne, "Status of Quantitative Genetic Theory," in *Proceedings of the Second International Conference on Quantitative Genetics*, ed. B.S. Weir (Sunderland, MA: Sinauer Associates, Inc., 1988), 721.

70 Ibid., 722–3.

71 Ibid., 741.

72 W.C. Hill, ed., *Quantitative Genetics: Part II, Selection* (New York: Van Nostrand Reinhold Company, 1984), 1, 2, 10, 11, 12.

73 See S. Wright, "Mendelian Analysis of the Pure Bred Breeds of Livestock, Part 1: The Measurement of Inbreeding and Relationship," *Journal of Heredity* 14 (1923): 339–48; and "Mendelian Analysis of the Pure Bred Breeds of Livestock, Part 2: The Duchess Family of Shorthorns as Bred by Thomas Bates," *Journal of Heredity* 14 (1923): 405–22.

74 S. Wright, "The Relationship of Livestock Breeding to Theories of Evolution," *Journal of Animal Science* 46 (1978): 1196–9.

75 See J.F. Crow, "Sewall Wright's Place in Twentieth-Century Biology," *Journal of the History of Biology* 23 (1990): 58, 62–6.

76 S. Wright, "The Relationship of Livestock Breeding to Theories of Evolution," 1196.

77 W. Provine, *Sewall Wright and Evolutionary Biology* (Chicago: University of Chicago Press, 1986), 156.

78 W.C. Hill, ed., *Quantitative Genetics: Part I, Explanation and Analysis of Continuous Variation* (New York: Van Nostrand Reinhold Company, 1984), 16.

79 Crow, "Sewall Wright's Place in Twentieth-Century Biology," 61.

80 Ibid., 63.

81 Provine, *Sewall Wright and Evolutionary Biology*, 138–9, 140, 141; and *Origins of Theoretical Population Genetics*, 160–1.

82 Quoted in Provine, *Sewall Wright and Evolutionary Biology*, 156. See as well, S. Wright, "The Effects of Inbreeding and Crossbreeding on Guinea Pigs," in Bulletin 1090 (United States Department of Agriculture, 1922); and *Systems of Mating and Other Papers* (Ames: Iowa State College Press, 1958), which contains reprints of "Systems of Mating," *Genetics* 6 (1921): 111–78, "Evolution in Mendelian Populations," *Genetics* 16 (1931): 97–159, "Correlation and Causation," *Journal of Agricultural Research* 20 (1921): 557–585, and "The Method of Path Coefficients," *Annals of Mathematical Statistics* 5 (1934): 161–215.

83 There are countless examples of this. A few are E.B. Babcock and R.E. Clausen, *Genetics in Relation to Agriculture* (New York: McGraw-Hill Book Company, Inc., 1918); C. Wriedt, *Heredity in Live Stock* (London: Macmillan and Co., 1930).

84 See A.R. Hallauer, "History, Contribution, and Future of Quantitative Genetics in Plant Breeding: Lessons from Maize," *Crop Science* 47 (2007), from International Plant Breeding Symposium, 2007, S4–19, for how population genetics related to quantitative genetics.

85 Provine, *The Origins of Theoretical Population Genetics*,164.

86 G.E. Dickerson, "Inbreeding and Heterosis in Animals," in *Proceedings of the Animal Breeding and Genetic Symposium in Honor of Dr. Jay L. Lush* (Blacksburg, Virginia: American Society of Animal Science and American Dairy Science Association, 1972), 54.

87 See Wright, "Mendelian Analysis of the Pure Bred Breeds of Livestock, Part 2," 405–22.

88 Freeman, "Genetic Statistics in Animal Breeding," in *Proceedings of the Animal Breeding and Genetic Symposium in Honor of Dr. Jay L. Lush* (Blacksburg, Virginia: American Society of Animal Science and American Dairy Science Association, 1972), 3.

89 Ibid., 4

90 A.B. Chapman, A.B. "Jay Lawrence Lush 1896–1982: A Brief Biography," *Journal of American Science* 69 (1991): 2673; Freeman, "Genetic Statistics in Animal Breeding," 5.

91 Crawford, ed., *Poultry Keeping and Genetics*, 956.

92 C.F. McClary, "Reciprocal Recurrent Selection Response in Poultry, Other Animals and Plants," Eighteenth Annual Session, National Poultry Breeders' Roundtable, 1969, 121.

93 D. Gianola, et al., ed., *Advances in Statistical Methods for Genetic Improvement of Livestock* (Berlin: Springer-Verlag, 1990), 4.

94 Ann Millán Gasca, "The Biology of Numbers: the Correspondence of Vito Volterra on Mathematical Biology," in *The Biology of Numbers: The Correspondence of Vito Volterra on Mathematical Biology*, ed. G. Israel and Ann Millán Gasca, vol. 26, *Science Networks. Historical Studies* (Berlin: Birkauser Verlag, 2002), 23–4.

95 See Millán. "The Biology of Numbers" and Kingsland, "Mathematical Figments," 235–56.

96 D.C. Warren, in *American Poultry History, 1823–1973*, ed. O.A. Hanke (Madison, Wisconsin: American Poultry Historical Society, 1974), 273.

97 Freeman, "Genetic Statistics in Animal Breeding," 6.

98 G.B. Havenstein, "Performance Changes in Livestock Following 50 Years of Genetic Selection," *Lohmann Information* 41 (December 2006): 30. See also B. Theunissen, "Breeding Without Mendelism: Theory and Practice of Dairy Cattle Breeding in the Netherlands, 1900–1950," *Journal of the History of Biology* 41 (2008): 637–76 for the effects of Lush on Dutch geneticists and British quantitative geneticists as well.

99 Provine, *Sewall Wright and Evolutionary Biology*, 321–26.

100 Freeman, "Genetic Statistics in Animal Breeding," 6.

101 L. Ollivier, L. "Jay Lush: Reflections on the Past," *Lohmann Information* 43 (2008): 3–12.

102 Chapman, "Jay Lawrence Lush 1896–1982," 2674.

103 J. Lush, "Family Merit and Individual Merit as Bases for Selection," *American Naturalist* 81 (1947): 241–61.

104 See, for example, J. Lush, *Animal Breeding Plans* (Ames: Collegiate Press, Inc., 1937).

105 J. Marie, "For Science, Love and Money: The Social Worlds of Poultry and Rabbit Breeding in Britain, 1900–1940," *Social Studies of Science* 38 (2008): 922–98.

106 For more on Punnett see F.A.E. Crew, "Reginald Crundall Punnett," *Genetics* 58 (1968): 1–7.

107 Marie, "For Science, Love and Money," 926–7.

108 See M.E. Derry, *Art and Science in Breeding: Creating Better Chickens* (Toronto: University of Toronto Press, 2012).

109 See C. Wreidt, "Formalism in Breeding of Livestock in Relation to Genetics," *Journal of Heredity* 16 (1925): 19–24; and his "The Inheritance of Butterfat Percentage in Crosses of Jerseys with Red Danes," *Journal of Heredity* 22 (1930): 45.

110 See Theunissen, "Breeding Without Mendelism," 659, 660, 664, 668.

111 A.L. Hagedoorn and G. Sykes, *Poultry Breeding: Theory and Practice* (London: Crosby Lockwood & Son Ltd., 1953), 14.

112 B. Theunissen, "Connecting Genetics, Evolutionary Theory and Practical Animal Breeding: Arend L. Hagadoorn (1885–1953)," (unpublished manuscript, 2012), 17.
113 Ibid., 18–21.
114 Ibid., 20.
115 Ibid., 24.
116 Hagedoorn, *Animal Breeding*, 10.
117 D.C. Warren, "A Half Century of Advances in the Genetics and Breeding Improvement of Poultry," *Poultry Science* 37 (1958): 6.
118 Ibid., 14.
119 "Better Plants and Animals," *Yearbook of Agriculture*, Book 1 (USDA, 1936), 985–7.
120 J. Lush, "Improving one Character by Breeding for Another," in *Fact Finding Conference of the Institute of American Poultry Industries, 1958–9* (unpublished, 1958–9), 186.
121 C. Bonneuil, "Mendelism, Plant Breeding and Experimental Cultures: Agriculture and the Development of Genetics in France," *Journal of the History of Biology* 39 (2006): 294.
122 Quoted in "Better Plants and Animals," *Yearbook of Agriculture*, Book I (USDA, 1936), 131.
123 See ibid., 832.

3 Practical Breeding

1 Bakewell to Culley, 15 December 1791, in, *Robert Bakewell Pioneer Livestock Breeder*, H.C. Pawson (London: Crosby Lockwood, 1957), 164.
2 *American Poultry Journal*, April 1911, 749; May 1912, 971; July 1912, 1165; December 1913, 1518, 1520–1, 1542.
3 See, for example, *American Poultry Journal*, May 1911, 1006.
4 See Ibid., April 1909, 426, 428, 430; October 1915, 1229, 1243; November 1915, 1315–16; July 1922, 737.
5 Ibid., April 1909, 426, 428, 430; October 1915, 1229, 1243; November 1915, 1315–16; July 1922, 737.
6 Ibid., August 1910, 976–7.
7 W.D. Termohlen, "Past History and Future Developments," *Poultry Science* 47 (1968): 12.
8 *American Poultry Journal*, August 1911, 1268. A particularly clear description of Felch's inbreeding strategies and his chart appeared in an agricultural circular published in 1911 in Alberta. "Practical poultry Keeping," Poultry Bulletin 2, Alberta Department of Agriculture, 1911, 73.

9 *Farmer's Advocate*, 1 May 1902, 348.

10 Ibid., December 1868, 185.

11 *Farming*, 18 January 1898, 156.

12 See, for example, *Canadian Poultry Chronicle*, February 1871, 118–119; October 1981, 49–51.

13 Ontario Legislature, Sessional Paper 1, 1872–3, Appendix F, 446; *Advocate*, 16 September 1895, 367.

14 *Canadian Poultry Chronicle*, April 1871, 149–50; May 1921, 567; G. Sawyer, *The Agribusiness Poultry Industry: A History of its Development* (New York: Exposition Press, 1971), 18; O.A. Hanke, edited by, *American Poultry History, 1823–1973* (Madison, Wisconsin: American Poultry History Society, 1974), 35–6.

15 An interesting discussion of the effects of shows on livestock breeding can be found in E.A. Heaman, *The Inglorious Acts of Peace: Exhibitions in Canadian Society during the Nineteenth Century* (Toronto: University of Toronto Press, 1999). See as well H. Ritvo, "Pride and Pedigree: The Evolution of the Victorian Dog Fancy," *Victorian Studies* 29 (1986): 227–53; and her *The Animal Estate* (Cambridge: Harvard University Press, 1987); K. White, "Victorian and Edwardian Dogs," *Veterinary History* 7 (1992): 72–8; and J. Blunt Lytton (Lady Wentworth), *Toy Dogs and Their Ancestors* (London: Duckworth & Co., 1911).

16 Quoted in *Farmer's Advocate*, June 1888, 178.

17 *Farmer's Advocate*, 1 March 1900, 128; 15 March 1900, 160.

18 Ibid.

19 See, for example, *Canadian Poultry Review*, January 1894, 22–3, and *American Poultry Journal*, August 1925, 756.

20 *Poultry Review*, June 1892, 86.

21 *Farmer's Advocate*, 15 June 1903, 559.

22 *American Poultry Journal*, September 1907, 690.

23 *Farmer's Advocate*, 12 January 1905, 50; 6 November 1919.

24 Ibid., 8 February 1912.

25 Ibid., 20 November 1920, 2035.

26 *American Poultry Journal*, November 1915, 1321–2.

27 Ibid., November 1925, 938–9.

28 For examples see *American Poultry Journal*, February 1912, 298; June 1921, 642.

29 M.E. Derry, *Ontario's Cattle Kingdom: Purebred Breeders and their World, 1870–1920* (Toronto: University of Toronto Press, 2001), 73–83; *Bred for Perfection: Shorthorn Cattle, Collies, and Arabian Horses Since 1800* (Baltimore: The Johns Hopkins University Press, 2003), 36–44.

30 *Farmer's Advocate*, 7 August 1919, 1429–30; "Official Record of Performance for Poultry," *Agricultural Gazette of Canada*, 1919, 796.
31 See *American Poultry Journal*, June 1921, 642; June 1922, 672–3, 674; April 1926, 466, 468.
32 Hanke, ed., *American Poultry History*, 703, 704.
33 A.L. Hagedoorn, and G. Sykes, *Poultry Breeding: Theory and Practice*, (London: Crosby Lockwood & Son Ltd., 1953), 217–220; Hanke, ed., *American Poultry History*, 702, 703, 704.
34 *American Poultry Journal*, August 1923, 893–4.
35 D.C. Warren, "A Half Century of Advances in the Genetics and Breeding Improvement of Poultry," *Poultry Science* 37 (1958): 13.
36 Sawyer, *The Agribusiness Poultry Industry*, 26; *American Poultry Journal*, December 1910, 1446.
37 See M. Derry, *Art and Science in Breeding: Creating Better Chickens* (Toronto: University of Toronto Press, 2012), 128–53.
38 See, for example, *Farmer's Advocate*, 26 August 1920, 1482; 20 February 1930, 267; 24 January 1953, 37; *The Globe*, 6 January 1915, 9; *Proceedings of the Select Committee of the House of Commons into Agricultural Conditions*, vol. 1 (Ottawa: Government of Canada, 1924), 521.
39 See, for example, *Farmer's Advocate*, 9 September 1937, 521; 24 January 1953, 37.
40 Ibid., 4 January 1906, 14; 11 April 1912, 688–9.
41 *American Poultry Journal*, April 1915, 693.
42 Ibid., January 1927, 11, 88, 90- 2, 94–7. See as well D. Fitzgerald, *Every Farm a Factory* (New Haven: Yale University Press, 2003), 106, 115.
43 *American Poultry Journal*, December 1925, 1032, 1038.
44 Hanke, *American Poultry History*, 253.
45 Walter Van Riper, "Aesthetic Notions in Animal Breeding," *Quarterly Review of Biology* 7 (1932): 84–7; review of *American Dairy Cattle: Their Past and Future* in *The Journal of Economic History* 2 (1942), 227–9; E. Parmelee Prentice, "Food for America, 1980-: The Supply of Animal Proteins: The Agricultural Colleges," *Political Science Quarterly* 66 (1951): 483; Hanke, ed., *American Poultry History*, 260.
46 *American Poultry Journal*, October 1913, 1278.
47 Ibid., September 1924, 902–3. See as well, Derry, *Art and Science in Breeding*, 121–5.
48 Sawyer, *The Agribusiness Poultry Industry*, 112; E.L. Schapsmeier and F.H. Schapsmeier, *Henry A. Wallace of Iowa: The Agrarian Years, 1910–1940* (Ames: The Iowa State University Press), 21, 27, 28.

49 See F.A. Hayes and G.T. Klein, *Poultry Breeding Applied* (Mount Morris, Illinois: Watt Publishing Company, 1953), 192–5; D.C. Warren, "Techniques of Hybridization of Poultry," *Poultry Science* 29 (1950): 60; A.L. Hagedoorn and G. Sykes, *Poultry Breeding: Theory and Practice* (London: Crosby Lockwood & Son Ltd., 1953), 184–5.

50 P.B. Seigel et al., "Genetic Selection Strategies – Population Genetics," *Poultry Science* 76 (1997): 1064.

51 Felch, *Breeding and Management of Poultry*, 20, 22.

52 H.W. Hawthorne, "A Five-Year Farm Management Survey in Palmer Township, Washington County, Ohio, 1912–1916," *USDA Bulletin* 716 (1916): 27, 29–31.

53 Hanke, ed., *American Poultry History*, 218.

54 *Farmer's Advocate*, 11 January 1958, 37; 8 February 1958; 14 February 1958, 3; 9 January 1960, 35.

55 See, for example, Ibid., 11 January 1958, 37; 8 February 1958, 5; 14 February 1959, 3; 9 January 1960, 35.

56 E.S. Synder, *A History of the Poultry Science Department at the Ontario Agricultural College, 1894–1968* (unpublished manuscript, 1970), 289.

57 *Farmer's Advocate*, 26 June 1941, 426.

58 Ibid., December 1868, 185.

59 H.L. Shrader, "The Chicken-of-Tomorrow Program: Its Influence on 'Meat-Type' Poultry Production," *Poultry Science* 31 (1952): 7–8.

60 J. Sapp, "The Struggle for Authority in the Field of Heredity, 1900–1932: New Perspectives on the Rise of Genetics," *Journal of the History of Biology* 16 (1983), 338.

61 F.B. Hutt, "Whither Poultry Genetics?," *Poultry Science* 21 (1965): 55.

62 See J. Blackman, "The Cattle Trade and Agrarian Change on the Eve of the Railway Age," *Agriculture History Review* 23 (1975): 48–62; P.W. Bidwell and J.L. Falconer, *History of Agriculture in the Northern United States, 1620–1860* (Washington: Carnegie Institute, 1925); R. Jones, *History of Agriculture in Ohio to 1880* (Kent: Kent State University Press, 1983); P. Henlein, "Cattle Droving from the Ohio Country, 1800–1850," *Agricultural History* 28 (1954): 83–94; A. Bogue, *From Prairie to Corn Belt: Farming on the Illinois and Iowa Prairies in the Nineteenth Century* (Chicago: University of Chicago Press, 1963); T. Jordan, *North American Cattle Ranching Frontiers: Origins, Diffusion, and Differentiation, Histories of the American Frontier* (Albuquerque: University of New Mexico Press, 1993); M.E. Derry, *Ontario's Cattle Kingdom: Purebred Breeders and Their World, 1870 to 1920* (Toronto: University of Toronto Press, 2001).

63 Series 1, Box 11, file 4, Shaver Collection, Archival and Special Collections, University of Guelph.

64 See "Better Plants and Animals," in *USDA Yearbook of Agriculture*, Part 1 (USDA, 1936), 849.
65 Series 1, Box 11, File 5, Shaver Collection, Archival and Special Collections, University of Guelph.
66 Ibid.
67 "Animal Breeding Practice," in *Evolution and Animal Breeding: Reviews on Molecular and Quantitative Approaches in Honour of Alan Robertson*, ed. W.C. Hill and T.F.C. Mackay (Wallingford, U.K.: C.A.B. International, 1989), 195, 196, 197.
68 See C.A. Narrod and K.O. Fuglie, "Private Investment in Livestock Breeding and Implications for Public Research Policy," *Agribusiness* 16 (2000): 457–70; E. Lutaaya et al., "Genetic Parameter Estimates from Joint Evaluation of Purebreds and Crossbreds in Swine Using the Crossbred Model," *Journal of Animal Science* 73 (2001): 3002–7; J.P. Cassady et al., "Heterosis and Recombination Effects on Pig Growth and Carcass Traits," *Journal of Animal Science* 80 (2002): 2286–2302; N. Ibanez-Esciche et al., "Genetic Evaluation Combining Purebred and Crossbred Data in a Pig Breeding Scheme," *Journal of Animal Science* 89 (2011): 3881–89.
69 I.M. Lerner and H. Donald, *Modern Developments in Animal Breeding* (London and New York: Academic Press, 1966), 170, 171, 178–9, 185.

4 New Directions: Artificial Insemination Technology

1 R. Vishwanath, "Artificial Insemination: the State of the Art," *Theriogenology* 59 (2003): 571, 572.
2 Auzias-Turenne, *Percherons and Normans, The Haras National Co. of Montreal* (Montreal: 1891), 23, 25, 26; G.H. Quetton, *On the Breeding of Horses and Other Domesticated Animals in Canada: Principally Crosses of Thoroughbreds and Large Mares* (Toronto: Williamson & Co., 1891), 14; Parliament of Canada, Sessional Paper, (henceforth known as SP, Canada) 15b, 1913, 219; Committee of the House of Lords, Britain, *Report from the Select Committee*, Appendix B, 1873, 336–338.
3 *Report from the Select Committee*, Appendix B, 1873, 339–30, 342, 348, 349.
4 Auzias-Turenne, *Percherons and Normans*, 26.
5 *Third Report of the Royal Commission on Horse Breeding*, 1890, 3, 4, 46–7, 89–91, 109, 111.
6 Ibid., 8.
7 Ibid., 19, 35.
8 For a more thorough discussion of the 1890 report of the British Horse Commission, see M.E. Derry, *Horses in Society*, 159–71.

9   *A Bill to Regulate the Use of Stallions for Stud Purposes in Ireland* (Bill 151), 1913.
10 C.C. Glenn, "Stallion Legislation and the Horse-Breeding Industry," in *USDA Yearbook of Agriculture*, Agriculture, United States, 1916, 290, 296. *Breeder's Gazette*, 2 March 1916, 495. For a more thorough discussion about stallion enrolment in the United States, see Derry, *Horses in Society*, 172–201; *Farmer's Advocate*, 1 August 1918, 1265; *Farmer's Advocate*, 18 July 1912, 1279; *Farm and Dairy*, 15 August 1918, 887; *Agricultural Gazette* 7 (1920): 413, 490; *Agricultural Gazette* 1 (1914): 187; *Agricultural Gazette* 6 (1919): 1057.
11 Ontario Legislature, Sessional Paper (henceforth known as SP), 1922, 39.
12 S. Wilmot, "From Public Service to Artificial Insemination: Animal Breeding Science, 1890–1951," Full Research Report, ESRC End of Award Report, 2007, Reference No. Res-000–23–0390, 19.
13 Ibid., 20.
14 S. Wilmot, "From 'Public Service' to Artificial Insemination: Animal Breeding Science and Reproductive Research in Early 20th Century Britain," *Studies in History and Philosophy of Biological and Biomedical Sciences* 38 (2007): 415.
15 "Better Bulls," Bulletin 281, Ontario Agricultural College, Ontario Department of Agriculture, 1920, 5.
16 H.A. Herman, *Improving Cattle by the Millions: NAAB and the Development and Worldwide Application of Artificial Insemination* (Columbia: University of Missouri Press, 1981), 7, 8.
17 Parliament of Canada, Sessional Paper (henceforth known as SP) 15a, 1909, 48.
18 "Better Bulls," 6–10; SP 39, Ontario, 1920, 53–63; SP 39, Ontario, 1922, 43–6; SP 15, Canada, 1923, 72.
19 SP 15, Canada, 1920, 21; SP 15, Canada, 1923, 72.
20 See B. Orland, "Turbo-Cows: Producing a Competitive Animal in the Nineteenth and Early Twentieth Centuries," in *Industrializing Organisms: Introducing Evolutionary History*, ed. S.R. Schrepfer and P. Scranton (London: Routledge, 2004) and T. Theunissen, "Breeding Without Mendelism: Theory and Practice of Dairy Cattle Breeding in the Netherlands, 1900–1950," *Journal of the History of Biology* 41 (2008): 637–76. See P.W. Bidwell and J.L. Falconer, *History of Agriculture in the Northern United States* (Washington: Carnegie Institute, 1925), 396; F. Bateman, "Improvement in American Dairy Farming, 1850–1910: A Quantitative Analysis," *Journal of Economic History* 28 (1968): 266–7; M. Derry, *Ontario's Cattle Kingdom*, 94–6; S. McMurry, *Transforming Rural America: Dairying Families and Agricultural Change, 1820–1885* (Baltimore: The Johns Hopkins University Press, 1995), 17, 18, 19; E.E. Lampard, *The Rise of the*

Dairy Industry in Wisconsin: A Study of Agricultural Change, 1820–1920 (Madison: State Historical Society of Wisconsin, 1963), 170, 171, 175; and J. Atack, and F. Bateman, *To Their Own Soil: Agriculture in the Antebellum North* (Ames: Iowa State University Press, 1987), 147. See as well SP 5, Ontario, 1869, 143; *Farmer's Advocate*, 1 February 1900, 64; 6 May 1909, 753.

21 J.L. Lush, "Genetics and Animal Breeding," in *Genetics in the 20th Century*, ed. L.C. Dunn (New York: The MacMillan Company, 1951), 500.

22 Canada, Department of Agriculture, *Publications of the International Agricultural Institute*, 1911–12, 41.

23 For a comprehensive history of the Friesian-Holstein breed in North America, see E.Y. Morwick, *The Chosen Breed: A Tale of Men, Women and the Canadian Holstein*, 2 vols. (Hamilton, Ontario: Seldon Griffin Graphics Inc., 2002).

24 Canada, Department of Agriculture, *Publications*, 1911–12, 41.

25 G.E. Reaman, *History of the Holstein-Friesian Breed in Canada* (Toronto: Collins, 1946), 320, 324–5, 340, 341, 343; Canada, Department of Agriculture, *Publications*, 1911–12, 41.

26 United States Department of Agriculture. "Better Plants and Animals," *Yearbook of Agriculture* (1936), 1010.

27 L.W. Morley, "Dairy Cattle Breed Associations," *Journal of Dairy Science* 39 (1956): 713.

28 Ibid.

29 B. Orland, "Turbo-Cows: Producing a Competitive Animal in the Nineteenth and Early Twentieth Centuries," in *Industrializing Organisms: Introducing Evolutionary History*, ed. S.R. Schrepfer and P. Scranton (London: Routledge, 2004), 182; Theunissen, "Breeding Without Mendelism," 657; Canada, Department of Agriculture, *Publications*, 1911–12, 41.

30 Orland, "Turbo-Cows," 182.

31 Theunissen, "Breeding without Mendelism," 657.

32 F.J. Arnold, "Fifty Years of DHIA Work," *Journal of Dairy Science* 39 (1956): 792; D.E. Voelker, "Dairy Herd Improvement Associations," *Journal of Dairy Science* 64 (1981): 1269.

33 United States Department of Agriculture. *Better Plants and Animals* (1936), 1010.

34 J.L. Lush, "Genetics and Animal Breeding" in *Genetics in the 20th Century*, ed. L.C. Dunn (New York: The MacMillan Company, 1951), 500.

35 Arnold, "Fifty Years of DHIA Work," 793; Voelker, "Dairy Herd Improvement Associations," 1275.

36 The description of dairy crossbreeding experiments has been derived from R.W. Touchberry, "Crossbreeding Effects in Dairy Cattle: The Illinois

Experiment, 1949–1969," *Journal of Dairy Science* 75 (1992): 640–667. See also J.L. Lush, "Dairy Cattle Genetics," *Journal of Dairy Science* 39 (1956): 693–4.

37 J.M. White et al., "Dairy Cattle Improvement and Genetics," *Journal of Dairy Science* 64 (1981): 1310, 1311.

38 R.C. Laben et al., "Some Effects of Inbreeding and Evidence of Heterosis Through Outcrossing in a Holstein-Friesian Herd," *Journal of Dairy Science* 38 (1955): 525–535.

39 F.R. Allaire et al., "Specific Combing Abilities among Dairy Sires," *Journal of Dairy Science* 48 (1965): 1096, 1099.

40 R.C. Beckett et al., "Specific and General Combining Abilities for Production and Reproduction among Lines of Holstein Cattle," *Journal of Dairy Science* 62 (1979): 613, 619.

41 J.M. White et al., "Dairy Cattle Improvement and Genetics," *Journal of Dairy Science* 64 (1981): 1310.

42 G.M. Trout, "Fifty Years of the American Dairy Science Association," *Journal of Dairy Science* 39 (1956): 625–7; Morley, "Dairy Cattle Breed Associations," 712.

43 United States Department of Agriculture, *Better Plants and Animals*, 150, 151.

44 Ibid., 997, 998, 999, 1004.

45 Ibid., 1005, 1008.

46 Herman, *Improving Cattle by the Millions*, 2–3; R.H. Foote, "The History of Artificial Insemination: Selected Notes and Notables," *Journal of Animal Science* 80 (2002): 2.

47 Herman, *Improving Cattle by the Millions*, 2–5.

48 Ibid., 4.

49 Ibid., 7; Foote, "The Artificial Insemination Industry," 14; and his "The History of Artificial Insemination: Selected Notes and Notables," *Journal of Animal Science* 80 (2002): 2, 3.

50 Herman, *Improving Cattle by the Millions*, 7, 10, 11.

51 H.D. Norman et al., "Overview of Progeny-Test Programs of Artificial-Insemination Organizations in the United States," *Journal of Dairy Science* 84 (2001): 1911.

52 Herman, *Improving Cattle by the Millions*, 13, 85, 87, 88, 135, 141.

53 T. Theunissen, "Breeding for Nobility or Production? Culture of Dairy Cattle Breeding in the Netherlands, 1945–1995," *Isis* 103 (2012): 278–309.

54 Foote, "The Artificial Insemination Industry," 24–5.

55 Herman, *Improving Cattle by the Millions*, 13, 85, 87, 88, 135, 141.

56 Morley, "Dairy Cattle Breed Associations," 714.

57 D.A. Funk, "Major Advances in Globalization and Consolidation of the Artificial Insemination Industry," *Journal of Dairy Science* 89 (2006): 1363.

58 Theunissen, "Breeding for Nobility or Production," 278–82.

59 Foote, "The History of Artificial Insemination," 4.

60 Herman, *Improving Cattle by the Millions,* 16, 17.

61 J.W.B. King, "Alan Robertson's Contributions to Theory and Application of Animal Improvement," in *Evolution and Animal Breeding: Reviews on Molecular and Quantitative Approaches in Honour of Alan Robertson,* eds. W.C. Hill and T.F.C. Mackay (Wallingford, U.K.: C.A.B. International, 1989), 158.

62 Wilmot, "From 'Public Service' to Artificial Insemination," *Studies in History and Philosophy of Biological and Biomedical Sciences* 38 (2007): 413, 419, 426, 427, 428.

63 Ibid., 428.

64 D.S. Falconer, "Quantitative Genetics in Edinburgh, 1947–1980," *Genetics* 133 (1993): 139.

65 Ibid., 139–40.

66 F.W. Nicholas, "Incorporation of New Reproductive Technology in Genetic Improvement Programs," in *Evolution and Animal Breeding: Reviews on Molecular and Quantitative Approaches in Honour of Alan Robertson,* eds. W.C. Hill and T.F.C. Mackay (Wallingford, U.K.: C.A.B. International, 1989), 203.

67 Ibid., 203.

68 A. Robertson and J.M. Rendel, "The Use of Progeny Testing with Artificial Insemination in Dairy Cattle," *Journal of Genetics* 50 (1950): 21–31. See as well Nicholas, "Incorporation of New Reproductive Technology," 203.

69 R. Frankham, "Alan Robertson's Contributions to Quantitative Genetics" in *Evolution and Animal Breeding: Reviews on Molecular and Quantitative Approaches in Honour of Alan Robertson,* eds. W.C. Hill and T.F.C. Mackay (Wallingford, U.K.: C.A.B. International, 1989), 83; and T.F.C. Mackay, "Alan Robertson (1920–1989)," *Genetics* 125 (1990): 2.

70 Nicholas, "Incorporation of New Reproductive Technology," 204.

71 Voelker, "Dairy Herd Improvement Associations," 1273; United States Department of Agriculture, *Better Plants and Animals,* 1005.

72 For a comprehensive discussion about breeding values and the impact of segregation on it see D.S. Falconer, *Introduction to Quantitative Genetics* (Edinburgh: Oliver and Boyd, 1960), 117–120.

73 L.N. Hazel, "The Genetic Basis for Constructing Selection Indexes," *Genetics* 28 (1943): 476, 488.

74 C.R. Henderson, "Historical Overview," in *Advances in Statistical Methods for Genetic Improvement of Livestock,* eds. D. Gianola et al. (Berlin: Springer-Verlag, 1990), 5, 6, 8.

75 P.L. Powell et al., "Major Advances in Genetic Evaluation Techniques," *Journal of Dairy Science* 89 (2006): 1337.

76 J.L. Lush, "Genetics and Animal Breeding" in *Genetics in the 20th Century*, ed. L.C. Dunn (New York: The MacMillan Company, 1951), 504, 505.

77 C.R. Henderson, "Young Sire Selection for Commercial Artificial Insemination," *Journal of Dairy Science* 47 (1964): 439.

78 Quoted in W.C. Hill, ed., *Quantitative Genetics: Part II, Selection* (New York: Van Nostrand Reinhold Company, 1984), 116. From Henderson's "General Flexibility of Linear Model Techniques for Sire Evaluation," *Journal of Dairy Science* 57 (1974): 964–972.

79 Herman, *Improving Cattle by the Millions*, 21.

80 C.R. Henderson, "Sire Evaluation and Genetic Trends," in *Proceedings of the Animal Breeding and Genetic Symposium in Honor of Dr. Jay L. Lush* (Blacksburg, Virginia: American Society of Animal Science and American Dairy Science Association, 1972), 10.

81 See, for example, C.R. Henderson, "Young Sire Selection," 439–46.

82 See, for example, C.R. Henderson, "General Flexibility of Linear Model Techniques for Sire Evaluation," *Journal of Dairy Science* 57 (1974): 963–72.

83 W.C. Hill, ed., *Quantitative Genetics: Part I, Explanation and Analysis of Continuous Variation* (New York: Van Nostrand Reinhold Company, 1984), 264.

84 J. Crow, review of *Proceedings of the Second International Conference on Quantitative Genetics*, by B.S. Weir et al., *Science* 242 (1988): 1449.

85 M. Lynch and B. Walsh, *Genetics and Analysis of Quantitative Traits* (Sunderland, MA: Sinauer Associates, Inc., 1998), 5, 7, 390.

86 J.L. Lush, "Genetics and Animal Breeding," in *A Short History of Genetics, The Development of Some of the Main Lines of Thought, 1864–1939*, ed. L.C. Dunn (New York: McGraw-Hill, 1965), 504. See also Lush, introduction to *Systems of Mating and Other Papers*, by S. Wright (Ames: Iowa State College Press, 1958). [Reprint of "Systems of Mating," *Genetics* 6 (1921): 111–178; "Evolution in Mendelian Populations," *Genetics* 16 (1931): 97–159; "Correlation and Causation," *Journal of Agricultural Research* 20 (1921): 557–585; "The Method of Path Coefficients," *Annals of Mathematic Statistics* 5 (1934): 161–215.]

87 I.M. Lerner and L.N. Hazel, "Population Genetics of a Poultry Flock under Artificial Selection," *Genetics* 32 (1947): 325, 339.

88 F. Pirchner, *Population Genetics in Animal Breeding* (San Francisco: W.H. Freeman and Company, 1969), 224, 225.

89 See S. Wright, "Experimental Results and Evolutionary Deductions," in *Evolution and the Genetics of Populations*, vol. 3 (Chicago: University of Chicago Press, 1977).

90 S. Wright, *Evolution and the Genetics of Populations*, vol. 3, 552, 553.
91 E. Mayr and W.B. Provine, *The Evolutionary Synthesis: Perspectives on the Unification of Biology* (Cambridge: Harvard University Press, 1980), 67.
92 W.C. Hill, ed., *Quantitative Genetics: Part I*, 150.
93 O. Kempthorne, introduction to *Proceedings of the International Conference on Quantitative Genetics*, eds. E.J. Pollak et al., (Ames: Iowa State University Press, 1977), 10, 11.
94 E.Y. Morwick, *The Chosen Breed: A Tale of Men, Women and the Canadian Holstein*, 2 vols. (Hamilton: Seldon Griffin Graphics Inc., 2002), 478–9, 707, 5067.
95 "The story of Starbuck's discovery," www.ciaq/ciaq/history/the-legend-of-starbuck/the-early-stages-of-the-starbuck-saga.html.
96 Theunissen, "Breeding Without Mendelism," 641, 643; P.O. Grothe, *Holstein Friesian: A Global Breed* (Netherlands: Misset Press, 1993 [English version 1994]), 15; S. Wilmot, "From 'Public Service' to Artificial Insemination," *Studies in History and Philosophy of Biological and Biomedical Sciences* 38 (2007): 414.
97 Theunissen, "Breeding Without Mendelism," 641, 654.
98 Morwick, *The Chosen Breed*, 13, 14, 15, 16, 19–22.
99 Government of Canada, *Report of the Committee on Agricultural Conditions, Part II*, 1924, 461–75; *Farmer's Advocate* 27 July 1916, 1252.
100 Grothe, *Holstein Friesian*, 15, 49, 50.
101 R.H. Mansfield, *Progress of the Breed: The History of U.S. Holsteins* (Holstein-Friesian World, Inc., 1985), 62.
102 Mansfield, *Progress of the Breed*, 125; and G. Simm, *Genetic Improvement of Cattle and Sheep* (Ipswich, U.K.: Farming Press, 1998), 219, 221, 222.
103 Simm, *Genetic Improvement of Cattle and Sheep*, 219, 221, 222.
104 D.A. Funk, "Major Advances in Globalization and Consolidation of the Artificial Insemination Industry," *Journal of Dairy Science* 89 (2006): 1363; M. Stolzman et al., "Friesian Cattle in Poland, Preliminary Results of Testing Different Strains," *World Animal Review* 38 (1981): 9.
105 H.D. Zarnecki et al., "Heterosis for Growth and Yield Traits from Crosses of Friesian Strains," *Journal of Dairy Science* 76 (1993): 1662, 1668.
106 Funk, "Major Advances," 1363.
107 Theunissen, "Breeding Without Mendelism," 637–76.
108 See Theunissen, "Breeding for Nobility or Productivity."
109 Funk, "Major Advances," 1364; P.L. Powell et al., "Progeny Testing and Selection Intensity for Holstein Bulls in Different Countries," *Journal of Dairy Science* 86 (2003): 3386. See also A. Torsell, "Prospects of Performing Multiple-Country Comparison of Dairy Sires for Countries not Participating in Interbull International Generic Evaluations"

(master's Thesis, Swedish University of Agricultural Sciences, Uppsala, 2007), accessed 29 December 2010, www.ex-epsilon.slu.se:8080/archive/00001885.

110 J.M. DeJarnette et al., "Sustaining the Fertility of Artificially Inseminated Dairy Cattle: The Role of the Artificial Insemination Industry," *Journal of Dairy Science* 87 (2004): E93.

111 C.W. Young et al., "Estimates of Inbreeding and Relationship Among Registered Holstein Females in the United States," *Journal of Dairy Science* 79 (1996): 502–5; Stachowicz, K. et al., "Rates of Inbreeding and Genetic Diversity in Canadian Holstein and Jersey Cattle," *Journal of Dairy Science* 94 (2011): 5160–75.

112 R. Mrode et al., "Genetic Relationships between the Holstein Cow Populations of Three European Dairy Countries," *Journal of Dairy Science* 92 (2009): 5760, 5764.

113 L.B. Hansen et al., "Is Crossbreeding the Answer for Reproductive Problems of Dairy Cattle?" *Proceedings of the Southwest Nutrition Conference*, 2005, 113.

114 G. Rogers, "Variability by Crossbreeding," *Dairy Crossbreeding*, posted by *GENO Global*, accessed 27 December 2010, http://www.dairycrossbreeding.com.

115 A.J. McAllister, "Is Crossbreeding the Answer to Questions of Dairy Breed Utilization?," *Journal of Dairy Science* 85 (2002): 2352.

116 M.C. Lucy, "Reproductive Loss in High-Producing Dairy Cattle: Where will it End," *Journal of Dairy Science* 84 (2001): 1278.

117 H.D. Norman, "Reproductive Status of Holstein and Jersey Cows in the United States," *Journal of Dairy Science* 92 (2009): 3519.

118 A. Sewalem et al., "Analysis of Inbreeding and its Relationship with Functional Longevity in Canadian Dairy Cattle," *Journal of Dairy Science* 89 (2006): 2210–2215.

119 See K. Ericson et al., "Crossbreeding Effects between Two Swedish Dairy Breeds for Reproductive Traits," *Journal of Animal Breeding and Genetics* 105 (1988): 441–51; B. Heins, "Impact of an Old Technology on Profitable Dairying in the 21st Century," in *Proceedings of the Fourth W.E. Petersen Symposium* (Minneapolis: University of Minnesota, 2007), 7–19; B.G. Cassell, "Mechanisms of Inbreeding Depression and Heterosis in Profitable Dairying," in *Proceedings of the Fourth W.E. Petersen Symposium* (Minneapolis: University of Minnesota, 2007), 1–6; P. Bijma et al., "Genetic Gain of Pure Line Selection and Combined Crossbred Purebred Selection with Constrained Inbreeding," *Animal Science* 72 (2001): 225–232; R. Prendiville, "A Comparison between Holstein-Friesian and Jersey Dairy Cows and Their F 1 Cross with Regard to Milk Yield,

Somatic Cell Score, Mastitis, and Milking Characteristics under Grazing Conditions," *Journal of Dairy Science* 93 (2009): 2741–50; B. Murray, "Dairy Crossbreeds – The Rare Breed," Ontario Ministry of Agriculture, Food and Rural Affairs, 2002, accessed 15 June 2009 and 27 December 2010, http://www.omafra.bov.on.ca/english/livestock/dairy/facts/ info_breed.htm; P.M. VanRaden et al., "Economic Merit of Crossbred and Purebred U.S. Dairy Cattle," *Journal of Dairy Science* 86 (2003): 1036–44; B.G. Cassell, "Dairy Crossbreeding: Why and How," Publication 404–093 (Blacksburg, Virginia: Virginia Polytechnic and State University, 2009); C. Maltecca, "Changes in Conception Rate, Calving Performance, and Calf Health and Survival from the Use of Cross-bred Jersey x Holstein Sires as Mates for Holstein Dams," *Journal of Dairy Science* 89 (2006): 2747–54; B.J. Heins et al., "Calving Difficulty and Stillbirths of Pure Holsteins Versus Crossbreds of Holstein with Normande, Montbeliarde, and Scandinavian Red," *Journal of Dairy Science* 89 (2006): 2805–20; and "Production of Pure Holsteins Versus Crossbreds of Holstein with Normande, Montbeliarde, and Scandinavian Red," *Journal of Dairy Science* 89 (2006): 2799–2804.

120 B.J. Heins and al., "Production of Pure Holsteins Versus Crossbreds of Holstein with Normande, Montbeliarde, and Scandinavian Red," *Journal of Dairy Science* 89 (2006): 2799.

121 N. Lopez-Villalobos et al., "Effects of Selection and Crossbreeding Strategies on Industry Profit in the New Zealand Dairy Industry," *Journal of Dairy Science* 83 (2000): 164, 171; Cluny Exports, "Crossbreeding Holstein-Friesian x Jersey Cattle," accessed 28 December 2010, www .clunyexports.com.au/media/crossbreedjhfjv11.pdf. See also N. Lopez-Villalobos et al., "Possible Effects of 25 Years of Selection and Crossbreeding on the Genetic Merit and Productivity of New Zealand Dairy Cattle," *Journal of Dairy Science* 83 (2000): 154, 162.

122 P.M. VanRaden et al., "Economic Merit of Crossbred and Purebred U.S. Dairy Cattle," *Journal of Dairy Science* 86 (2003): 1043.

123 See "Crossbreeding – It is the Next Magic Pill?," Genex Cooperative, Inc., 2002, accessed 15 June 2009, http://www.bullsemen.com/cattle-breeding/bull-semen-article.php?arti; Select Sires, "Crossbreeding 101: What Every Producer Should Know," accessed 27 December 2010, www .selectsires.com/dairy/crossbreeding_101.pdf and "Crossbreeding 103: A Further Update on Crossbreeding Results," accessed 27 December 2010, www.selectsires.com/dairy/crossbreeding_103_aug2006.pdf.

124 "Crossbreeding – It is the Next Magic Pill?," 3, 4.

125 Select Sires, "Crossbreeding 101: What Every Producer Should Know," accessed 27 December 2010, www.selectsires.com/dairy/crossbreeding_101.pdf.

126 Ibid.

127 Ibid.

128 See B.G. Cassell, "Dairy Crossbreeding: Why and How," Publication 404–093 (Virginia Polytechnic and State University, 2009); and his "Mechanisms of Inbreeding Depression and Heterosis in Profitable Dairying," *Proceedings of the Fourth W.E. Petersen Symposium* (Minneapolis: University of Minnesota, 2007), 1–6.

129 *Hoard's Dairyman*, 10 January 2008, 10.

130 B.L. Golden et al., "Milestones in Beef Cattle Genetic Evaluations," *Journal of Animal Science* 87 (2009): E3–10; A.L. Eller, "A Look Back at BIF History," *Proceedings of Annual Beef Improvement Federation Symposium*, 2007 10–14.

131 R.M. Bourdon, "Shortcomings of Current Genetic Evaluation Systems," *Journal of Animal Science* 76 (1998): 2308–23.

132 P. Brassley, "Cutting Across Nature? The History of Artificial Insemination in Pigs in the United Kingdom," *Studies in History and Philosophy of Biological and Biomedical Sciences* 38 (2007): 458.

133 For more on pig breeding see Webb, "Animal Breeding Practice," in *Evolution and Animal Breeding: Reviews on Molecular and Quantitative Approaches in Honour of Alan Robertson*, eds. W.C. Hill and T.F.C. Mackay (Wallingford, U.K.: C.A.B. International, 1989), 195–7.

134 Brassley, "Cutting Across Nature?," 452.

135 Ibid., 459.

136 Ibid., 444–5.

137 DeJarnette et al., "Sustaining the Fertility," E101.

138 Ibid.

139 See Simm, *Genetic Improvement of Cattle and Sheep*, 207.

140 http://www.semex.com/i?lang=en&page=genomicfaq.shtml.

5 Molecular Genetics, the Rise of Genomics, and Livestock Breeding

1 E.P. Cunningham, *Quantitative Genetic Theory and Livestock Improvement* (University of New England, 1979), 9; *Proceedings of the Second International Conference on Quantitative Genetics*, ed. B.S. Weir (Sunderland, MA: Sinauer Associates, Inc., 1988), 536; M. Lynch and B. Walsh, *Genetics and Analysis of Quantitative Traits* (Sunderland, MA: Sinauer Associates, Inc., 1998), 390.

2 D.A. Funk, "Major Advances in Globalization and Consolidation of the Artificial Insemination Industry," *Journal of Dairy Science* 89 (2006): 1365.

3 F.W. Nicholas, "Genetic Improvement through Reproductive Technology," *Animal Reproduction Science* 42 (1996): 205–14.

4 M.B. Wheeler, "Production of Transgenic Livestock: Promise Fulfilled," *Journal of Animal Science* 81 (2003): 32–7; V.G. Pursel et al., "Status of Research with Transgenic Farm Animals," *Journal of Animal Science* 71 (1993 – Supplement 3): 10–19; C.G. Van Reenen, et al., "Transgenesis May Affect Farm Animal Welfare: A Case for Systematic Risk Assessment," *Journal of Animal Science* 79 (2001): 1763–79.

5 M. Maxmen, "Politics Holds Back Animal Engineers," *Nature* 490 (2012): 318–9.

6 Bovine, equine, and porcine cloning service is offered by companies such as ViaGen, headquartered in Texas, which deals as well in gene banking and other genetic services. The company acquired the rights to the Roslin technique and after taking over ProLinia of Athens, Georgia in 2003.

7 V.K. Mcelheny, *Watson and DNA: Making a Scientific Revolution* (Persus Publishing, 2003), 216, 217, 218–9, 221, 223, 226, 232, 236, 243, 244, 246.

8 P.E. Smouse, review of *Proceedings of the International Conference on Quantitative Genetics*, ed. E. Pollack et al., *American Journal of Human Genetics* 31 (1979): 754; E. Reeve, review of *Proceedings of the Second International Conference on Quantitative Genetics*, ed. B.S. Weir et al., *Genetical Research* 53 (1989): 239; B.S. Weir, "The Third International Conference of Quantitative Genetics," *Genetica* 136 (2009): 211.

9 C. Haley, "Human and Livestock Genetics: Parallel Evolution and Horizontal Exchange," *Journal of Animal Breeding and Genetics* 126 (2009): 413–14.

10 *Proceedings of Conference on Quantitative Genetics*, 1977, 719, 720.

11 Ibid., 717, 718.

12 Ibid., 56.

13 Ibid.

14 Ibid., 588.

15 Smouse, review of *Proceedings of the International Conference on Quantitative Genetics*, 754.

16 *Proceedings of Conference on Quantitative Genetics*, 1977, 50.

17 Ibid., 794.

18 See A. Robertson, "Molecular Biology and Animal Breeding," *Annales de Génétique et de Sélection animale* 2 (1970): 393–402.

19 J.I. Weller, Quantitative Trait Loci Analysis in Animals (New York: CABI Publishing, 2001), 6, 7, 8. See also Robertson, "Molecular Biology and Animal Breeding."

20 See J.L. Williams, "The Use of Marker-Assisted Selection in Animal Breeding and Biotechnology," *Revue Scientifique et Technique – Office International* 24 (2005): 381, 383.

21 B.J. Hayes, "QTL Mapping, MAS, and Genomic Selection," Short Course called Animal Breeding and Genetics, Department of Animal Science, Iowa State University, 2007, 55.

22 T.F.C. Mackay, "Alan Robertson (1920–1989)," *Genetics* 125 (1990): 3.

23 See, for example, J.N. Thompson, "Quantitative Variation and Gene Number" *Nature* 258 (1975): 665–8.

24 *Proceedings of Second Conference on Quantitative Genetics*, 1988, 243, 244, 249.

25 Weller, *Quantitative Trait Loci Analysis*, 6, 7, 8.

26 *Proceedings of Second Conference on Quantitative Genetics*, 1988, 161.

27 See C.W. Beattie, "Development of Detailed Microsatellite Linkage Maps in Livestock," in *Biotechnology's Role in the Genetic Improvement of Farm Animals*, ed. R.H. Miller et al. (Beltsville: Beltsville Symposia in Agricultural Research, American Society of Animal Science, 1996), 52, 57; J. I. Weller, "Introduction to QTL Detection and Marker-Assisted Selection," in *Biotechnology's Role in the Genetic Improvement of Farm Animals*, ed. R.H. Miller et al. (Beltsville: Beltsville Symposia in Agricultural Research, American Society of Animal Science, 1996), 259, 261, 261.

28 See W. Barendse, "Development and Commercialization of a Genetic Marker for Marbling of Beef in Cattle: A Case Study," in *Intellectual Property Rights in Animal Breeding and Genetics,* ed. M. Rothschild and S. Newman (New York: CABI Publishing, 2002), 197–212.

29 Sale catalogue of Shadybrook Farm, 28 September 2002, 8. This pattern can also be seen in many issues of Shorthorn Country over 2001/2.

30 A.L. Van Eenennaam et al., "Validation of Commercial DNA Tests for Quantitative Beef Quality Traits," *Journal of Animal Science* 85 (2007): 891–7.

31 Ibid., 899.

32 R.W. Fairhill, "The Use of Markers in Poultry Improvement," in *Biotechnology's Role in the Genetic Improvement of Farm Animals*, ed. R.H. Miller et al. (Beltsville: Beltsville Symposia in Agricultural Research, American Society of Animal Science, 1996), 297.

33 See F.S. Collins et al., "A Vision for the Future of Genomics Research," *Nature* 422 (2003): 835–47. See Toronto International Data Release Workshop Authors, "Publication Data Sharing," *Nature* 461 (2000): 168–70. International Human Genome Sequencing Consortium, "Initial Sequencing and Analysis of the Human Genome," *Nature* 409 (2001): 860–921. See H. Nebel et al., "Patenting and Sequencing the Genome," in *Intellectual Property Rights in Animal Breeding and Genetics*, ed. Rothschild and Newman, 119–146. Mcelheny, *Watson and DNA*, 267.

J.C. Venter, "The Sequence of the Human Genome," *Science* 291 (2001): 1304–51. J.C. Venter, *A Life Decoded: My Genome, My Life* (New York: Viking, 2007), 189, 190, 197. For more on Venter and Celera see R. Preston, "The Genome Warrior," *The New Yorker,,* 12 June 2000, 66–83; and J. Shreeve, *The Genome War: How Craig Venter Tried to Capture the Code of Life and Save the World* (New York: Alfred A. Knopf, 2004). Anonymous, "The International HapMap Project," *Nature* 426 (2003): 791.

34 Venter, *A Life Decoded*, 197.
35 T.M. Powledge, "U.S. Genome Sequencing Priorities Decided," *Genome Biology* 3 (2002). http://genomebiology.com/2002/3/5/spotlight-2020913–02.
36 J.E. Fulton, "Molecular Genetics in a Modern Poultry Breeding Organization," *World's Poultry Science* 64 (2008): 171–6.
37 Bovine HapMap Consortium, http://genomes.arc.georgetown.edu/drupal/bovine/?q=hapmap_funding; Bovine HapMap Consortium, "Genome-Wide Survey of SNP Variation Uncovers the Genetic Structure of Cattle Breeds," *Science* 324 (2009): 528–32; Bovine Genome Sequencing and Analysis Consortium et al., "The Genome Sequence of Taurine Cattle: A Window to Ruminant Biology and Evolution," *Science* 324 (2009): 522–28; H.A. Lewin, "It's a Bull Market," *Science* 324 (2009): 478–9.
38 Bin Fan et al., "Development and Application of High-Density SNP Arrays in Genomic Studies of Domestic Animals," *Asian-Australasian Journal of Animal Science* 23 (2010): 834, 835.
39 Anonymous, "The International HapMap Project," 791.
40 B. Walsh, "Quantitative Genetics, Version 3.0: Where Have We Gone Since 1987 and Where Are We Headed?," *Genetica* 136 (2009): 215.
41 J. van der Werf, "Animal Breeding and the Black Box of Biology," *Journal of Animal Breeding and Genetics* 124 (2007): 101.
42 S. Strauss, "Biotech Breeding Goes Bovine," *Nature Biotechnology* 28 (2010): 541. See M.R. Dentine, "Improving Dairy Cattle Using Marker Data," in *Biotechnology's Role in the Genetic Improvement of Farm Animals,* ed. R.H. Miller et al. (Beltsville: Beltsville Symposia in Agricultural Research, American Society of Animal Science, 1996), 282–7; J.W. Keele et al., "Databases and Information Systems Needed for Maps and Marker-Assisted Selection," in *Biotechnology's Role in the Genetic Improvement of Farm Animals,* ed. R.H. Miller et al. (Beltsville: Beltsville Symposia in Agricultural Research, American Society of Animal Science, 1996), 300.
43 S. Strauss, "Biotech Breeding Goes Bovine," *Nature Biotechnology* 28 (2010): 540–1.
44 J.I. Weller, "Introduction to QTL Detection and Marker-Assisted Selection," in *Biotechnology's Role in the Genetic Improvement of Farm Animals,* ed. R.H.

Miller et al. (Beltsville: Beltsville Symposia in Agricultural Research, American Society of Animal Science, 1996), 272.

45 Strauss, "Biotech Breeding Goes Bovine," 540.

46 D.A. Funk, "Major Advances in Globalization and Consolidation of the Artificial Insemination Industry," *Journal of Dairy Science* 89 (2006): 1365.

47 P. Schaeffer et al., "Strategy for Applying Genome-Wide Selection in Dairy Cattle," *Journal of Animal Breeding and Genetics* 123 (2006): 219.

48 P.L. Powell et al., "Progeny Testing and Selection Intensity for Holstein Bulls in Different Countries," *Journal of Dairy Science* 86 (2003): 3391.

49 T.H.E. Meuwissen et al., "Prediction of Total Genetic Value Using Genome-wide Marker Maps," *Genetics* 157 (2001): 1819–1829.

50 Ibid., 1828.

51 B.J. Hayes et al., "Genomic Selection in Dairy Cattle: Progress and Challenges," *Journal of Dairy Science* 92 (2009): 437.

52 P.M. VanRaden et al., "Reliability of Genomic Predictions for North American Holstein Bulls," *Journal of Dairy Science* 92 (2009): 20.

53 B.J. Hayes et al., "Genomic Selection in Dairy Cattle," 434.

54 J. van der Werf, "Animal Breeding and the Black Box of Biology," *Journal of Animal Breeding and Genetics* 124 (2007): 101.

55 *Holstein World*, September 2008, 40.

56 J. Perkel, "SNP Genotyping: Six Technologies that Keyed a Revolution," *Nature Methods* 5 (2008): 448.

57 T.H.E. Meuwissen "Genomic Selection: Marker Assisted Selection on a Genome Wide Scale," *Journal of Animal Breeding and Genetics* 124 (2007): 321.

58 Website, Accessed 20 July 2009.

59 Bin Fan et al., "Development and Application of High-Density SNP Arrays in Genomic Studies of Domestic Animals," *Asian-Australasian Journal of Animal Science* 23 (2010): 837.

60 Semex Alliance, Canada, http://www.semex.com.

61 *Holstein World*, September 2008, 40.

62 M.E. Goddard, "How can we best use DNA data in the selection of cattle?," in *Proceedings of the Beef Improvement Federation* (41st Annual Research Symposium, April 30 – May 3, 2009, Sacramento, California, USA), 87.

63 S.S. Moore, "The Bovine Genome Sequence – Will it Live up to the Promise?," *Journal of Animal Science and Genetics* 126 (2009): 257; G.R. Wiggans, "Selection of Single-Nucleotide Polymorphisms and Quality of Genotypes Used in Genomic Evaluation of Dairy Cattle in the United States and Canada," *Journal of Dairy Science* 92 (2009): 3431.

64 G.R. Wiggans, "Selection of Single-Nucleotide Polymorphisms," 3435.

65 Ibid.

66 Ibid.
67 *Hoard's Dairyman*, 10 February 2009, 85.
68 Strauss, "Biotech Breeding Goes Bovine," 540.
69 J.P. Chesnais, "How is the AI industry using Genomic tools in practice?," *Interbull Bulletin* 41 (2010): 60.
70 Strauss, "Biotech Breeding Goes Bovine," 540.
71 *Hoard's Dairyman*, 10 March 2011.
72 J.P. Chesnais, "How is the AI industry using Genomic tools in practice?," 59.
73 From P.M. VanRaden et al., "Reliability of Genomic Predictions for North American Holstein Bulls," *Journal of Dairy Science* 92 (2009): 16–24.
74 A. Loberg et al., "Interbull Survey on the Use of Genomic Evaluations," in *Interbull Bulletin* 38 (2009): 3, 4, 5, 6, 9.
75 B. Muir et al., "International Genomic Cooperation – North American Perspective," *Interbull Bulletin* 41 (2010): 71–6.
76 Ibid.
77 Ibid.
78 X. David et al., "International Genomic Cooperation: EuroGenomics significantly improves reliability of Genomic evaluations," *Interbull Bulletin* 41 (2010): 77–8. See as well, R. Dassonneville et al., "Effect of Imputing Markers from a Low-density Chip on the Reliability of Genomic Breeding Values in Holstein Populations," *Journal of Dairy Science* 94 (2011): 3679–86.
79 B. Muir et al., "International Genomic Cooperation – North American Perspective," 75; P.M. VanRaden et al., "International Genomic Evaluation Methods for Dairy Cattle," *Genetics Selection Evolution* 42 (2010): 7–15; M.A. Nilforooshan et al., "Validation of National Genomic Evaluations," *Interbull Bulletin* 42 (2010): 56–61.
80 See A. Torsell, "Prospects of Performing Multiple-Country Comparison of Dairy Sires for Countries not Participating in Interbull International Generic Evaluations" (master's thesis, Swedish University of Agricultural Sciences, Uppsala, 2007), accessed 29 December 2010, www.ex-epsilon .slu.se:8080/archive/00001885.
81 Examples abound. See Bin Fan et al., "Development and Application of High-Density SNP Arrays," 833–47; Suchocki, T. et al., "Testing Candidate Gene Effects on Milk Production in Dairy Cattle under Various Parameterizations and Modes of Inheritance," *Journal of Dairy Science* 93 (2010): 2703–17; and P.M. VanRaden et al., "Combining Different Marker Densities in Genomic Evaluation," *Interbull Bulletin* 42 (2010): 113–117.
82 N. McHugh et al., "Use of Female Information in Dairy Cattle Genomic Breeding Programs," *Journal of Dairy Science* 94 (2011): 4109.

83 USDA, "Use of Single Nucleotides Polymorphisms to Verify Parentage," Animal Improvement Programs, Report for 2008, accessed July 20, 2009, http://www.ars.usda.gov/research/projects.htm?ACCN_NO=.

84 Strauss, "Biotech Breeding Goes Bovine," 542.

85 G. Moser et al., "Accuracy of Direct Genomic Values in Holstein Bulls and Cows using Subsets of SNP Markers," *Genetics Selection Evolution* 42 (2010): 50.

86 P.M. VanRaden et al., "Combining Different Marker Densities in Genomic Evaluation," *Interbull Bulletin* 42 (2010): 116.

87 R.G. Wiggans et al., "Use of the Illumina Bovine3K BeadChip in Dairy Genomic Evaluation," *Journal of Dairy Science* 95 (2012): 1552–58.

88 See K.A. Weigel et al., "Potential Gains in Lifetime Net Merit from Genomic Testing of Cows, Heifers, and Calves on Commercial Dairy Farms," *Journal of Dairy Science* 95 (2012): 2215–25.

89 See, for example, B.L. Harris et al., "The Impact of High Density SNP chips on Genomic Evaluation in Dairy Cattle," *Interbull Bulletin* 42 (2010): 40–3.

90 G. Rincon et al, "Hot Topic: Performance of Bovine High-Density Genotyping Platforms in Holsteins and Jerseys," *Journal of Dairy Science* 94 (2011): 6116–21.

91 B.J. Hayes et al., "Accuracy of Genomic Breeding Values in Multi-Breed Dairy Cattle Populations," *Genetics Selection Evolution* 41 (2009): 41–2; N. Ibanez-Escriche et al., "Genomic Selection of Purebreds for Crossbred Performance," *Genetics Selection Evolution* 41 (2009): 12–21.

92 See J.B. Cole et al., "Breeding and Genetics Symposium: Really Big Data: Processing and Analysis of Very Large Data Sets," *Journal of Animal Science* 90 (2012): 723–33.

93 See A.P.W. De Roos et al., "Genomic Breeding Value Estimation using Genetic Markers, Inferred Ancestral Haplotypes, and the Genomic Relationship Matrix," *Journal of Dairy Science* 94 (2011): 4708–14.

94 Cole et al. "Breeding and Genetics Symposium," 724.

95 Ibid., 726–9

96 Ibid., 731.

97 Ibid.

98 *Ontario Beef Farmer*, February/March 2010, 4.

99 *Shorthorn Country*, July 2009, 122.

100 Ibid.

101 L.D. Leachman, "Combined Selection for the Beef Cattle Industry," *Proceedings of the Beef Improvement Federation*, 2009, 216.

102 A. Toosi et al., "Genomic Selection in Admixed and Crossbred Populations," *Journal of Animal Science* 88 (2010): 3246.

103 M. Tixier-Boichard et al., "A Century of Poultry Genetics," *World's Poultry Science* 68 (2012): 312–3.

104 J.E. Fulton, "Molecular Genetics in a Modern Poultry Breeding Organization," *World's Poultry Science* 64 (2008): 171–6.

105 See Tixier-Boichard et al., "A Century of Poultry Genetics," 316–7, for a short review of such joint efforts across Europe.

106 W.M. Muir et al., "Review of the Initial Validation and Characterization of a 3K Chicken SNP Array," *World's Poultry Science* 64 (2008): 219–25.

107 B.J. Hayes et al., "QTL Mapping, MAS, and Genomic Selection," 14.

108 See, for example, B. Abasht et al.,"Extent and Consistency of Linkage Disequilibrium and Identification of DNA Markers for Production and Egg Quality Traits in Commercial Layer Chicken Populations," *BMC Genomics* 10, Supplement 2 (2009): http://www.biomedcentral.com/1471–2164–10-S2-S2; and C. Andreescu et al., "Linkage Disequilibrium in Related Breeding Lines of Chickens," *Genetics* 177 (2007): 2161–9.

109 M.AM. Groenen et al., "The Development and Characterization of a 60K SNP Chip for Chickens," *BMC Genomics* 12 (2011): http://www.biomedcentral.com/1471–2164/12/274.

110 H-J. Megens et al., "Compariosn of Linkage Disequilibrium and Haplotype Diversity on Macro-and Microchromosomes in Chickens," *BMC Genetics* 10 (2009): http://www.biomedcentral.com/1471–2156/10/86. For more on genomics and chicken breeding see: C. Andreescu et al., "Linkage Disequilibrium in Related Breeding Lines of Chickens," *Genetics* 177 (2007): 2161–9; B. Abasht et al., "Extent and Consistency of Linkage Disequilibrium and Identification of DNA Markers for Production and Egg Quality Traits in Commercial Layer Chicken Populations," *BMC Genomics* 10, Supplement 2 (2009): http://www.biomedcentral.com/1471–2164–10-S2-S2; and P.B. Siegel, "Progress from Chicken Genetics to the Chicken Genome," *Poultry Science* 85 (2006): 2050–60.

111 F. Callaway, "Chicken project gets off the ground," *Nature* 509 (2014): 546.

112 I. Misztal, "Shortage of Quantitative Geneticists in Animal Breeding," *Journal of Animal Breeding and Genetics* 124 (2007): 255–56.

113 Ibid.

114 R.D. Green, "ASAS Centennial Paper: Future Needs in Animal Breeding and Genetics," *Journal of Animal Science* 87 (2009): 797.

115 Ibid., 798. See as well R.D. Green et al., "Identifying the Future Needs for Long-Term U.S.D.A. Efforts in Agricultural Animal Genomics," *International Journal of Biological Sciences* 3 (2007): 185–91.

116 Cole et al., "Breeding and Genetics Symposium," 730.

6 Biology, Industry Needs, and Morality

1 See, for example, M.A. Vyas, ed., *Issues in Ethics and Animal Rights* (New Delhi: Regency Publications (Kindle format), 2013).
2 *Farmer's Advocate*, April 1883, 103.
3 Ibid., 29 April 1915, 717.
4 Ibid., 26 February 1914, 366.
5 Ibid., April 1883, 102.
6 Ibid., May 1890, 142.
7 Ibid., 27 April 1905, 621.
8 Ibid., 12 August 1915, 1272; 25 November 1916, 1842; 1 March 1918, 365.
9 Ibid., 1 March 1918, 365.
10 T.L. Beauchamp et al., *The Human Use of Animals: Case Studies in Ethical Practice* (Oxford: Oxford University Press, 2008), 61, 63, 65, 66.
11 C.A. Wolf, "Custom Dairy Heifer Grower Industry Characteristics and Contract Terms," *Journal of Dairy Science* 86 (2003): 3016–22.
12 G.E. Seidel, "Overview of Sexing Semen," *Theriogenology* 68 (2007): 443–6; D.L. Garner et al., "History of Commercializing Sexed Semen for Cattle," *Theriogenology* 69 (2008): 886–95; G. Abdel-Azim et al., "Genetic Impacts of Using Female-Sorted Semen in Commercial and Nucleus Herds," *Journal of Dairy Science* 90 (2007): 1554–63; J. Fetrow, "Sexed Semen: Economics of a New Technology," *Proceedings of Western Dairy Management Conference*, 2007.
13 K.A. Weigel, "Exploring the Role of Sexed Semen in Dairy Production Systems," *Journal of Dairy Science* 87 (2004): E125.
14 D. Rath et al., "Improved Quality of Sex-sorted Sperm: A Prerequisite for Wider Commercial Application," *Theriogenology* 71 (2009): 22; J.M. DeJarnette, et al., "Evaluating the Success of Sex-Sorted Semen in the US Dairy Herds from on Farm Records," *Theriogenology* 71 (2009): 49–58; W.D. Hobenboken, "Applications of Sexed Semen in Cattle Production," *Theriogenology* 52 (1999): 1421–33; R.P. Amann, "Issues affecting Commercialization of Sexed Semen," *Theriogenology* 52 (1999): 1441–57.
15 *Hoard's Dairyman*, 10 March 2008, 174.
16 Ibid., April 2008, 272.
17 Ibid., 25 January 2008, 58.
18 Ibid., 25 March 2009, 205.
19 Ibid., 25 January 2008, 47.
20 Ibid.
21 E.F. Kaleta et al., "Approaches to Determine the Sex Prior to and after Incubation of Chicken Eggs and of Day-old-chicks," *World's Poultry*

Science 64 (2008): 391–9; "Avian Sex Determination and Sex Diagnosis," *World's Poultry Science* 59 (2003): 5–64; D. Martin, "Special Report: Avian Sex Determination and Sex Diagnosis," *World's Poultry Science* 59 (2003): 1.

22 "Avian Sex Determination and Sex Diagnosis," *World's Poultry Science* 59 (2003): 7

23 E.F. Kaleta et al., "Approaches to Determine the Sex Prior to and after Incubation of Chicken Eggs and of Day-old-chicks," *World's Poultry Science* 64 (2008): 397.

24 Sessional Paper (SP) 2, Ontario Legislature, 1893, *Report of the Ontario Commission on the Dehorning of Cattle* (1892): 29, 42, 69.

25 *The Canadian Breeder*, 29 October 1885, 659.

26 See ibid., 2 January 1885, 4, 5; *Advocate*, November 1882, 286.

27 *Canadian Live Stock and Farm Journal*, April 1886, 105.

28 Ibid., May 1886, 127.

29 *Report of the Commission on the Dehorning of Cattle*, 1892, 12.

30 Ibid., 18.

31 Ibid., 23, 24.

32 Ibid., 24, 25.

33 *Farmer's Advocate*, 15 March 1895, 117; 1 April 1895, 132; 15 July 1895, 283.

34 Ibid., 16 November 1896, 479.

35 *Farming World for Farmers and Stockmen*, 28 May 1901, 1010.

36 *Farmer's Advocate*, 11 May 1905, 697; 18 May 1905, 741.

37 Ibid., 15 May 1905, 374.

38 Ibid., 26 April 1917, 714; 19 March 1914, 527.

39 See for example, *Advocate*, 8 May 1919, 913; 6 May 1920, 869; 26 April 1917, 714; 14 May 1914, 957; 21 May 1914, 992; 11 June 1914, 1123; 30 March 1916, 555; 1 May 1913, 816; 19 March 1914, 527; 13 April 1916, 653.

40 M.A.G. von Keyserlingk et al., "Invited Review: The Welfare of Dairy Cattle – Key Concepts and the Role of Science," *Journal of Dairy Science* 92 (2009): 4105.

41 *Hoard's Dairyman*, 25 April 2008, 286.

42 Ibid., 10 February 2009, 83.

43 Ibid.

44 Beauchamp et al., ed., *The Human Use of Animals*, 8, 9; *Hoard's Dairyman*, December 2007, 801.

45 *The Canadian Breeder*, 23 January 1885, 53.

46 Ibid., 31 December 1885, 806.

47 Ibid., 19 November 1885, 708, 709; *The Canadian Live-Stock and Farm Journal*, April 1886, 105; June 1895, 122; *The Farming World and Canadian Farm and Home*, 15 August 1904, 579.

48 *Farmer's Advocate*, 20 March 1902, 213.
49 Ibid., 22 August 1918, 1357.
50 See *Farmer's Advocate*, October 1890, 331, February 1891, 39; S. Plimsoll, *Cattle Ships* (London: Kegan Paul, Trench, Truber, & Co. Ltd., 1890); G.H. Peters, *The Plimsoll Line: the Story of Samuel Plimsoll, Member of Parliament for Derby, 1868 to 1880* (Chichester: Rose, 1975); N. Jones, *The Plimsoll Sensation: the Great Campaign to Save Lives at Sea* (London: Little, Brown, 2006).
51 Harrison, R., *Animal Machines: The New Factory Farming Industry* (London: V. Stuart, 1964).
52 *Hoard's Dairyman*, December 2007, 801.
53 See M. Coulon et al., "Dairy Cattle Exploratory and Social Behaviors: Is There an Effect of Cloning?," *Theriogenology* 68 (2007): 1097–1103.
54 See M. Greger, "Trait Selection and Welfare of Genetically Engineered Animals in Agriculture," *Journal of Animal Science* 88 (2010): 811–14.
55 A.B. Lawrence, "Applied Animal Behaviour Science: Past, Present and Future Prospects," *Applied Animal Behaviour Science* 115 (2008): 3.
56 Ibid., 6–7.
57 See, for example, *Hoard's Dairyman*, 10 January 2008, 9; July 2007, 484.
58 Ibid., December 2007, 801.
59 Ibid.

Conclusions

1 I. Misztal, "Methods to approximate reliabilities in single-step genomic evaluations," *Journal of Dairy Science* 96 (2013): 647–54; O. Christensen, "Compatibility of pedigree-based and marker-based relationship matrices for single-step genetic evaluation," *Genetics Selection Evolution* 44 (2012): 37, and "Single-step methods for genomic evaluation of pigs," *Animal* 6 (2012): 1565–71; Cy Chen, "Genome-wide marker- assisted selection combining all pedigree phenotypic information with genotypic data in one-step: an example using Broiler chickens," *Journal of Animal Science* 89 (2010): 23–8; A Legarra, "A Relationship Matrix including Full Pedigree and Genomic Information," *Journal of Dairy Science* 92 (2009): 4656–64; S Forni et al., "Different genomic relationship matrices for single-step analysis using phenotypic, pedigree and genomic information," *Genetics Selection Evolution* 43 (2011): 1.
2 J.E. Fulton, "Genomic selection for poultry breeding," *Animal Frontiers* 2 (2012): 30–6.

3 O. Christensen, "Single-step methods for genomic evaluation of pigs," *Animal* 6 (2012): 1565–71; Cy Chen, "Genome-wide marker- assisted selection combining all pedigree phenotypic information with genotypic data in one-step: an example using Broiler chickens," *Journal of Animal Science* 89 (2010): 23–8.

4 See T.B. Kinney Jr., "Poultry Breeding Research in North America," *World's Poultry Science Journal* 30–1 (1974–5): 8–27; N.D. Bayley, "Is There a Future for Land-Grant College Research and Extension?.," *Poultry Science* 52 (1973): 5–15; Y.V. Thaxton et al., "The Decline of Academic Poultry Science in the United States of America," *World's Poultry Science* 59 (2003): 303–13; S.L. Pardue, "Education Opportunities and Challenges in Poultry Science: Impact of Resource Allocation and Industry Needs," *Poultry Science* 76 (1997): 938–43; W.E. Shaklee, "Federal-Grant Funds and Poultry Breeding Research in the United States," *World's Poultry Science* 29 (1972): 215–21.

5 F.B. Hutt, "Whither Poultry Genetics?," *Poultry Science* 21 (1965): 60.

6 See Kinney Jr., "Poultry Breeding Research in North America," 8–27; and N.D. Bayley, "Is There a Future for Land-Grant College Research and Extension?" *Poultry Science* 52 (1973): 5–15.

7 See, for example, C.A. Narrod and K.O. Fuglie, "Private Investment in Livestock Breeding with Implications for Public Research Policy,"· *Agribusiness* 16 (2000): 457–70.

8 T.B. Rodenburg et al., "The role of breeding and genetics in the welfare of farm animals," *Animal Frontiers* 2 (2012): 16–21; R. Poletto et al., "The Five Freedoms in the global agriculture market: Challenges and achievements as opportunities," *Animal Frontiers* 2 (2012): 22–30.

9 For the history of sheep breeding see M.L. Ryder, *Sheep and Man*, (London: Duckworth, 1983); his "The History of Sheep Breeds in Britain," *Agricultural History Review* 12 (1964): Part 1: 1–12, Part 2: 79–97; and G.G.S. Bowie, "New Sheep for Old-Changes in Sheep Farming in Hampshire, 1792–1879," *Agricultural History Review* 35 (1987): 15–23.

10 R.M. Bourdon, "Shortcomings of Current Genetic Evaluation Systems," *Journal of Animal Science* 76 (1998): 2308–23.

11 See, for example, Lady Wentworth, Lady (Judith Anne Blunt), *The Authentic Arabian Horse* (London: George Allen & Unwin Ltd., 1945); and her *Thoroughbred Racing Stock* (New York: Charles Scribner's Sons, 1938).

12 A.T. Bowling et al., "A Pedigree-Based Study of Mitochondrial d-Loop DNA Sequence Variation Among Arabian Horses," *Animal Genetics* 31 (2000): 1–7.

13 I. Glażewska et al., "A New View on Dam Lines in Polish Arabian Horses Based on mtDNA Analysis," *Genetics Selection Evolution* 39 (2007): 609–19.

14 ENCODE project consortium, "An Integrated Encyclopaedia of DNA Elements in the Human Genome," *Nature* 489 (2012): 57.

15 *Nature* 489 (2012): 52, 54.

16 Ibid., 496 (2013): 420.

Glossary

alleles Subunits that make up half the DNA in an individual. Alleles represent the contribution of one parent to the total DNA make-up of the offspring. See also double helix.

ancestry breeding Selecting animals for breeding on the basis of their parents or more remote ancestors.

ANOVA (Analysis of Variance) A statistical tool used to validate an explanation for observed data.

Bakewellian breeding principle Breeding via inbreeding and selection on the basis of the progeny test.

bioinformatics and functional genomics The study of the dynamics of the DNA in its entirety; that is, the functioning of so-called "junk DNA" with regions of DNA that code for protein (areas known to influence how characteristics are inherited). See also SNPs; gene.

biological lock Breeding that results in progeny that will not replicate themselves, compelling the buyer to go back to the breeder for the next generation.

biometry The statistical study of inheritance of variable traits within populations. A form of classical genetics. Supported Darwinism and the theory of continuous inheritance. See also traits as quantitative.

BLUP (best linear unbiased prediction) A statistical model designed to predict an animal's breeding potential under environmentally neutral conditions. See also sire indexing; MACE.

boar Male pig.

chromosome A stretch of DNA containing many genes and other nucleotide sequences. Species have different numbers of chromosomes. Sex chromosomes govern the sex of the offspring and other characteristics that are linked with sex inheritance. Autosome chromosomes carry the rest of genetic hereditary information.

classical genetics The assessment of inheritance on the basis of probabilities that are estimated by statistics. Classical genetics does not study the underlying genetic architecture of DNA. Normally associated with Mendelism, biometry, population genetics, and quantitative genetics.

cloning The replication, via recombinant DNA technology, of an individual using its DNA.

cockerel Young cock/rooster, or male chicken.

crossbreeding The mating of animals not related to each other. They can belong to the same breed or different breeds.

double helix Describes the shape of DNA molecules where two long chains of nucleotide subunits twist around each other, forming a right-handed helix. There are four bases in the chains: A (adenine), T (thymine), G (guanine), or C (cytosine). They join each other in the following manner: where one strand has an A, the other has a T, and where there is a G on a strand, its partner has a C. Species differ only by the sequence of the A, G, T, and C nucleotides. See also SNPs.

EPD (estimated progeny difference) An EPD is a prediction of an animal's likelihood of passing on a trait in relation to breed average for that trait.

ET and MOET (embryo transfer and multiple ovulation embryo transfer) Reproductive technologies that increase the production of female mammals. First the animal is made to super-ovulate. Her eggs are then fertilized by artificial insemination or natural service. After a short time, the fertilized eggs are flushed from the mother's uterus, and then transplanted into numerous recipient females.

gene A stretch of DNA that codes for protein, meaning that it specifies how the protein used to build cells in the body will develop and function.

genomics The study of DNA at the molecular level, but across all chromosomes in any given species. A form of molecular genetics.

genotype The genetic make-up of an individual. Not observable by assessing the physical animal.

gilt Young female pig.

GMACE MACE evaluations modified to include genomic data. See also MACE.

haplotype A combination of DNA sequences at different loci on the chromosome that are transmitted together from one generation to the next.

heifer A young female cow that has not yet calved.

heterosis An explanation for hybrid vigour, which is based on the theory that a return to heterozygosis from homozygosis brings about increased vigour.

heterozygosis If an animal inherits a trait in a dominant/recessive way, it is heterozygotic to that trait, meaning that while it might itself demonstrate one form of the trait, it can transmit another. Such an animal will not breed truly for a certain form of that trait. See also crossbreeding.

homozygosis A uniform inheritance of traits, on either a double dominant or double recessive basis. If an animal is homozygotic for a trait, it can pass only one form of that trait to its offspring. It will, therefore, breed truly in connection with that trait. (See inbreeding.)

hybrid vigour Found in progeny who are superior to their parents because of crossbreeding or hybridizing. Hybrid vigour occurs only in the first generation cross.

hybridizing or hybrid breeding The breeding of two individuals in an effort to gain hybrid vigour or superiority in the offspring. The offspring, however, will not breed truly to their improvement. Hybrid breeding can be done across breeds, inbred lines within a breed, or across species.

inbreeding The mating of animals closely related to each other. In and in breeding means intense inbreeding.

individual worth Selecting individual animals for breeding on the basis of their performance.

infinitesimal model of inheritance A theory that explains how quantitative traits are inherited. It states that traits are regulated by a large number of segregating Mendelian units acting together.

LD (linkage disequilibrium) The non-random association of nucleotides at two or more loci, meaning that more variation (polymorphism) in genetic markers (at loci) occurs in a population than would be expected in random formation. See also SNPs.

locus The point where the two alleles join.

MACE (multiple across-country evaluations) A statistical method, developed by quantitative geneticists, to evaluate the genetic worth of bulls for each dairy breed on an international basis. MACE suggested that by statistically neutralizing the effects of environmental conditions and husbandry practices, as well as national evaluation procedures – all of which mask true genetic worth – the breeding value of a bull, from an international perspective, could be known.

mass selection Selecting animals for breeding from a group on the basis of certain characteristics. The favouring of various individuals according to some set criteria (such as looks or production history).

Mendelism The study of heredity on the basis of Mendel's laws. A form of classical genetics.

microsatellites Repeating sequences of base pairs in DNA. They can be used as genetic markers to explain the presence of certain characteristics.

molecular genetics The study of inheritance at the DNA level; therefore, the study of genetic architecture.

next-generation sequencing A technology used in genomics, and the one that largely replaced shotgun sequencing. Next-generation sequencing allows for a high volume of DNA data to be sequenced more rapidly. See also shotgun sequencing.

outcrossing Effectively the same as crossbreeding.

path coefficient A way of calculating the level of shared genetic material that would result from different inbreeding systems – brother to sister, first cousins, double first cousins, half brother to half sister, and so on. An accurate quantitative way to measure the level of homozygosis resulting from any type of inbreeding.

phenotype – Observable characteristics in an animal – either what it looks like, or what it can produce.

population genetics The study of inheritance of qualitative traits within the framework of populations. A form of classical genetics. See also traits as qualitative.

progeny testing Selection of animals for breeding purposes on the basis of their progeny.

pullet Young hen, or female chicken.

purebred breeding The selection of individuals for breeding on the basis that both have registered pedigrees in a public herd book.

quantitative genetics The study of inheritance of quantitative traits in populations. A form of classical genetics. Closely related to population genetics. See also traits as quantitative.

QTL (quantitative trait loci) The loci of genetic material proven to relate to productive traits in livestock. The loci can be genes, or even simply stretches of DNA that do not code for protein (meaning they do not specify how the protein used to build cells in the body will develop and function) if these stretches are known to be close to the genes that regulate traits sought after.

recombinant DNA The transferring of sections of DNA from one organism into another, in order to study and manipulate that DNA. Recombinant DNA technology opened up the possibility of studying genetics at the DNA level, and laid the basis for genomics.

RFLPs (restriction fragment length polymorphisms) A technique used to exploit variations in DNA sequences. Largely obsolete today.

sire indexing A sire index for dairy bulls works on the assumption that the level of inheritance of a daughter's ability to produce milk is halfway between that of her sire and her dam, and that by knowing the milk producing level of dam and daughter, the sire's transmitting ability can be calculated. With no daughters yet in existence, a young male's breeding potential value can only be calculated on the basis of an average value between that of his two parents – namely, his mother's milking ability and his sire's ability to produce good daughters.

shotgun sequencing A methodology of sequencing an entire genome. DNA is sheared into random fragments, which are subsequently cloned into an appropriate vector (normally a bacteria using recombinant DNA technology). The clones are then sequenced at each end, and overlapping readings taken with other segments. A composite of sequences are then rejoined, a process that is continued until the entire genome had been put together.

SNPs (single nucleotide polymorphisms) An evident deviation in an individual or group from population-norm sequencing patterns of the four bases in the DNA molecule. The insertion or deletion of an AT or GC unit creates the variation. See also double helix.

standardbred breeding The selection of animals for breeding on the basis of meeting a recognized standard.

Thoroughbred breeding The name is applied only to the breeding of the Thoroughbred horse, which originated from the crossing of Arabians on local British horses, thereby making the animals "thoroughly" bred.

training population A group of animals whose profiles are used to establish a template for superior DNA profiles.

traits as qualitative Traits that separate one species from another. Early Mendelists supported the view that any variation in all traits was effectively qualitative and therefore could not be explained by inheritance factors. Since variation could not be inherited, it was always discontinuous. Today when traits are looked at qualitatively, they are studied under population genetic theory, which addresses evolution and the rise of speciation.

traits as quantitative Supports the view that variation and differences in traits are inheritable. Traits are inherited in a quantitative way; that is, on the basis of amount – more or less. The biometricians, in opposition to the Mendelists, believed all differences were primarily quantitative and therefore not only inheritable, but also continuous from generation to generation. Greater variation in quantitative traits could lead to qualitative traits that then separated one species from another, and therefore supports Darwin's contention that species evolve over time from other species. Today when traits are studied quantitatively the focus is on changes over short generational spans (such as better meat production and milk yields in cattle, or egg laying in chickens), and not on evolution. Traits in farm animals are looked at quantitatively under quantitative genetic theory, in order to find better ways to predict the outcome of breeding strategies.

transgenics The trans-positioning of certain genetic material from one organism into another.

true breeding The breeding of animals that will largely replicate themselves in the next generation.

unit character theory The theory that quantitative traits are inherited on the basis of a single gene.

Bibliography

Primary Sources

General

Note: A number of the entries below cite the Proceedings of the Beef Improvement Federation Conferences of various years. These articles are all on the available on the website, www. beefimprovement.org/proceedings.html. (If you get a 404 error, ignore it, go to "Convention" on the menu and press "Proceedings." The articles for all the conference years are there.)

Abasht, B. et al. "Extent and Consistency of Linkage Disequilibrium and Identification of DNA Markers for Production and Egg Quality Traits in Commercial Layer Chicken Populations." *BMC Genomics* 10, Supplement 2 (2009). http://www.biomedcentral.com/1471-2164-10-S2-S2.

Abdel-Azim, G. et al. "Genetic Impacts of Using Female-Sorted Semen in Commercial and Nucleus Herds." *Journal of Dairy Science* 90 (2007): 1554–63.

Abdel-Azim, G., and A.E. Freeman. "Superiority of QTL-assisted Selection in Dairy Cattle Breeding Schemes." *Journal of Dairy Science* 85 (2002): 1869–80.

Aerts, J. et al. "Extent of Linkage Disequilibrium in Chickens." *Cytogenetic and Genome Research* 117 (2007): 338–45.

Agriculture and Agri-Food Canada et al. *2011 Edition – Statistics of the Canadian Dairy Industry.*

Aguilar, I. et al. "Hot Topic: A Unified Approach to Utilize Phenotypic, Full Pedigree, and Genomic Information for Genetic Evaluation of Holstein Final Score." *Journal of Dairy Science* 93 (2010): 743–52.

Allaire, F.R. et al. "Specific Combing Abilities Among Dairy Sires." *Journal of Dairy Science* 48 (1965): 1096–1100.

Amann, R.P. "Issues affecting Commercialization of Sexed Semen." *Theriogenology* 52 (1999): 1441–57.

– "Treatment of Sperm to Predetermine Sex." *Theriogenology* 31 (1989): 49–60.

Amer, P.R. et al. "Implications of Avoiding Overlap Between Training and Testing Data Sets when Evaluating Genomic Predictions of Genetic Merit." *Journal of Dairy Science* 93 (2010): 3320–30.

Anderson, K. "Breed and Breeder Adaptation to Genome-Enhanced/Enabled Selection Information." In *Proceedings of the Beef Improvement Federation Conference* (2009), 107–111.

Andersson, L. "Genome-wide Association Analysis in Domestic Animals: a Powerful Approach for Genetic Dissection of Trait Loci." *Genetica* 136 (2009): 341–49.

– "Mapping Genes Influencing Meat Characteristics in Pigs." In *Biotechnology's Role in the Genetic Improvement of Farm Animals*, edited by R.H. Miller et al., 276-81. Beltsville Symposia in Agricultural Research, American Society of Animal Science, 1996.

Andreescu, C., et al. "Linkage Disequilibrium in Related Breeding Lines of Chickens." *Genetics* 177 (2007): 2161–9.

Arnold, F.J. "Fifty Years of DHIA Work." *Journal of Dairy Science* 39 (1956): 792–4.

Ashwell, M.S. et al. "A Genome Scan to Identify Quantitative Trait Loci Affecting Economically Important Traits in a US Holstein Population." *Journal of Dairy Science* 84 (2001): 2535–42.

Auldist, M.J. et al. "Comparative Reproductive Performance and Early Lactation Productivity of Jersey x Holstein Cows in Predominantly Holstein Herds in a Pasture-Based Dairying System." *Journal of Dairy Science* 90 (2007): 4856–62.

"Avian Sex Determination and Sex Diagnosis." *World's Poultry Science* 59 (2003): 5–64.

Babcock, E.B. and R.E. Clausen. *Genetics in Relation to Agriculture*. New York: McGraw-Hill Book Company, 1918.

Bagnato, A. and A Rosati. "From the Editors – Animal Selection: the Genomic Revolution." *Animal Frontiers* 2 (2012): 1-2.

Bailey, J. and G. Culley. *General View of the Agriculture of Northumberland, Cumberland and Westmoreland*. London: B. McMillan, 1805. Reprinted Newcastle upon Tyne: Frank Graham, 1972.

Bailey, M.T. et al. "'Breed Registries' Regulations on Artificial Insemination and Embryo Transfer." *Journal of Equine Veterinary Medicine* 15 (1995): 60–5.

Bajema, C.J., ed. *Artificial Selection and the Development of Evolutionary Theory*: *Benchmark Papers in Systematic and Evolutionary Biology 4*. Stroudsburg, Pa: Hutchinson Ross Publishing Company, 1982.

Bakewell Letters – Culley and British Museum Collections, Part II. In *Robert Bakewell: Pioneer Livestock Breeder*, edited by H.C. Pawson. London: Crosby Lockwood & Son, 1957.

Barker, J.S.F. "Quantitative Genetics, Ecology, and Evolution." In *Proceedings of the Second International Conference on Quantitative Genetics*, edited by B.S. Weir, 596-600. Sunderland, MA: Sinauer Associates, 1988.

Barrington, A. and K. Pearson. "On the Inheritance of Coat-Colour in Cattle: Part 1. Shorthorn Crosses and Pure Shorthorns." *Biometricka* 4 (1906): 427–64.

Bartlett, J.W. et al. "A Comparison of Inbreeding and Outcrossing Holstein-Friesian Cattle." *Vol. 712 of Bulletin (New Jersey Experiment Station)*. New Jersey: New Jersey Experiment Station, 1944.

Batra, T.R. et al. "Birth Weight and Gestation Period in Purebred and Crossbred Dairy Cattle." *Journal of Dairy Science* 57 (1974): 323–7.

Bayley, N.D. "Is There a Future for Land-Grant College Research and Extension?" *Poultry Science* 52 (1973): 5–15.

Baylor College of Medicine. "Bovine Genome Project." http://www.hgsc.bcm .tmc.edu/project-species-m-Bovine.hgsc?pageLoc

Beattie, C.W. "Development of Detailed Microsatellite Linkage Maps in Livestock." In *Biotechnology's Role in the Genetic Improvement of Farm Animals*, edited by R.H. Miller et al., 51–60. Beltsville Symposia in Agricultural Research, American Society of Animal Science, 1996.

Beckett, R.C. et al. "Specific and General Combining Abilities for Production and Reproduction among Lines of Holstein Cattle." *Journal of Dairy Science* 62 (1979): 613–20.

Bell, A.E. "Heritability in Retrospect." *Journal of Heredity* 68 (1977): 297–300.

Bell, A.E. et al. "Systems of Breeding Designed to Utilize Heterosis in the Domestic Fowl." *Poultry Science* 31 (1952): 11–22.

Bereskin, B. et al. "Crossbreeding Dairy Cattle. III. First Lactation Production." *Journal of Dairy Science* 49 (1966): 659–67.

– "Crossbreeding Dairy Cattle. IV. Effects of Breed Group, Lactation Production and Pregnancy on Body Growth. Generation 1." *Journal of Dairy Science* 50 (1967): 876–83.

Bichard, M. "Changes in Quantitative Genetic Technology in Animal Breeding." In *Proceedings of the Second International Conference on Quantitative Genetics*, edited by B.S. Weir, 145-9. Sunderland, MA: Sinauer Associates, 1988.

Biely, J., H.C. Gasperdone, and W.H. Pope. "Broiler Production: 25 Years of Progress (Canada verses U.S.A.)." *World's Poultry Science Journal* 27 (1971): 241–61.

Bijma, P. et al. "Genetic Gain of Pure Line Selection and Combined Crossbred Purebred Selection with Constrained Inbreeding." *Animal Science* 72 (2001): 225–32.

Birchler, J.A. et al. "In Search of the Molecular Basis of Heterosis." *The Plant Cell* 15 (2003): 2236–9.

– "Unraveling the Genetic Basis of Hybrid Vigor." *Proceedings of the National Academy of Sciences of the USA* 103 (2006): 12957–8.

Blakeslee, A.F. "Fancy Points vs. Utility." *Journal of Heredity* 6 (1915): 175–81.

Boichard, D. et al. "Where is Dairy Cattle Breeding Going? A Vision for the Future." *Interbull Bulletin* 41 (2010): 63–7.

Boichard, D. et al. "New Phenotypes for New Breeding Goals in Dairy Cattle." *Animal* 6 (2012): 544–50.

Botstein, D. et al. "Construction of a Genetic Linkage Map in Man using Restriction Fragment Length Polymorphisms." *American Journal of Human Genetics* 32 (1980): 314–31.

Bourdon, R.M. *Understanding Animal Breeding*. 2nd ed. Upper Saddle River, NJ: Prentice Hall, 2000.

– "Shortcomings of Current Genetic Evaluation Systems." *Journal of Animal Science* 76 (1998): 2308–23.

Bouwman, A.C. et al. "Imputation of non-genotyped individuals based on genotyped relatives: assessing the imputation accuracy of a real case scenario in dairy cattle." *Genetics Selection Evolution* 46 (2014): 6.

Bovine Genome Sequencing and Analysis Consortium et al. "The Genome Sequence of Taurine Cattle: A Window to Ruminant Biology and Evolution." *Science* 324 (2009): 522–8.

Bovine HapMap Consortium. http://genomes.arc.georgetown.edu/drupal/bovine/?q=hapmap_funding.

– "Genome-Wide Survey of SNP Variation Uncovers the Genetic Structure of Cattle Breeds." *Science* 324 (2009): 528–32.

Bowling, A.T. et al. "A Pedigree-Based Study of Mitochondrial d-Loop DNA Sequence Variation Among Arabian Horses." *Animal Genetics* 31 (2000): 1–7.

Bowling, G.A. "Tempest in a Teapot." *Guernsey Breeder's Journal* (November 1946): 1417.

Brah, G.S. et al. "Cross-line Productivity in Relation to Selection in Pure-line White Leghorns." *Journal of Animal Breeding and Genetics* 104 (1987): 391–6.

Brandt, G.W. et al. "Effects of Crossbreeding on Production Traits in Dairy Cattle." *Journal of Dairy Science* 49 (1966): 1249–53.

Brandt, H. et al. "Estimation of Genetic and Crossbreeding Parameters for Preweaning Traits in German Angus and Simmental Beef Cattle and the Reciprocal Crosses." *Journal of Animal Science* 88 (2010): 80–6.

Brewbaker, J.L. *Agricultural Genetics*. Englewood Cliffs, NJ: Prentice-Hall, 1964.

Brown, E. *Poultry Breeding and Production*. London: Ernest Benn, 1929.

Brown, W.A. "The Poultry Industry in Maine." *O.A.C. Review* 22 (1910): 239–45.

Burnham, G. *Burnham's New Poultry Book*. Boston: Lee & Shepard, 1877.

– *The History of the Hen Fever*. San Diego: Frank E. Marcy, 1935. Reprint of the 1855 edition.

Cain, J., and I. Layland. "The Situation in Genetics: Dunn's 1927 Russian Tour." *Mendel Newsletter* 12 (2003): 10–15.

Callaway, E. "Chicken Project Gets off the Ground." *Nature* 509 (2014): 546.

Calus, M.P.L. "Accuracy of Breeding Value when Using and Ignoring the Polygenic Effect in Genome Breeding Value Estimation with a Marker Density of one SNP per cM." *Journal of Animal Breeding and Genetics* 124 (2007): 362–8.

Calus, M.P.L. et al. "Accuracy of Genomic Selection Using Different Methods to Define Haplotypes." *Genetics* 178 (2008): 553–61.

Campbell, M.H. "Inheritance of Black and Red Colors in Cattle." *Genetics* 9 (1924): 419–41.

Canada. Department of Agriculture. *Publications of the International Agricultural Institute*. Ottawa: Department of Agriculture, 1911–12.

Cassady, J.P. et al. "Heterosis and Recombination Effects on Pig Growth and Carcass Traits." *Journal of Animal Science* 80 (2002): 2286–302.

Cassell, B.G. "Dairy Crossbreeding: Why and How." *Publication 404–093*. Virginia Polytechnic and State University, 2009.

– "Mechanisms of Inbreeding Depression and Heterosis in Profitable Dairying." In *Proceedings of the Fourth W.E. Petersen Symposium*, 1-6. Minneapolis: University of Minnesota, 2007.

Cassell, B.G. et al. "Birth Weights, Mortality, and Dystocia in Holsteins, Jerseys, and Their Reciprocal Crosses in the Virginia Tech and Kentucky Crossbreeding Project." Abstract. *Journal of Dairy Science* 88 (2005): 92.

Cassell, B.G. et al. "Genetic and Phenotypic Relationships among Type Traits in Holstein-Friesian Cattle." *Journal of Dairy Science* 56 (1973) 1171–7.

Castle, W.E. *Heredity in Relation to Evolution and Animal Breeding*. New York: D. Appleton and Company, 1911.

– "Inheritance of Quantity and Quality of Milk Production in Dairy Cattle." *Proceedings of the National Academy of Science, U.S.A.* 5 (1919): 428–34.

– "Pure Lines and Selection." *Journal of Heredity* 5 (1914): 93–7.

– "Some Biological Principles of Animal Breeding." *American Breeders' Magazine* 3 (1912): 270–82.
– "The Mutation Theory of Organic Evolution, from the Standpoint of Animal Breeding." *Science* 21 (1905): 521–5.
Cerchiaro, I. et al. "A Field Study on Fertility and Purity of Sex-Sorted Cattle Semen." *Journal of Dairy Science* 90 (2007): 2538–42.
Chebel, R.C. et al. "Sex-Sorted Semen for Dairy Heifers: Effects on Reproductive and Lactational Performance." *Journal of Dairy Science* 93 (2010): 2496–507.
Chen, C. et al. "Genome-wide Marker-assisted Selection Combining all Pedigree Phenotypic Information with Genotypic Data in One-step: an Example using Broiler Chickens." *Journal of Animal Science* 89 (2010): 23-8.
Chesnais, J.P. "How is the AI industry using Genomic tools in practice?" *Interbull Bulletin* 41 (2010): 59–62.
Christensen, O. "Compatibility of Pedigree-based and Marker-based Relationship Matrices for Single-step Genetic Evaluation." *Genetics Selection Evolution* 44 (2012): 37.
– "Single-step Methods for Genomic Evaluation in Pigs." *Animal* 6 (2012): 1565-71.
Christensen, O., et al. "Genomic Evaluation of both Purebred and Crossbred Performances." *Genetics Selection Evolution* 46 (2014): 23
Cole, L.J. et al. "Inheritance in Crosses of Jersey and Holstein-Friesian with Aberdeen Angus Cattle. III. Growth and Body Type, Milk Yield, and Butterfat Percentage." *American Naturalist* 82 (1948): 265–80.
– "Biological Sciences Needed by a Student Specializing in Animal Husbandry." *Journal of Animal Science* (1936): 179–84.
Cole, J.B. "Distribution and Location of Genetic Effects for Dairy Traits." *Journal of Dairy Science* 92 (2009): 2931–46.
Cole, J.B. et al. "Visualization of Results from Genomic Evaluations." *Journal of Dairy Science* 93 (2010): 2727–40.
– "Breeding and Genetics Symposium: Really Big Data: Processing and Analysis of Very Large Data Sets." *Journal of Animal Science* 90 (2012): 723–33.
Colleau, J.J. "Genetic Improvement by Embryo Transfer within Selection Nuclei in Dairy Cattle." *Genetic Selection Evolution* 17 (1985): 499–538.
Colley, A. et al. "Single Bovine Sperm Sex Typing by Amelogenin Nested PCR." *Theriogenology* 70 (2008): 978–83.
Collins, F.S. et al. "A Vision for the Future of Genomics Research." *Nature* 422 (2003): 835–47.
"Compassion in World Farming. "Beyond Calf Exports Stakeholder Forum." Accessed 29 January 2009. http://www.ciwf.org.uk/what_we_do/calves/beyond_calf_exports.aspx.

Coulon, M. et al. "Dairy Cattle Exploratory and Social Behaviors: Is There an Effect of Cloning?" *Theriogenology* 68 (2007): 1097–103.

Crittenden, L.B. et al. "Insertion of Retroviruses into the Avian Germ Line." In *Proceedings of the Second International Conference on Quantitative Genetics*, edited by B.S. Weir, 207-14. Sunderland, MA: Sinauer Associates, 1988.

Croquet, C. et al. "Inbreeding Depression for Global and Partial Economic Indexes, Production, Type and Functional Traits." *Journal of Dairy Science* 89 (2006): 2257–67.

"Crossbreeding – It is the Next Magic Pill?" Genex Cooperative Inc. Accessed 15 June 2009. http://www.bullsemen.com/cattle-breeding/bull-semen-article .php?arti.

Crow, J.R. Review of *Proceedings of the Second International Conference on Quantitative Genetics*, by B.S. Weir et al. *Science* 242 (1988): 1449–50.

Culley, G. *Observations on Live Stock: Containing Hints for Choosing and Improving the Best Breeds of the Most Useful Kinds of Domestic Animals*. London: D. Longworth, 1804.

Cunningham, E.P. "Structure of Dairy Cattle Breeding in Western Europe and Comparisons with North America." *Journal of Dairy Science* 66 (1983): 1579–87.

Daetwyler, H.D. et al. "Inbreeding in Genome-Wide Selection." *Journal of Animal Breeding and Genetics* 124 (2007): 369–76.

Dal Zotto, R. et al. "Use of Crossbreeding with Beef Bulls in Dairy Herds: Effects on Age, Body Weight, Price, and Market Value of Calves sold at Livestock Auctions." *Journal of Dairy Science* 87 (2009): 3053–9.

Daley, D.A. "Crossbreeding – Back to the Future." In *Proceedings of the Beef Improvement Federation Conference* (2009), 132–36.

Damme, K. et al. "Fattening Performance, Meat Yield and Economic Aspects of Meat and Layer Hybrids." *World's Poultry Science* 59 (2003): 50–3.

Damou, R.A. Jr. et al. "Genetic Analysis of Crossbreeding Beef Cattle." *Journal of Animal Science* 20 (1961): 849–57.

Darwin, C. *On the Origin of Species*. London: John Murray, 1859.

– *The Variation of Animals and Plants under Domestication*. 1883. Reprint of the first edition, with a forward by Harriet Ritvo. Baltimore: The Johns Hopkins University Press, 1998.

Darwin Correspondence Project. "Darwin C.R. to Gray, Asa." University of Cambridge. Accessed 14 October 2012. www.darwinproject.ac.uk/ entry-2125.

Dassonneville, R. et al. "Effect of Imputing Markers from a Low-density Chip on the Reliability of Genomic Breeding Values in Holstein Populations." *Journal of Dairy Science* 94 (2011): 3679–86.

Davenport, C.B. "The Relationship of the Association to Pure Research." *American Breeders' Magazine* 1 (1911): 66–7.

David, X. et al. "International Genomic Cooperation: EuroGenomics significantly improves reliability of Genomic evaluations." *Interbull Bulletin* 41 (2010): 77–8.

Davoli, R. et al. "Molecular Approaches in Pig Breeding to Improve Meat Quality." *Briefings in Functional Genomics* 6 (2007): 313–21.

De Boer, I.J.M. et al. "An Ethical Evaluation of Animal Biotechnology: The Case of Using Clones in Dairy Cattle." *Animal Science* 61 (1995): 453–63.

DeJarnette, J.M. et al. "Sustaining the Fertility of Artificially Inseminated Dairy Cattle: The Role of the Artificial Insemination Industry." *Journal of Dairy Science* 87 (2004): E93–104.

DeJarnette, J.M. et al. "Evaluating the Success of Sex-Sorted Semen in the US Dairy Herds from on Farm Records." *Theriogenology* 71 (2009): 49–58.

Dekkers, J.C.M. "Commercial Application of Marker and Gene Assisted Selection in Livestock: Strategies and Lessons," *Journal of Animal Science* 82 (2004): E313–328.

– "Marker-Assisted Selection for Commercial Crossbred Performance." *Journal of Animal Science* 85 (2007a): 2104–14.

– "Prediction of Response to Marker-Assisted and Genomic Selection Using Selection Index Theory." *Journal of Animal Breeding and Genetics* 124 (2007): 331–41.

– "Structure of Breeding Programs to Capitalize on Reproductive Technology for Genetic Improvement," *Journal of Dairy Science* 75 (1992): 2880–91.

Dempster, E.R., and I.M. Lerner. "Heritability of Threshold Characters." *Genetics* 35 (1950): 212–37.

Dentine, M.R. "Improving Dairy Cattle Using Marker Data." In *Biotechnology's Role in the Genetic Improvement of Farm Animals*, edited by R.H. Miller et al., 282–7. Beltsville Symposia in Agricultural Research, American Society of Animal Science, 1996.

Department of Agriculture. "Better Plants and Animals," Part 1, *Yearbook of Agriculture*. United States: Department of Agriculture, 1936.

De Roos, A.P.W. et al. "Genomic Breeding Value Estimation using Genetic Markers, Inferred Ancestral Haplotypes, and the Genomic Relationship Matrix." *Journal of Dairy Science* 94 (2011): 4708–14.

– "Linkage Disequilibrium and Persistence of Phase in Holstein Friesian, Jersey and Angus Cattle." *Genetics* 179 (2008): 1503–1512.

De Vries, A. "Sexed Semen Economics." Milkproduction.com, 2008. Accessed 24 March 2009. http://www.milkproduction.com/Library/Articles/Sexed _semen_econo.

De Vries, A. et al. "Exploring the Impact of Sexed Semen on the Structure of the Dairy Industry." *Journal of Dairy Science* 91 (2008): 847–56.

Dickinson, F.N. et al. "Liveability of Purebred Versus Crossbred Dairy Cattle." *Journal of Dairy Science* 44 (1961): 879–87.

Dürr, J. et al. "International cooperation: the Pathway for cattle genomics." *Animal Frontiers* 2 (2012): 16-21.

East, E.M. "Heterosis." *Genetics* 21 (1936): 375–97.

Edress, V. et al. "The Effect of Using Geneology-based Haplotypes for Genomic Selection." *Genetics Selection Evolution* 45 (2013): 5.

Eggen, A. "The Development and Application of Genomic Selection as a new Breeding Paradigm." *Animal Frontiers* 2 (2012): 10-15.

Ellegren, H. "Hens, Cocks and Avian Sex Determination." *EMRO (European Molecular Biology Organization) Reports* 2 (2001): 192-6.

Ellendorff, F. et al. "Current Knowledge on Sex Determination and Sex Diagnosis: Potential Solutions." *World's Poultry Science* 59 (2003): 7; in "Avian Sex Determination and Sex Diagnosis" review, 5–64.

Eller, A.L. "A Look Back at BIF History." In *Proceedings of Annual Beef Improvement Federation Symposium*, 10–14. 2007.

Ellinger, T. "The Variation and Inheritance of Milk Characters." *Proceedings of the National Academy of Sciences, U.S.A.* 9 (1923): 111–16.

Emmerson, D.A. "Commercial Approaches to Genetic Selection for Growth and Feed Conversion in Domestic Poultry." *Poultry Science* 76 (1997): 1121–5.

ENCODE Project Consortium, "An Integrated Encyclopaedia of DNA Elements in the Human Genome." *Nature* 489 (2012): 52, 54, 57–74.

Enfield, R.D. "New Sources of Variation." In *Proceedings of the Second International Conference on Quantitative Genetics*, edited by B.S. Weir, 215-18. Sunderland, MA: Sinauer Associates, 1998.

Ericson, K. et al. "Crossbreeding Effects between Two Swedish Dairy Breeds for Reproductive Traits." *Journal of Animal Breeding and Genetics* 105 (1988): 441–51.

Fact Finding (and later *Proceedings*) *of the Annual National Poultry Breeders' Roundtable*, Poultry Breeders of America, unpublished, 1958–1969.

Fahrenkrug, S.C. et al. "Precision Genetics for Complex Objectives in Animal Agriculture." *Journal of Animal Science* 88 (2010): 2530–9.

Fairhill, R.W. "The Use of Markers in Poultry Improvement." In *Biotechnology's Role in the Genetic Improvement of Farm Animals*, edited by R.H. Miller et al., 289–99. Beltsville Symposia in Agricultural Research, American Society of Animal Science, 1996.

Falconer, D.S. "The Problem of Environment and Section." *American Naturalist* 86 (1952): 293–8.

Fan, Bin et al. "Development and Application of High-Density SNP Arrays in Genomic Studies of Domestic Animals." *Asian-Australasian Journal of Animal Science* 23 (2010): 833–47.

Felch, I.K. *The Breeding and Management of Poultry*. Hyde Park, NY: Norfolk County Press, 1877.

– *Poultry Culture: How to Raise, Mate and Judge Thoroughbred Fowls*. Chicago: Donohue, Henneberry, 1902.

Fetrow, J. "Sexed Semen: Economics of a New Technology." In *Proceedings of Western Dairy Management Conference*. 2007. www.wdmc.org/proceed.htm.

Ferdosi, M.H. et al. "Detection of Recombination Events, Haplotype Reconstruction and Imputation of Sires using Half-sib Genotypes." Genetics Selection Evolution 46 (2014): 11.

Fisher, R.A. "The Correlation between Relatives on the Supposition of Mendelian Inheritance." *Transactions of the Royal Society of Edinburgh* 52 (1918): 399–433.

– *The Theory of Inbreeding*. London: Oliver and Boyd, 1949.

Fohrman, M.H. "A Cross-breeding Experiment with Dairy Cattle." BDIM-INF-30. USDA, Bureau of Dairy Industry, 1946.

– "Cross-Breeding Dairy Cows." In *Yearbook of Agriculture*, Part 1, 177–84. USDA, Department of Agriculture, 1943-47.

Fohrman, M.H. et al. "A Cross-breeding Experiment with Dairy Cattle." *USDA Technical Bulletin* 1074. 1954. www.nal.usda.gov/ref/USDA/tb.htm.

Foote, R.H. "The Artificial Insemination Industry." In *New Technologies in Animal Breeding*, edited by B.J. Brackett et al., 13-39. New York: Academic Press, 1981.

– "The History of Artificial Insemination: Selected Notes and Notables." *Journal of Animal Science* 80 (2002): 1–10.

Forni, S. et al. "Different Genomic Relationship Matrices for Single-step Analysis using Phenotypic, Pedigree and Genomic Information." Genetics Selection Evolution 43 (201): 1.

Forum: Genomics. "Encode Explained." *Nature* 489 (2012): 52–5.

Frankham, R. "Exchanges in the rRNA Multigene Family as a Source of Genetic Variation." In *Proceedings of the Second International Conference on Quantitative Genetics*, edited by B.S. Weir, 236-42. Sunderland, MA: Sinauer Associates, 1998.

Fraser, A. *Animal Husbandry Heresies*, London: Crosby Lockwood & Son, 1960.

Frijters, A.C.J. et al. "What Affects Fertility of Sexed Bull Semen More, Low Sperm Dosage or the Sorting Process?" *Theriogenology* 71 (2009): 64–7.

Fulton, J.E. "Molecular Genetics in a Modern Poultry Breeding Organization." *World's Poultry Science* 64 (2008): 171–6.

– "Genomic Selection for Poultry Breeding." *Animal Frontiers* 2 (2012): 30-6.
Funk, D.A. "Major Advances in Globalization and Consolidation of the Artificial Insemination Industry." *Journal of Dairy Science* 89 (2006): 1362–8.
Garcia, E.M. et al. "Improving the Fertilizing Capacity of Sex Sorted Boar Spermatozoa." *Theriogenology* 68 (2007): 771–8.
Garner, D.L. et al. "History of Commercializing Sexed Semen for Cattle." *Theriogenology* 69 (2008): 886–95.
Garrick, D. et al. "Consequences of Genomic Prediction in Cattle." *Interbull Bulletin* 41 (2010): 51–8.
– "The Nature and Scope of Some Whole Genome Analyses in US Beef Cattle." In *Proceedings of the Beef Improvement Federation, 41st Annual Research Symposium* (2009), 92–102.
Georges, M. et al. "Mapping Quantitative Trait Loci Controlling Milk Production in Dairy Cattle by Exploiting Progeny Testing." *Genetics* 139 (1995): 907–20.
Gerken, M. et al. "Growth, Behaviour and Carcass Characteristics of Egg-type Cockerels Compared to Male Broilers." *World's Poultry Science* 59 (2003): 46–9.
Gianola, D. et al., eds. *Advances in Statistical Methods for Genetic Improvement of Livestock*. Berlin: Springer-Verlag, 1990.
Gill, J.L. et al. "Association of Selected SNP with Carcass and Taste Panel Assessed Meat Quality Traits in a Commercial Population of Aberdeen Angus-sired Beef Cattle. *Genetics Selection Evolution* 41 (2009): 36–47.
Givens, M.D. et al. "Relative Risks and Approaches to Biosecurity in the Use of Embryo Technologies in Livestock." *Theriogenology* 68 (2007): 298–307.
Glażewska, I. et al. "A New View on Dam Lines in Polish Arabian Horses Based on mtDNA Analysis." *Genetics Selection Evolution* 39 (2007): 609–19.
Gobin, A. "Mendelism in Animal Breeding as Developed by Professor Leopold Frateur, Louvain (1877–1946)." *Argos* 23 (2000): 111–18.
Goddard, M.E. "Genomic Selection." *Journal of Animal Breeding and Genetics* 124 (2007): 323–30.
– "How can we best use DNA data in the Selection of Cattle?" In *Proceedings of the Beef Improvement Federation Conference* (2009), 81–8.
– "Implementation of Genomic Evaluation in the Dairy Industry." In *Proceedings of the Beef Improvement Federation Conference* (2009).
– "View of the Future: Could Genomic Evaluation Become the Standard?" *Interbull Bulletin* 39 (2009): 83–7.
Goddard, M.E. et al. "Mapping Genes for Complex Traits in Domestic Animals and Their Use in Breeding Programmes." *Nature Reviews Genetics* 10 (2009): 381–91.

Golden, B.L. et al. "Milestones in Beef Cattle Genetic Evaluations." *Journal of Animal Science* 87 (2009): E3–10.

Gordon, K. "DNA Companies Share Vision for Whole-Genome Future." Paper presented at the Beef Improvement Federation Conference, Sacramento, California, May 2009.

Goto, E., and A.W. Nordskog. "Heterosis in Poultry: Estimation of Combining Ability Variance From Diallel Crosses of Inbred Lines in the Fowl." *Poultry Science* 38 (1959): 1381–8.

Gowe, R.S. "Comments on the Conference." In *Proceedings of the International Conference on Quantitative Genetics*, edited by E. Pollak et al., 799-801. Ames: Iowa State University Press, 1977.

Gowen, J.W. "Studies in Inheritance of Certain Characters of Crosses Between Dairy and Beef Breeds of Cattle." *Journal of Agricultural Research* 15 (1918): 1–57.

– "Inheritance in Crosses of Dairy and Beef Breeds of Cattle. II. On the Transmission of Milk Yield to the First Generation." *Journal of Heredity* 11 (1920): 300–16.

– "Inheritance in Crosses of Dairy and Beef Breeds of Cattle. III. Transmission of Butterfat Percentage of the First Generation." *Journal of Heredity* 11 (1920): 365–77.

Graaf, S.P. et al. "Sperm Sexing in Sheep and Cattle: The Exception and the Rule." *Theriogenology* 71 (2009): 89–97.

Graham, W.R. "Modern Poultry Tendencies." *O.A.C. Review* 39 (1926): 55–6.

Grantham, J.A. et al. "Genetic Relationships between Milk Production and Type in Holsteins." *Journal of Dairy Science* 57 (1974): 1483–8.

Green, R.D. "ASAS Centennial Paper: Future Needs in Animal Breeding and Genetics." *Journal of Animal Science* 87 (2009): 793–800.

Green, R.D. et al. "Identifying the Future Needs for Long-Term USDA Efforts in Agricultural Animal Genomics." *International Journal of Biological Sciences* 3 (2007): 185–91.

Greger, M. "Trait Selection and Welfare of Genetically Engineered Animals in Agriculture." *Journal of Animal Science* 88 (2010): 811–14.

Groenen, M.A.M. et al. "The Development and Characterization of a 60K SNP Chip for Chickens." *BMC Genomics* 12 (2011). http://www.biomedcentral.com/1471-2164/12/274.

Gupta, P.K. "Quantitative Genetics on the Rise." *Current Science* 93 (2007): 1051–2.

Gura, S. *Livestock Genetics Companies: Concentration and Proprietary Strategies of an Emerging Power in the Global Food Economy*. Ober-Ramstadt, Germany: League for Pastoral Peoples and Endogenous Livestock Development, 2007.

Gurkiewicz, M. et al. "Too Much Quantification Hinders Creativity." *Science* 324 (2009): 1515.

Hagedoorn, A.L. *Animal Breeding*. 6th ed. London: Crosby Lockwood & Son, 1962.

Hagedoorn, A.L., and G. Sykes. *Poultry Breeding: Theory and Practice*, London: Crosby Lockwood & Son, 1953.

Hagger, C. "Genetic Effects Estimated from Crosses and Backcrosses of Two Related Lines of White Leghorn Chickens." *Journal of Animal Breeding and Genetics* 106 (1989): 241–8.

Haley, C. "Human and Livestock Genetics: Parallel Evolution and Horizontal Exchange." *Journal of Animal Breeding and Genetics* 126 (2009): 413–14.

Hallauer, A.R. "History, Contribution, and the Future of Quantitative Genetics in Plant Breeding: Lessons from Maize." *Crop Science* 47 (2007): S4-19.

Hansen, L.B. "Consequences of Selection for Milk Yields from a Geneticist's Point of View." *Journal of Dairy Science* 83 (2000): 1145–50.

Hansen, L.B. et al. "Is Crossbreeding the Answer for Reproductive Problems of Dairy Cattle?" In *Proceedings of the Southwest Nutrition Conference*, 113-18. 2005.

Hardy, I.C.W. "Factors Influencing Avian Sex Ratios." *World's Poultry Science* 59 (2003): 19–24.

Hare, E. "Duration of Herd Participation in Dairy Herd Improvement Milk Recording in the United States." *Journal of Dairy Science* 87 (2004): 2743–7.

Harlizius, B. et al. "A Single Nucleotide Polymorphism Set for Paternal Identification to Reduce the Costs of Trait Recording in Commercial Pig Breeding." *Journal of Animal Science* 89 (2011): 1661–8.

Harper, M.H. *Breeding of Farm Animals*, New York: Orange Judd Company, 1920.

Harris, B.L. et al. "Genomic Predictions for New Zealand Dairy Bulls and Integration with National Genetic Evaluation." *Journal of Dairy Science* 93 (2010): 1243–52.

Harris, B.L. et al. "The Impact of High Density SNP chips on Genomic Evaluation in Dairy Cattle." *Interbull Bulletin* 42 (2010): 40–3.

Harris, D.L. "Past, Present and Potential Contributions of Quantitative Genetics to Applied Animal Breeding." In *Proceedings of the International Conference on Quantitative Genetics*, edited by E. Pollak et al., 587-609. Ames: Iowa State University Press, 1977.

– "What's Different about Chickens?" In *Proceedings of the International Conference on Quantitative Genetics*, edited by E. Pollak et al., 793-8. Ames: Iowa State University Press, 1977.

Harris, D.L. et al. "Breeding for Profit: Synergism between Genetic Improvement and Livestock Production (A Review)." *Journal of Animal Science* 72 (1994): 2178–200.

Hasler, J.F. "Current Status and Potential for Embryo and Reproductive
   Technology in Dairy Cattle." *Journal of Dairy Science* 83 (1992): 1145–50.
Hawken, R.J. et al. "Whole Genome Amplification of DNA Extracted from
   Cattle Semen Samples." *Journal of Dairy Science* 89 (2006): 2217–21.
Hayes, B.J. et al. "Accuracy of Genomic Breeding Values in Multi-Breed Dairy
   Cattle Populations." *Genetics Selection Evolution* 41 (2009): 41–9.
Hayes, B.J. et al. "Genomic Selection in Dairy Cattle: Progress and
   Challenges." *Journal of Dairy Science* 92 (2009): 433–443.
Hayes, B.J. et al. "The Distribution of the Effects of Genes Affecting Quantitative
   Traits in Livestock." *Genetic Selection and Evolution* 33 (2001): 209–29.
–   "QTL Mapping, MAS, and Genomic Selection." Short Course called Animal
   Breeding and Genetics, Department of Animal Science, Iowa State University,
   2007.
Hays, F.A., and G.T. Klein. *Poultry Breeding Applied*. Mount Morris, Illinois: Watt
   Publishing Company, 1952.
Hawley, R.J. "Genetic Modification of Pigs by Nuclear Transfer."
   Xenotransplantation 9 (2002): 159–63.
Hazel, L.N. "The Genetic Basis for Constructing Selection Indexes." *Genetics*
   28 (1943): 476–90.
Hazel, L.N. et al. "The Selection Index – Then, Now, and for the Future." *Journal
   of Dairy Science* 77 (1994): 3236–51.
Heape, W.H. *The Breeding Industry: Its Value to the Country, and Its Needs*.
   Cambridge: Cambridge University Press, 1906.
Heinrichs, A.J. "Raising Dairy Replacement Heifers to Meet the Needs of the
   21st Century." *Journal of Dairy Science* 76 (1993): 3179–87.
Heins, B. "Impact of an Old Technology on Profitable Dairying in the 21st
   Century." In *Proceedings of the Fourth W.E. Petersen Symposium*, 7–19.
   Minneapolis: University of Minnesota, 2007.
Heins, B.J. et al. "Calving Difficulty and Stillbirths of Pure Holsteins Versus
   Crossbreds of Holstein with Normande, Montbeliarde, and Scandinavian
   Red." *Journal of Dairy Science* 89 (2006): 2805–20.
Heins, B.J. et al. "Production of Pure Holsteins Versus Crossbreds of Holstein
   with Normande, Montbeliarde, and Scandinavian Red." *Journal of Dairy
   Science* 89 (2006): 2799–804.
Henderson, C.R. *Applications of Linear Models in Animal Breeding*. Guelph,
   Ontario: University of Guelph, 1984. cgil.uoguelph.ca/pub/Henderson.html.
–   "Young Sire Selection for Commercial Artificial Insemination." *Journal of
   Dairy Science* 47 (1964): 439–46.
–   "General Flexibility of Linear Model Techniques for Sire Evaluation." *Journal
   of Dairy Science* 57 (1974): 963–72.

Hill, E.W. et al. "History and Integrity of Thoroughbred Dam Lines Revealed in Equine mtDNA Variation." *Animal Genetics* 33 (2002): 287–94.

Hill, W.C. "Can more be learned from selection experiments of value in animal breeding programmes? Or is it time for an obituary?" *Journal of Animal Breeding and Genetics* 128 (2011): 87–94.

– "Variation in Response to Selection." In *Proceedings of the International Conference on Quantitative Genetics*, edited by E. Pollak et al., 343-63. Ames: Iowa State University Press, 1977.

– "Understanding and Using Quantitative Genetic Variation." *Philosophical Transactions of the Royal Society B: Biological Sciences* 365 (2010): 73–85.

Hill, W.C., ed. *Quantitative Genetics: Part I, Explanation and Analysis of Continuous Variation*. New York: Van Nostrand Reinhold Company, 1984.

– *Quantitative Genetics: Part II, Selection*. New York: Van Nostrand Reinhold Company, 1984.

Hill, W.C., and T.F.C. Mackay, eds. *Evolution and Animal Breeding: Reviews on Molecular and Quantitative Approaches in Honour of Alan Robertson*. Wallingford, UK: C.A.B. International, 1989.

Hiroyuki, H. "Systems Approaches to Beef Cattle Production Systems Using Modeling and Simulation." *Animal Science Journal* 81 (2010): 411–24.

Hobenboken, W.D. "Applications of Sexed Semen in Cattle Production." *Theriogenology* 52 (1999): 1421–33.

Hocking, P.M. et al. "Factors Affecting Length of Herdlife in Purebred and Crossbred Dairy Cattle." *Journal of Dairy Science* 71 (1988): 1011–24.

Housman, W. *The Improved Shorthorn, Notes and Reflections upon Some Facts in Shorthorn History, with Remarks upon Certain Principles of Breeding*. London: Ridgeway, 1876.

Hutchison, J.L. et al. "Characterization and Usage of Sexed Semen from US Field Data." Abstract. *Theriogenology* 71 (2000): 48.

Hu, X. et al. "Advanced Technologies for Genomic Analysis in Farm Animals and Its Application for QTL Mapping." *Genetica* 136 (2009): 371–86.

Hutt, F.B., and B.A. Rasmusen. *Animal Genetics*. 2nd ed. New York: John Wiley & Sons, 1982.

Hutt, F.B. and R.K. Cole. "Heterosis in an Interstrain Cross of White Leghorns." *Poultry Science* 31 (1952): 365–374.

Ibanez-Escriche, N. et al. "Genomic Selection of Purebreds for Crossbred Performance." *Genetics Selection Evolution* 41 (2009): 12–21.

Ibanez-Escriche, N. et al. "Genetic Evaluation Combining Purebred and Crossbred Data in a Pig Breeding Scheme." *Journal of Animal Science* 89 (2011): 3881–9.

Illumina. "BovineSNP50 Genotyping BeadChip." Accessed 20 July 2009. http://www.illumina.com/pages.ilmn?ID=3.

Interbull. "Interbull Summary." Accessed 17 June 2009. http://www-interbull
.slu.se/summary/framesida-summary.htm.

"International HapMap Project, The." *Nature* 426 (2003): 789–96.

International Human Genome Sequencing Consortium, "Initial Sequencing
and Analysis of the Human Genome." *Nature* 409 (2001): 860–921.

Jakubec, V. et al. "Crossbreeding in Farm Animals. I Analysis of Complete
Diallel Experiments by Means of Three Models with Application to
Poultry." *Journal of Animal Breeding and Genetics* 104 (1987): 283–94.

Jasper, A.W. "The Farmer and the Poultry Industry." *World's Poultry Science
Journal* 27 (1971): 43–57.

– "Genetic Effects of Heterosis in F 1 and Backcrosses of Inbred Lines of
White Leghorns." *Journal of Animal Breeding and Genetics* 103 (1986): 26–32.

Joerg, H. et al. "Validating Bovine Sexed Semen Samples Using Quantitative
PCR." *Journal of Animal Breeding and Genetics* 121 (2004): 209–15.

Johannsen, W. "The Genotype Conception of Heredity." *American Naturalist*
45 (1911): 129–59.

Johnson, L.A. et al. "A Study of Nicking in Jersey Cattle." *Journal of Dairy Science*
23 (1940): 709–18.

Johnston, D.J. et al. "New Developments in Beef Cattle Genetic Evaluation in
Australia." *Interbull Bulletin* 37 (2007): 8–11.

Jones, D.F. "Dominance of Linked Factors as a Means of Accounting for
Heterosis." *Genetics* 2 (1917): 466–79.

– *Genetics in Plant and Animal Improvement*. New York: John Wiley & Sons, 1925.

Joshi, A. "Sir R.A. Fisher and the Evolution of Genetics." *Resonance* 2 (1997):
27–31.

Jull, A. "Inbreeding and Crossbreeding in Poultry." *Journal of Heredity* 24
(1933): 93–101.

– *Poultry Breeding*. 3rd ed. New York: John Wiley & Sons, 1952.

Jull, A. et al. "The Poultry Industry." In *Yearbook of Agriculture*, 377-456.
USDA, Department of Agriculture, 1924.

Kagami, H. "Sex Reversal in Chicken." *World's Poultry Science* 59 (2003): 15–18.

Kaleta, E.F. et al. "Approaches to Determine the Sex Prior to and after
Incubation of Chicken Eggs and of Day-old-chicks." *World's Poultry Science*
64 (2008): 391–9.

Kavar, T. et al. "History of Lipizzan Horse Maternal Lines as Revealed by
mtDNA Analysis." *Genetics Selection Evolution* 34 (2002): 635–48.

Keele, J.W. et al. "Databases and Information Systems Needed for Maps and
Marker-Assisted Selection." In *Biotechnology's Role in the Genetic Improvement
of Farm Animals*, edited by R.H. Miller et al., 300-6. Beltsville Symposia in
Agricultural Research, American Society of Animal Science, 1996.

Kemmer, D. et al. "Whose Genome is Next?" *Genome Biology* 3 (2002). http://
genomebiology.com/2002/3/12/reports/4037.

Kemp, R. et al. "Collection and Application of Genetic Information from a
Canadian Perspective." In *Proceedings of the Beef Improvement Federation
Conference* (2008), 31-9.

Kempthorne, O. "An Overview of the Field of Quantitative Genetics." In
Proceedings of the Second International Conference on Quantitative Genetics, edited
by B.S. Weir, 47-56. Sunderland, MA: Sinauer Associates, 1988.

Kinney, T.B. Jr. "Poultry Breeding Research in North America." *World's Poultry
Science Journal* 30–31 (1974–5): 9–27.

Kizilkaya, K. et al. "Genomic Prediction of Simulated Multibreed and
Purebred Performance Using Observed Fifty Thousand Single Nucleotide
Polymorphism Genotypes." *Journal of Animal Science* 88 (2010): 544–51.

Klein, S. et al. "Management of Newly Hatched Male Layer Chicks – Current
Knowledge on Sex Determination and Sex Diagnosis in Chickens: Potential
Solutions." *World's Poultry Science* 59 (2003): 62–4.

Klein, S. et al. "Sexing the Freshly Laid Egg – Development of Embryos after
Manipulation; Analytical Approach and Localization of the Blastoderm in
the Intact Egg." *World's Poultry Science* 59 (2003): 39–45.

Kolbehdari, D. et al. "A Whole-Genome Scan to Mao Quantitative Trait Loci
for Conformation and Functional Traits in Canadian Holstein Bulls." *Journal
of Dairy Science* 91 (2008): 2844–56.

Komender, P. "Crossbreeding in Farm Animals. III. A General Method of
Comparing Models to Estimate Crossbreeding Parameters with an Application
to Diallel Crossbreeding Experiments." *Journal of Animal Breeding and Genetics*
105 (1988): 362–71.

König, A. et al. "Economic Evaluation of Genomic Breeding Programs."
Journal of Dairy Science 92 (2009): 382–91.

Kuehn, L.A. et al. "Across-Breed EPD Tables for the Year 2009 Adjusted
to Breed Differences for Birth Year of 2007." In *Proceedings of the Beef
Improvement Federation, 41st Annual Research Symposium* (2009), 160–83.

Kyriazakis, I., and C.T. Wittemore. *Whittemore's Science and Practice of Pig
Production*. 3rd ed. Oxford: Blackwell Printing, 2007.

Laben, R.C. et al. "Some Effects of Inbreeding and Evidence of Heterosis
Through Outcrossing in a Holstein-Friesian Herd." *Journal of Dairy Science*
38 (1955): 525–35.

Lande, R. et al. "The Efficiency of Marker Assisted Selection in Dairy Cattle
Breeding Schemes." *Genetics* 124 (1990): 743–53.

Lasley, J.F. *Genetics of Livestock Improvement*, Englewood Cliffs, New Jersey:
Prentice-Hall, 1963.

Laurie, D.F. "Poultry Breeding in South Australia." *American Breeders' Magazine* 1 (1911): 52–60.

Lawrence, A.B. "Applied Animal Behaviour Science: Past, Present and Future Prospects." *Applied Animal Behaviour Science* 115 (2008): 1–24.

Leachman, L.D. "Combined Selection for the Beef Cattle Industry." In *Proceedings of the Beef Improvement Federation* (2009), 216–28.

Legarra, A. et al. "A Relationship Matrix Including Full Pedigree and Genomic Information." *Journal of Dairy Science* 92 (2009): 4656–63.

Lerner, I.M. "Lethal and Sublethal Characters in Farm Animals." *Journal of Heredity* 35 (1947): 219–24.

– *Population Genetics and Animal Improvement*. Cambridge: Cambridge University Press, 1950.

– *The Genetic Basis of Selection*. New York; John Wiley & Sons, 1958.

Lerner, I.M., and L.N. Hazel. "Population Genetics of a Poultry Flock under Artificial Selection." *Genetics* 32 (1947): 325–39.

Lerner, I.M., and H. Donald. *Modern Developments in Animal Breeding*. London and New York: Academic Press, 1966.

Lerner, I.M., and E.R. Dempster. "Some Aspects of Evolutionary Theory in the Light of Recent Work on Animal Breeding." *Evolution* 2 (1948) 19–28.

Levine, A.S. "'Agriculture' is not a Dirty Word." *Science* 324 (2009): 1140–1140.

Lewin, H.A. "It's a Bull Market." *Science* 324 (2009): 478–9.

Lewontin, R.C. "The Relevance of Molecular Biology to Plant and Animal Breeding." In *Proceedings of the International Conference on Quantitative Genetics*, edited by E. Pollak et al., 55–62. Ames: Iowa State University Press, 1977.

Lin, C.Y. et al. "Reproductive Performance of Crossline and Pureline Dairy Heifers." *Journal of Dairy Science* 71 (1984): 2420–8.

Liu, Y. et al. "Accounting for Heterogeneity of Variances to Improve the Precision of QTL Mapping in Dairy Cattle." *Animal Science Journal* 78 (2007): 371–7.

Loberg, A. et al. "Interbull Survey on the Use of Genomic Information." *Interbull Bulletin* 39 (2009): 3–14, Uppsala, Sweden.

Lopez-Villalobos, N. et al. "Effects of Selection and Crossbreeding Strategies on Industry Profit in the New Zealand Dairy Industry." *Journal of Dairy Science* 83 (2000): 164–72.

– "Profitabilities of Some Mating Systems for Dairy Herds in New Zealand." *Journal of Dairy Science* 83 (2000): 144–53.

– "Possible Effects of 25 Years of Selection and Crossbreeding on the Genetic Merit and Productivity of New Zealand Dairy Cattle." *Journal of Dairy Science* 83 (2000): 154–63.

Lucy, M.C. "Reproductive Loss in High-Producing Dairy Cattle: Where will it End." *Journal of Dairy Science* 84 (2001): 1277–93.

Lush, J. *Animal Breeding Plans*. Ames: Collegiate Press, 1937.

– "Dairy Cattle Genetics." *Journal of Dairy Science* 39 (1956): 693–4.

– "Family Merit and Individual Merit as Bases for Selection." *American Naturalist* 81 (1947): 241–61.

– "Genetics and Animal Breeding." In *Genetics in the 20th Century*, edited by L.C. Dunn, 493–525. New York: The MacMillan Company, 1951.

– *Notes on Animal Breeding*. Unpublished manuscript, 1933.

Lutaaya, E. et al. "Genetic Parameter Estimates from Joint Evaluation of Purebreds and Crossbreds in Swine Using the Crossbred Model." *Journal of Animal Science* 73 (2001): 3002–7.

Lynch, M., and B. Walsh. *Genetics and Analysis of Quantitative Traits*. Sunderland, MA: Sinauer Associates, 1998.

Lyons, L.A. "Comparative Genomics: The Next Generation." In *Biotechnology's Role in the Genetic Improvement of Farm Animals*, edited by R.H. Miller et al., 114–21. Beltsville Symposia in Agricultural Research, American Society of Animal Science, 1996.

MacArthur, D. "Face up to False Positives." *Nature* 487 (2012): 427–8.

Macciotta, N.P.P. et al. "Using Eigenvalues as Variance Priors in the Prediction of Genomic Breeding Values by Principal Component Analysis." *Journal of Dairy Science* 93 (2010): 2765–74.

Mackay, T.F.C. "Alan Robertson (1920–1989)." *Genetics* 125 (1990): 1–7.

– "Transposable Element-Induced Quantitative Genetic Variation in Drosophila." In *Proceedings of the Second International Conference on Quantitative Genetics*, edited by B.S. Weir, 219–35. Sunderland, MA: Sinauer Associates, 1988.

MacLeod, I.M. et al. "Power of a Genome Scan to Detect and Locate Quantitative Trait Loci in Cattle using Dense Single Nucleotide Polymorphisms." *Journal of Animal Breeding and Genetics* 127 (2010): 133–42.

Madalena, F.E. et al. "The Value of Sexed Bovine Semen." *Journal of Animal Breeding and Genetics* 121 (2004): 253–9.

Maher, B. "The Human Encyclopaedia." *Nature* 489 (2012): 46–8.

Mäki-Tanila, A. "Animal Breeding Further Ameliorated." *Journal of Animal Breeding and Genetics* 124 (2007): 1–2.

– "Seeing the Genomic Value of Markers." *Journal of Animal Breeding and Genetics* 123 (2006): 217.

Maltecca, C. et al. "Changes in Conception Rate, Calving Performance, and Calf Health and Survival from the Use of Cross-bred Jersey x Holstein Sires as Mates for Holstein Dams." *Journal of Dairy Science* 89 (2006): 2747–54.

Mardis, E.R. "A Decade's Perspective on DNA Sequencing Technology." *Nature* 470 (2011): 198–203.

Marie, J. "The Situation in Genetics II: Dunn's 1927 European Tour." *Mendel Newsletter* 13 (2004): 1–8.

Martin, D. "Special Report: Avian Sex Determination and Sex Diagnosis." *World's Poultry Science* 59 (2003): 1.

Matukumalli, L.K. et al. "Development and Characterization of a High Density SNP Genotyping Assay for Cattle." *PLoS One* 4 (2009). www.plosone.org.

Maxmen, M. "Politics Holds Back Animal Engineers." *Nature* 490 (2012): 318–19.

McAllister, A.J. "Are Today's Dairy Cattle Breeding Programs Suitable for Tomorrow's Production Requirements?" *Canadian Journal of Animal Science* 60 (1980): 253–64.

– "Is Crossbreeding the Answer to Questions of Dairy Breed Utilization?" *Journal of Dairy Science* 85 (2002): 2352–57.

McDowell, R.E. et al. "Interbreed Matings in Dairy Cattle. II. Herd Health and Viability." *Journal of Dairy Science* 51 (1968): 1275–83.

McHugh, N. et al. "Use of Female Information in Dairy Cattle Genomic Breeding Programs." *Journal of Dairy Science* 94 (2011): 4109–18.

McLaren, D.G. "Biotechnology Transfer: A Pig Breeding Company Perspective." *Animal Biotechnology* 3 (1992): 37–54.

Medrano, J.F. et al. "Dairy Cattle Breeding Simulation Program: A Simulation Program to Teach Animal Breeding Principles and Practices." *Journal of Dairy Science* 93 (2010): 2816–26.

Megens, H.J. et al. "Comparison of Linkage Disequilibrium and Haplotype Diversity on Macro-and Microchromosomes in Chickens." *BMC Genetics* 10 (2009). http://www.biomedcentral.com/1471-2156/10/86.

Meinert, T.R. "Usability for Genetic Evaluation of Records from Herds Participating in Progeny Test Programs of Artificial Insemination Organizations." *Journal of Dairy Science* 80 (1997): 2599–605.

Merilä, J. "It's the Genotype, Stupid!" *Journal of Animal Breeding and Genetics* 126 (2009): 1–2.

Merks, J.W.M., et al. "New Phenotypes for New Breeding Goals in Pigs." *Animal* 6 (2012): 535–43.

Meuwissen, T.H.E. "Genomic Selection: Marker Assisted Selection on a Genome Wide Scale." *Journal of Animal Breeding and Genetics* 124 (2007): 321–2.

Meuwissen, T.H.E. et al. "Prediction of Total Genetic Value Using Genome-wide Marker Maps." *Genetics* 157 (2001): 1819–29.

Meuwissen, T.H.E. et al. "Maximizing the Response of Selection with a Predefined Rate of Inbreeding: Overlapping Generations." *Journal of Animal Science* 76 (1998): 2575–83.

Miglior, F. et al. "Inbreeding of Canadian Holstein Cattle." *Journal of Dairy Science* 78 (1995): 1163–7.

Miglior, F. et al. "Overview of Different Breeding Objectives in Various Countries." In *Proceedings of the 11th World Holstein Friesian Federation, Session 4*, 7-11. 2005.

Miglior, F. et al. "Production Traits of Holstein Cattle: Estimation of Nonadditive Genetic Variance Components and Inbreeding Depression." *Journal of Dairy Science* 78 (1995): 1174–80.

Miller, P.D. "Artificial Insemination Organizations." *Journal of Dairy Science*, 64 (1981): 1283–7.

– "Implementation Technology for Genetic Improvement: Industry's View." *Journal of Dairy Science* 71 (1988): 1967–1971.

Mingay, E., ed. *Arthur Young and His Times*. London: The Macmillan Press, 1975.

Misztal, I. "Shortage of Quantitative Geneticists in Animal Breeding." *Journal of Animal Breeding and Genetics* 124 (2007): 255–6.

Misztal, I., et al. "Methods to Approximate Reliabilities in Single-step Genomic Evaluation." Journal of Dairy Science 93 (2013): 647-54.

Moore, S.S. "The Bovine Genome Sequence – Will it Live up to the Promise?" *Journal of Animal Science and Genetics* 126 (2009): 257–8.

Moreira, H.L.M. "Inaugural Editorial: Genomics and Quantitative Genetics, a new Journal." *Genomics and Quantitative Genetics* 1 (2010): 1–3.

Morin, P.A. et al. "SNPs in Ecology, Evolution, and Conservation." *Trends in Ecology and Evolution* 19 (2004): 208–16.

Morley, L.W. "Dairy Cattle Breed Associations." *Journal of Dairy Science* 39 (1956): 712–14.

Morse, G.B. "Poultry Pathology: Its Place in the Curriculum." *O.A.C. Review* 22 (1910): 246–51.

Morwick, E.Y. *The Chosen Breed: A Tale of Men, Women and the Canadian Holstein.* 2 vols. Hamilton: Seldon Griffin Graphics, 2002.

Moser, G. et al. "Accuracy of Direct Genomic Values in Holstein Bulls and Cows using Subsets of SNP Markers." *Genetics Selection Evolution* 42 (2010): 37–51.

Moser, G. et al. "A Comparison of Five Methods to Predict Genomic Breeding Values of Dairy Bulls from Genome-Wide SNP Markers." *Genetics Selection Evolution* 41 (2009): 56–61.

Mrode, R. et al. "Preliminary Evaluations of Production Traits in Mixed Breed Populations in the United Kingdom." *Interbull Bulletin* 38 (2008): 8–12, Uppsala, Sweden.

Mrode, R. et al. "Genetic Relationships between the Holstein Cow Populations of Three European Dairy Countries." *Journal of Dairy Science* 92 (2009): 5760–4.

Mrode, R. et al. "Understanding Genomic Evaluations from Various Evaluation Methods and GMACE." *Interbull Bulletin* 42 (2010): 52–5.

Muir, B. et al. "International Genomic Cooperation – North American Perspective." *Interbull Bulletin* 41 (2010): 71–6.

Muir, W.M. "Comparison of Genomic and Traditional BLUP-Estimated Breeding Value Accuracy and Selection Response under Alternative Trait and Genomic Parameters." *Journal of Animal Breeding and Genetics* 124 (2007): 342–55.

Muir, W.M. et al. "Review of the Initial Validation and Characterization of a 3K Chicken SNP Array." *World's Poultry Science* 64 (2008): 219–25.

Murray, B. "Dairy Crossbreeds – The Rare Breed." *Ontario Ministry of Agriculture, Food and Rural Affairs*. Reviewed June 2010. http://www.omafra .bov.on.ca/english/livestock/dairy/facts/info_breed.htm.

– *Dairy Genomic Selection*, Factsheet 12-023, Ministry of Agriculture, Food, and Rural Affairs, Ontario. 2012.

Nandi, S. et al. "Sex Diagnosis and Sex Determination." *World's Poultry Science* 59 (2003) 8–14.

Narrod, C.A., and K.O. Fuglie. "Private Investment in Livestock Breeding and Implications for Public Research Policy." *Agribusiness* 16 (2000): 457–70.

National Agricultural Statistics Service. "Overview of the United States Dairy Industry." USDA, 2010.

Nicholas, F.W. "Genetic Improvement through Reproductive Technology." *Animal Reproduction Science* 42 (1996): 205–14.

Nilforooshan, M.A. et al. "Validation of National Genomic Evaluations." *Interbull Bulletin* 42 (2010): 56–61.

Nomura, T. et al. "Optimization of Selection and Mating Schemes in Closed Broiler Lines." *Animal Science Journal* 73 (2002): 435–43.

Nordskog, A.W., L.T. Smith, and R.E. Philips. "Heterosis in Poultry: Crossbreds versus Top-Crossbreds." *Poultry Science* 38 (1959): 1372–80.

Norman, H.D. et al. "Overview of Progeny-Test Programs of Artificial-Insemination Organizations in the United States." *Journal of Dairy Science* 84 (2001): 1899–1912.

Norman, H.D. et al. "Reproductive Status of Holstein and Jersey Cows in the United States." *Journal of Dairy Science* 92 (2009): 3517–28.

Norman, H.D. et al. "Timeliness and Effectiveness of Progeny Testing Through Artificial Insemination." *Journal of Dairy Science* 86 (2003): 1513–25.

Ødegård, J. et al. "Identity-by-descent Genomic Selection using Selective and Sparse Genotyping", Genetics Selection Evolution 46 (2014): 3

Ollivier, L. "Current Principles and Future Prospects in Selection in Farm Animals." *Proceedings of the Second International Conference on Quantitative*

Genetics, edited by B.S. Weir, 438-49. Sunderland, MA: Sinauer Associates, 1988.

Olynk, N.J. et al. "Expected Net Present Value of Pure and Mixed Sexed Semen Artificial Insemination." *Journal of Dairy Science* 90 (2007): 2569–76.

Ontario Agricultural College, Ontario Department of Agriculture. "Better Bulls." *Bulletin 281*, 1920.

Ontario Agricultural College. "The Department of Poultry Husbandry." *News Bulletin*, 1959.

Orde, A., ed. *Mathew and George Culley: Travel Journals and Letters, 1765–1798*. Oxford: Oxford University Press, 2002.

Orozco, F., and J.L. Campo. "A Comparison of Purebred and Crossbred Genetic Parameters in Layers." *World's Poultry Science Journal* 30–31 (1974–75): 149–53.

Pace, M.M. "Implications of Sexed Semen on the Beef Industry." In *Proceedings of the 39th Annual Florida Beef Cattle Short Course*. 1990.

Pardue, S.L. "Educational Opportunities and Challenges in Poultry Science: the Impact of Resource Allocation and Industry Needs." Poultry Science 76 (1997): 938-43.

Pearl, R. "Inheritance of Hatching Quality of Eggs in Poultry." *American Breeders' Magazine* 1 (1911): 129–33.

Pearson, K. "A Mendelian's View of the Law of Ancestral Inheritance." *Biometrika* 3 (1904): 109–12.

– "On the Fundamental Conceptions of Biology." *Biometrika* 1 (1902): 320–44.

– "On the Systematic Fitting of Curves to Measurements and Observations: Part I." *Biometrika* 1 (1902): 265–303.

– "On the Systematic Fitting of Curves to Measurements and Observations: Part II." *Biometrika* 2 (1902): 1–23.

Pearson, P.B., and J.L. Lush. "A Linebreeding Program for Horse Breeding." *Journal of Heredity* 24 (1933): 185–91.

Pérez-Enciso, M. "Population and Quantitative Genetics for a Cause." *Journal of Animal Breeding and Genetics* 125 (2008): 217–18.

Perkel, J. "SNP Genotyping: Six Technologies that Keyed a Revolution." *Nature Methods* 5 (2008): 447–453.

Pims, K. "Experience with Crossbreeding – From Headaches to Happiness" In *Proceedings of the Fourth W.E. Petersen Symposium*, 20–2. Minneapolis: University of Minnesota, 2007.

Pirchner, F. *Population Genetics in Animal Breeding*. San Francisco: W.H. Freeman and Company, 1969.

– "Finding Genes Affecting Quantitative Traits in Domestic Animals." In *Proceedings of the Second International Conference on Quantitative Genetics*, edited by B.S. Weir, 243-9. Sunderland, MA: Sinauer Associates, 1988.

Phelps, P. et al. "Automated Identification of Male Layer Chicks Prior to Hatch." *World's Poultry Science* 59 (2003): 33–8.

Philipsson, J. et al. "Present and Future Use of Selection Index Methodology in Dairy Cattle." *Journal of Dairy Science* 77 (1994): 3252–61.

Plimsoll, S. *Cattle Ships*. London: Kegan Paul, Trench, Truber, & Co., 1890.

Poletto, R. et al. "The Five Freedoms in the Global Agriculture Market: Challenges and Achievements as Opportunities." *Animal Frontiers* 2 (2012): 22–30.

Pollak, E.J. "Keeping the 'Genetic Doors' Open between Canada and the U.S." In *Proceedings of the Beef Improvement Federation Conference* (2008), 1–3.

Pollak, E.J. et al., eds. *Proceedings of the International Conference on Quantitative Genetics*. Ames: Iowa State University Press, 1977.

Porto-Neto, L.R. et al. "The Extent of Linkage Disequilibrium in Beef Cattle Breeds using High-Gensity SNP Genotypes." Genetics Selection Evolution 46 (2014): 22.

Poultry Breeders of America. *Fact Finding* (and later *Proceedings*) *of the Annual National Poultry Breeders' Roundtable*. Unpublished documents, 1958–1969.

Powell, P.L. et al. "Accuracy of Foreign Dairy Bull Evaluations in Predicting United States Evaluations for Yield." *Journal of Dairy Science* 87 (2004): 2621–6.

Powell, P.L. et al. "International Dairy Bull Evaluations Expressed in National, Subglobal, and Global Scales." *Journal of Dairy Science* 85 (2002): 1863–8.

Powell, P.L. et al. "Major Advances in Genetic Evaluation Techniques." *Journal of Dairy Science* 89 (2006): 1337–48.

Powell, P.L. et al. "Progeny Testing and Selection Intensity for Holstein Bulls in Different Countries." *Journal of Dairy Science* 86 (2003): 3386–93.

Powledge, T.M. "U.S. Genome Sequencing Priorities Decided." *Genome Biology* 3 (2002). http://genomebiology.com/2002/3/5/spotlight-20020524-02.

– "High-Priority Genomes Announced." *Genome Biology* 3 (2002). http://genomebiology.com/2002/3/9/spotlight-20020524-02.

Preisinger, R. "Sex Determination in Poultry – A Primary Breeder's View." *World's Poultry Science* 59 (2003): 54–8.

Prendiville, R. et al. "A Comparison between Holstein-Friesian and Jersey Dairy Cows and Their F 1 Cross with Regard to Milk Yield, Somatic Cell Score, Mastitis, and Milking Characteristics under Grazing Conditions." *Journal of Dairy Science* 93 (2009): 2741–50.

Proceedings of the Animal Breeding and Genetic Symposium in Honor of Dr. Jay L. Lush. American Society of Animal Science and American Dairy Science Association, Blacksburg, Virginia, 1972.

Pursel, V.G. et al. "Status of Research with Transgenic Farm Animals." *Journal of Animal Science* 71 (1993 – Supplement 3): 10–19.

Rath, D. et al. "Improved Quality of Sex-sorted Sperm: A Prerequisite for Wider Commercial Application." *Theriogenology* 71 (2009): 22–9.

Rathje, T.A. "Strategies to Manage Inbreeding Accumulation in Swine Breeding Company Nuclear Herds: Some Case Studies." In *Proceedings of the Midwestern Section, American Society of Animal Science*, 1–8. 2000.

Reed, O.E. "Is the Crossbred Dairy Cow on Her Way?" *Country Gentleman*, June 1946.

Reeve, E. Review of *Proceedings of the Second International Conference on Quantitative Genetics*, by B.S. Weir et al. *Genetical Research* 53 (1989): 239–40.

Ren, W. et al. "Improved Flow Cytometric Sorting of X- and Y- Chromosome Bearing Sperm: Substantial Increase in Yield of Sexed Semen." *Molecular Reproduction and Development* 52 (1999): 50–6.

Rendel, J.M., and A. Robertson. "Estimation of Genetic Gain in Milk Yield by Selection in a Closed Herd of Dairy Cattle." *Journal of Genetics* 50 (1950): 1–8.

Report of the Committee on Agricultural Conditions. Government of Canada, 1924.

Report of the Ontario Commission on the Dehorning of Cattle. Ontario Government, 1892.

Report of the 245th Session of the Intergroup on the Welfare and Conservation of Animals. European Parliament, July 2008.

Research Committee on Animal Breeding. "Live-Stock Genetics." *Journal of Heredity* 6 (1915): 21–31.

Reynolds, L.P. "Perspectives: The Decline of Domestic Animal Research in Agriculture and Biomedicine." *Journal of Animal Science* 87 (2009): 4181–2.

Rice, J. "Report of the Committee on Breeding Poultry: Some Principles of Poultry Breeding." In *American Breeders Association*, 376–9. 1909.

Rice, V.A. "Crossbreeding – Blood from Turnips?" *Hoard's Dairyman*, 10 September 1946, 669.

Riesenbeck, A. et al. "Review of International Trade in Boar Semen." *Reproduction in Domestic Animals* 46 (2011): 1–3.

Rincon, G. et al. "Hot Topic: Performance of Bovine High-Density Genotyping Platforms in Holsteins and Jerseys." *Journal of Dairy Science* 94 (2011): 6116–21.

Roberts, R.M. "Farm Animal Research in Crisis." *Science* 324 (2009): 468–9.

Robertson, A. "Crossbreeding Experiments with Dairy Cattle (A Review)." *Animal Breeding Abstracts* 17 (1949): 201.

– "Inbreeding in Artificial Selection Programmes." *Genetic Research* 2 (1961): 189–94.

– "Molecular Biology and Animal Breeding." *Annales de Génétique et de Sélection animale* 2 (1970): 393–402.

Robertson, A., and J.M. Rendel. "The Use of Progeny Testing with Artificial Insemination in Dairy Cattle." *Journal of Genetics* 50 (1950): 21–31.

Robinson, O.W. et al. "Genetic Parameters of Body Size in Purebred and Crossbred Dairy Cattle." *Journal of Dairy Science* 63 (1980): 1887–99.

Rodenburg, T.B. et al. "The Role of Breeding and Genetics in the Welfare of Farm Animals." *Animal Frontiers* 2 (2012): 16–21.

Rodriguez, H. et al. "Semen Technologies in Domestic Animal Species." *Animal Frontiers* 3 (2013): 26-33.

Roff, D.A. "A Centennial Celebration for Quantitative Genetics." *Evolution* 61 (2007): 1017–32.

Rogers, G. "Variability by Crossbreeding." *Dairy Crossbreeding*, posted by *GENO Global*. Last modified 15 December 2008. http://www.dairycrossbreeding .com.

Sanders, A. *Red, White and Roan*. Chicago: American Shorthorn Breeders' Association, 1936.

– *Short-Horn Cattle, a Series of Historical Sketches, Memoirs and Records of the Breed and Its Development in the United States and Canada*. Chicago: Sanders Publishing, 1900.

Sanders, J.H. *The Breeds of Live Stock, and the Principles of Heredity*. Chicago: J.H. Sanders Publishing Company, 1887.

Schaeffer, P. et al. "Strategy for Applying Genome-Wide Selection in Dairy Cattle." *Journal of Animal Breeding and Genetics* 123 (2006): 218–23.

Schefers, J.M. et al. "Genomic Selection in Dairy Cattle: Integration of DNA Testing into Breeding Programs." *Animal Frontiers* 2 (2012): 4-9.

Schenkel, F.S. et al. "Reliability of Genomic Evaluation of Holstein Cattle in Canada." *Interbull Bulletin* 39 (2009): 51–8, Uppsala, Sweden.

Schuler, J. "Inserting Genes Affecting Quantitative Traits." In *Proceedings of the Second International Conference on Quantitative Genetics*, edited by B.S. Weir, 198–9. Sunderland, MA: Sinauer Associates, 1988.

Seath, D.M., and J.L. Lush. "'Nicking' in Dairy Cattle." *Journal of Dairy Science* 23 (1940): 103–13.

Sebright, Sir J.S. *The Art of Improving the Breeds of Domestic Animals*. London: John Harding, 1809.

Seeman, G. "Organizational Framework for Hatcheries." *World's Poultry Science* 59 (2003): 59–61.

Seidel, G.E. "Overview of Sexing Semen." *Theriogenology* 68 (2007): 443–6.

———. "Sperm Sexing Technology – The Transition to Commercial Application. An Introduction to the Symposium 'Update on Sexing Mammalian Sperm'." *Theriogenology* 71 (2009): 1–3.

Sessional Papers, Parliament of Canada.

Sessional Papers, Legislature of Ontario.

Sewalem, A. et al. "Analysis of Inbreeding and its Relationship with Functional Longevity in Canadian Dairy Cattle." *Journal of Dairy Science* 89 (2006): 2210–16.

"Sexed Semen Symposium: Applying Sexed Semen in Cattle." *Journal of Animal Science* 88 (2010) E-Supplement: 783–4.

Shaklee, W.E. "Federal-Grant Funds and Poultry Breeding Research in the United States." World's Poultry Science 29 (1972): 215-21.

Shook, G.E. et al. "Major Advances in Determining Appropriate Selection Goals." *Journal of Dairy Science* 89 (2006): 1349–61.

Shreffler, D.C. et al. "Effects of Crossbreeding on Rate of Growth in Dairy Cattle." *Journal of Dairy Science* 42 (1959): 607–20.

Shull, G.H. "What is 'Heterosis'?" *Genetics* 33 (1948): 439–46.

Siegel, P.B. "Progress from Chicken Genetics to the Chicken Genome." *Poultry Science* 85 (2006): 2050–60.

Shaw, T. *Animal Breeding*. Chicago: Orange Judd Company, 1901.

– *The Study of Breeds*. Chicago: Orange Judd Company, 1900.

Shrader, I.I.L. "The Chicken-of-Tomorrow Program: Its Influence on 'Meat-Type' Poultry Production." *Poultry Science* 31 (1952): 3–10.

Skinner, J.L. "150 Years of the Poultry Industry." *World's Poultry Science Journal* 30–31 (1974–5): 27–31.

Slate, J. et al. "Gene Mapping in the Wild with SNPs: Guidelines and Future." *Genetica* 136 (2009): 97–107.

Slocum, R.R. "Poultry Breeding." *Journal of Heredity* 6 (1915): 483–7.

– "The Value of the Poultry Show." In *Report*, Bureau of Animal Industry, 357–63. USDA, United States Department of Agriculture, 1908.

Smith, C. "Potential for Animal Breeding, Current and Future." In *Proceedings of the Second International Conference on Quantitative Genetics*, edited by B.S. Weir, 150–60. Sunderland, MA: Sinauer Associates, 1988.

Smith, C. et al. "The Use of Genetic Polymorphisms in Livestock Improvement." *Journal of Animal Breeding and Genetics* 103 (1986): 205–17.

Smith, G.C. et al. "Trends in Value-Added Marketing: Challenges and Similarities Among Branded Beef Products; Importance of Animal Welfare." In *Proceedings of the Beef Improvement Federation Conference* (2008), 18–20.

Smith, L.A. et al. "The Effects of Inbreeding on the Lifetime Performance of Dairy Cattle." *Journal of Dairy Science* 81 (1998): 2729–37.

Smouse, P. Review of *Proceedings of the International Conference on Quantitative Genetics*, by E. Pollak et al. *American Journal of Human Genetics* 31 (1979): 754–5.

Snelling, W.M. et al. "Genome-Wide Association Study of Growth in Crossbred Beef Cattle." *Journal of Animal Science* 88 (2010): 837–48.

Snyder, E.S. *A History of the Poultry Science Department at the Ontario Agricultural College, 1894–1968*. Unpublished manuscript, 1970.

Soller, M. "Marker-Assisted Selection – An Overview." *Animal Biotechnology* 5 (1994): 193–207.

Soller, M. et al. "Genomic Genetics and the Utilization for Breeding Purposes of Genetic Variation Between Populations." In *Proceedings of the Second International Conference on Quantitative Genetics*, edited by B.S. Weir, 161–81. Sunderland, MA: Sinauer Associates, 1988.

Sorensen, A.C. et al. "Udder Health Shows Inbreeding Repression in Danish Holsteins." *Journal of Dairy Science* 89 (2006): 4077–82.

Sorensen, D. "Developments in Statistical Analysis in Quantitative Genetics." *Genetica* 136 (2009): 319–32.

Sorensen, M.K. "Crossbreeding – An Important Part of Sustainable Breeding in Dairy Cattle and Possibilities for Implementation." In *Proceedings of the Fourth W.E. Petersen Symposium*, 29–40. Minneapolis: University of Minnesota, 2007.

Speidel, S. "Genetic Analysis of Longitudinal Data in Beef Cattle." In *Proceedings of the Beef Improvement Federation, 41st Annual Research Symposium* (2009), 229–49.

Springer, N.M., and R.M. Stupar. "Allelic Variation and Heterosis in Maize: How do Two Haves Make More than a Whole?" *Genome Research* 17 (2007): 264–75.

Stachowicz, K. et al. "Rates of Inbreeding and Genetic Diversity in Canadian Holstein and Jersey Cattle." *Journal of Dairy Science* 94 (2011): 5160–75.

Stella, A. et al. "Strategies for Continual Application of Marker-Assisted Selection in an Open Nucleus Population." *Journal of Dairy Science* 85 (2002): 2358–67.

Stinchcombe, J.R. et al. "Combining Population Genomics and Quantitative Genetics: Finding the Genes Underlying Ecologically Important Traits." *Heredity* 100 (2008): 158–70.

Stolzman, M. et al. "Friesian Cattle in Poland, Preliminary Results of Testing Different Strains." *World Animal Review* 38 (1981): 9–15.

Stoltzus, A. and K. Cable. "Mendelism-Mutationsim: the Forgotten Evolutionary Synthesis." Journal of the History of Biology (May 9, 2014). http://link.springer.com/article/10.1007%2Fs10739-014-9383-2.

Strauss, S. "Biotech Breeding Goes Bovine." *Nature Biotechnology* 28 (2010): 540–3.

Sturtevant, A.H. *A History of Genetics*. New York: Cold Spring Harbor Laboratory Press, 2001. First published 1964 by Harper and Row.

Suchocki, T. et al. "Testing Candidate Gene Effects on Milk Production in Dairy Cattle under Various Parameterizations and Modes of Inheritance." *Journal of Dairy Science* 93 (2010): 2703–17.

Swan, A.A. et al. "Evaluation and Exploitation of Crossbreeding in Dairy Cattle." *Journal of Dairy Science* 75 (1992): 624–39.

Tallman, R.M. et al. "Estimation of the Proportion of Genetic Variation Accounted for by DNA Tests." In *Proceedings of the Beef Improvement Federation, 41st Annual Research Symposium* (2009), 184–209. Tallman, R.M. et al. "High-Density SNP Scan of Production and Product Quality Traits in Beef Cattle." *Journal of Animal Science* 87 (2009): E-Supplement.

Tallman, R.M. et al. "Proposed Strategy for Selection against Recessive Genetic Defects through a Combination of Inbreeding and DNA Markers." In *Proceedings of the Beef Improvement Federation Conference* (2009), 137–44.

Taylor, J.F. et al. "Net Present Value and Economic Merit of Sexed Semen and Splitting Units of Semen for Australian Holsteins." *Journal of Dairy Science* 71 (1988): 3100–11.

Tellam, R.L. "Unlocking the Bovine Genome." *BMC Genomics* 10 (2009). http://biomedcentral.com/1471-2164/10/193.

Termohlen, W.D. "Past History and Future Developments." *Poultry Science* 47 (1968): 6–22.

Tess, M. "Emerging Technologies in Genetic Improvement – Convergence of Quantitative and Molecular Tools." In *Proceedings of the Beef Improvement Federation Conference* (2008), 4–10.

Thaxton, Y.V. et al. "The Decline of Academic Poultry Science in the United States of America." World's Poultry Science 59 (2003): 303-13.

Tier, B. "Genomic Selection: Promises and Priority." *Journal of Animal Breeding and Genetics* 127 (2010): 169–70.

Tiersch, T.R. "Identification of Sex in Chickens by Flow Cytometry." *World's Poultry Science* 59 (2003): 25–32.

Tixier-Boichard, M. et al. "A Century of Poultry Genetics." *World's Poultry Science* 68 (2012): 307–21.

Toosi, A. et al. "Genomic Selection in Admixed and Crossbred Populations." *Journal of Animal Science* 88 (2010): 32–46.

Toronto International Data Release Workshop Authors. "Publication Data Sharing." *Nature* 461 (2000): 168–70.

Torsell, A. "Prospects of Performing Multiple-Country Comparison of Dairy Sires for Countries not Participating in Interbull International Generic Evaluations." MSc thesis, Swedish University of Agricultural Sciences, Uppsala, 2007. www.ex-epsilon.slu.se:8080/archive/00001885.

Touchberry, R.W. "A Comparison of the General Merits of Purebred and Crossbred Dairy Cattle Resulting from Twenty Years (Four Generations) of Cows Breeding." *National Breeders Roundtable*, Nineteenth Annual Session, 1970, 18–63.

– "Crossbreeding Effects in Dairy Cattle: The Illinois Experiment, 1949-1969." *Journal of Dairy Science* 75 (1992): 640–667.

Touchberry, R.W. et al. "Body Weight Changes in Lactating Purebred and Crossbred Cattle." *Journal of Dairy Science* 59 (1976): 733–43.

Tran, H.T. et al. "An Estrogen Sensor for Poultry Sex Sorting." *Journal of Animal Science* 88 (2010): 1358–64.

Trout, G.M. "Fifty Years of the American Dairy Science Association." *Journal of Dairy Science* 39 (1956): 625–50.

Troyer, A.F. "Adaptedness and Heterosis in Corn and Mule Hybrids." *Crop Science* 46 (2006): 528–43.

Tyler, W.J. et al. "Growth and Production of Inbred and Outbred Holstein-Friesian Cattle. *Journal of Dairy Science* 32 (1940): 247–56.

USDA. "Use of Single Nucleotides Polymorphisms to Verify Parentage." In *Animal Improvement Programs, Report for 2008*. Accessed 20 July 2009. http:// www.ars.usda.gov/research/projects.htm?ACCN_NO=.

VanDoormaal, B. "Genetic Evaluation of Dairy Cattle in Canada." Canadian Dairy Network, 2007.

– "Value of DHI: Today and in the Future." Canadian Dairy Network, 2011.

Van Eenennaam, A.L. et al. "Validation of Commercial DNA Tests for Quantitative Beef Quality Traits." *Journal of Animal Science* 85 (2007): 891–900.

VanRaden, P.M. "Accounting for Inbreeding and Crossbreeding in Genetic Evaluation of Large Populations." *Journal of Dairy Science* 75 (1992): 3136–44.

– "Integrating Use of New Markers into the Real World." In *Biotechnology's Role in the Genetic Improvement of Farm Animals*, edited by R.H. Miller et al., 307–14, Beltsville Symposia in Agricultural Research, American Society of Animal Science, 1996.

– "Merits of Purebred and Crossbred Dairy Cows for Productive Life." Abstract. *Journal of Dairy Science* 78, Supplement 1 (1995): 159.

– "Why we Don't Patent." *Journal of Animal Breeding and Genetics* 126 (2009): 91.

VanRaden, P.M. et al. "Combining Different Marker Densities in Genomic Evaluation." *Interbull Bulletin* 42 (2010): 113–117, Uppsala, Sweden.

– "Economic Merit of Crossbred and Purebred U.S. Dairy Cattle." *Journal of Dairy Science* 86 (2003): 1036–44.

– "Reliability of Genomic Predictions for North American Holstein Bulls." *Journal of Dairy Science* 92 (2009): 16–24.

– "Genetic Evaluation for Cow Fertility." *Journal of Dairy Science* 87 (2004): 2285–92.

– "Genetic Evaluation Using Combined Data from All Breeds and Crossbred Cows." In *Proceedings of the Fourth W.E. Petersen Symposium*, 23–8. Minneapolis: University of Minnesota, 2007.

- "Genomic Measures of Relationship and Inbreeding." *Interbull Bulletin* 37 (2007): 33–36. http://journal.interbull.org/index.php/ib/issue56.
- "Selection and Mating Considering Expected Inbreeding of Future Progeny." *Journal of Dairy Science* 82 (1999): 2771–8.
- "Genetic Evaluations for Mixed-Breed Populations." *Journal of Dairy Science* 90 (2007): 2434–41.
- "International Genomic Evaluation Methods for Dairy Cattle." *Genetics Selection Evolution* 42 (2010): 7–15.
- "Combining Different Marker Densities in Genomic Evaluation." *Interbull Bulletin* 42 (2010): 113–18, Uppsala, Sweden.
Van Reenen, C.G. et al. "Transgenesis May Affect Farm Animal Welfare: A Case for Systematic Risk Assessment." *Journal of Animal Science* 79 (2001): 1763–79.
Van Tassell, C.P. et al. "SNP Discovery and Allele Frequency Estimation by Deep Sequencing of Reduced Representation Libraries." *Nature Methods* 5 (2008): 247–52.
VanVleck, L.D. "Observations on Selection Advances in Dairy Cattle." In *Proceedings of the Second International Conference on Quantitative Genetics*, edited by B.S. Weir, 433–7. Sunderland, MA: Sinauer Associates, 1988.
- "Potential Impact of Artificial Insemination, Sex Selection, Embryo Transfer, Cloning, and Selfing in Dairy Cattle." In *New Technologies in Animal Breeding*, edited by B.G. Brackett et al., 221–42. New York: Academic Press, 1981.
- "Sampling the Young Sire in Artificial Insemination." *Journal of Dairy Science* 47 (1964): 441–6.
VanVleck, L.D. et al. "Genetic Value of Sexed Semen to Produce Dairy Heifers." *Journal of Dairy Science* 59 (1976): 1802–7.
Venter, J.C. "The Sequence of the Human Genome." *Science* 291 (2001): 1304–51.
Verley, F.A., et al. "Effects of Crossbreeding on Reproductive Performance of Dairy Cattle." *Journal of Dairy Science*, 44 (1961): 2058–67.
Verley, F.A. et al. "Reproductive Performance of Crossbred and Purebred Dairy Cows." *Journal of Dairy Science* 69 (1986): 518–526.
Vignal, A. et al. "A Review of SNP and other Types of Molecular Markers and Their Use in Animal Genetics." *Genetics Selection Evolution* 34 (2002): 275–305.
Vishwanath, R. "Artificial Insemination: the State of the Art." *Theriogenology* 59 (2003): 571–84.
Visscher, P.M. "Whole Genome Approaches to Quantitative Genetics." *Genetica* 136 (2009): 351–8.
Voelker, D.E. "Dairy Herd Improvement Associations." *Journal of Dairy Science* 64 (1981): 1269–77.

von Keyserlingk, M.A.G. et al. "Invited Review: The Welfare of Dairy Cattle – Key Concepts and the Role of Science." *Journal of Dairy Science* 92 (2009): 4101–11.

Wahlberg, P. "Chicken Genomics – Linkage and QTL Mapping." Digital Comprehensive Summaries of Uppsala Dissertations from the Faculty of Medicine 409, 2009.

Wall, E.S. et al. "Effects of Including Inbreeding, Heterosis and Recombination Loss in Precition of Breeding Values for Fertility Traits." *Interbull Bulletin* 31 (2003): 17–21, Uppsala, Sweden.

Wall, R.J. et al. "Are Animal Models as Good as We Think?" *Theriogenology* 69 (2008): 2–9.

Walsh, B. "Quantitative Genetics, Version 3.0: Where Have We Gone Since 1987 and Where Are We Headed?" *Genetica* 136 (2009): 213–223.

Walsh, B. et al. *Advanced Topics in Breeding and Evolution*. Vol. 2 of *Genetics and Analysis of Quantitative Traits*. Last modified May 15, 2011. http://nitro .biosci.arizona.edu/zbook/NewVolume_2/newvol2.html.

– "Quantitative Genetics in the Age of Genomics." *Theoretical Population Biology* 59 (2001): 175–84.

Warren, D.C. "A Half Century of Advances in the Genetics and Breeding Improvement of Poultry." *Poultry Science* 37 (1958): 3–20.

– *Practical Poultry Breeding*, New York: The Macmillan Company, 1953.

– "Techniques of Hybridization of Poultry." *Poultry Science* 29 (1950): 59–63.

Watson, J.D., and F.H.C. Crick. "A Structure for Deoxyribose Nucleic Acid." *Nature* 171 (1953): 737–8.

Weigel, K.A. "Exploring the Role of Sexed Semen in Dairy Production Systems." *Journal of Dairy Science* 87 (2004): E120–30.

– "Crossbreeding: A Dirty Word or an Opportunity?" *Western Dairy Management Conference*, 189–2002. 2005.

Weigel, K.A. et al. "Potential Gains in Lifetime Net Merit from Genomic Testing of Cows, Heifers, and Calves on Commercial Dairy Farms." *Journal of Dairy Science* 95 (2012): 2215–25.

Weigel, K.A. et al. "Results of a Producer Survey Regarding Crossbreeding on U.S. Dairy Farms." *Journal of Dairy Science* 86 (2003): 4148–54.

Weigel, K.A. et al. "Prediction of Unobserved Single Nucleotide Polymorphism Genotypes of Jersey Cattle Using Reference Panels and Population-Based Imputation Algorithms." *Journal of Dairy Science* 93 (2010): 2229–38.

Weir, B.S. "Quantitative Genetics in 1987." *Genetics* 117 (1987): 601–2.

Weir, B.S. et al., eds. *Proceedings of the Second International Conference on Quantitative Genetics*. Sunderland, MA: Sinauer Associates, 1988.

Weir, B.S. et al. "The Third International Conference of Quantitative Genetics."
    Genetica 136 (2009): 211–12.
Weldon, W.F.R. "Mendel's Laws of Alternative Inheritance in Peas." *Biometrika*
    1 (1902): 228–54.
Weller, J.I. "Introduction to QTL Detection and Marker-Assisted Selection."
    In *Biotechnology's Role in the Genetic Improvement of Farm Animals*, edited
    by R.H. Miller et al., 259–75. Beltsville: Beltsville Symposia in Agricultural
    Research, American Society of Animal Science, 1996.
–   *Quantitative Trait Loci Analysis in Animals.* New York: CABI Publishing, 2001.
Weller, J.I. et al. "Power of Daughter and Grandmother Designs for
    Determining Linkage Between Marker Loci and Quantitative Trait Loci in
    Dairy Cattle." *Journal of Dairy Science* 73 (1990): 2525–37.
Weller, J.I. et al. "A Simple Method for Genomic Selection of Moderately Sized
    Dairy Cattle Populations." *Animal* 6 (2012): 193–202.
Wentworth, Lady (Judith Anne Blunt). *The Authentic Arabian Horse.* London:
    George Allen & Unwin, 1945.
–   *Thoroughbred Racing Stock.* New York: Charles Scribner's Sons, 1938.
Werf, van der, J. "Animal Breeding and the Black Box of Biology." *Journal of
    Animal Breeding and Genetics* 124 (2007): 101.
White, J.M. et al. "Dairy Cattle Improvement and Genetics." *Journal of Dairy
    Science* 64 (1981): 1305–17.
Wiggans, G.R. "Selection of Single-Nucleotide Polymorphisms and Quality of
    Genotypes Used in Genomic Evaluation of Dairy Cattle in the United States
    and Canada." *Journal of Dairy Science* 92 (2009): 3431–6.
Wiggans, G.R. et al. "The Genomic Evaluation System in the United States:
    Past, Present, Future." *Journal of Dairy Science* 94 (2011): 3202–11.
Wiggans, G.R. et al. "Selection and Management of DNA Markers for Use in
    Genomic Evaluation." *Journal of Dairy Science* 93 (2010): 2287–92.
Wiggans, G.R. et al. "Use of the Illumina Bovine3K BeadChip in Dairy
    Genomic Evaluation." *Journal of Dairy Science* 95 (2012): 1552–8.
Willham, R.L., et al. "Theory of Heterosis." *Journal of Dairy Science* 68 (1985):
    2411–17.
Williams, A.R. et al. "Two-, Three- and Four-Breed Rotational Crossbreeding of
    Beef Cattle: Reproductive Traits." *Journal of Animal Science* 68 (1990): 1536–46.
Williams, J.L. "The Use of Marker-Assisted Selection in Animal Breeding and
    Biotechnology." *Revue Scientifique et Technique – Office International* 24 (2005):
    379–91.
Williams, J.L. et al. "Estimation of Breed and Heterosis Effects for Growth and
    Carcass Traits in Cattle using Established Crossbreeding Studies." *Journal of
    Animal Science* 88 (2010): 460–6.

Wheeler, M.B. "Production of Transgenic Livestock: Promise Fulfilled." *Journal of Animal Science* 81 (2003): 32–7.

Wolf, C.A. "Custom Dairy Heifer Grower Industry Characteristics and Contract Terms." *Journal of Dairy Science* 86 (2003): 3016–22.

Wolfova, M. et al. "Selection for Profit in Cattle: II. Economic Weights for Dairy and Beef Sires in Crossbreeding Systems." *Journal of Dairy Science* 90 (2007): 2456–67.

Woodward, T.E. et al. "Results of Inbreeding Grade Holstein-Friesian Cattle." In *Technical Bulletin* 927. USDA, 1946.

World Holstein Friesian Federation. *Newsletter* 3 (2010). www.whff.info/_pdf/newsletter/WHFF_newsletter03_2010-06.pdf.

Wriedt, C. *Heredity in Live Stock*. London: Macmillan and Co., 1930.

– "Formalism in Breeding of Livestock in Relation to Genetics." *Journal of Heredity* 16 (1925): 19–24.

– "The Inheritance of Butterfat Percentage in Crosses of Jerseys with Red Danes." *Journal of Heredity* 22 (1930): 45–63.

Wright, S. "Coefficients of Inbreeding and Relationship." *American Naturalist* 56 (1922): 330–8.

– "Experimental Results and Evolutionary Deductions." In *Evolution and the Genetics of Populations*, vol.3. Chicago: University of Chicago Press, 1977.

– "Mendelian Analysis of the Pure Bred Breeds of Livestock, Part 1: The Measurement of Inbreeding and Relationship." *Journal of Heredity* 14 (1923): 339–48.

– "Mendelian Analysis of the Pure Bred Breeds of Livestock, Part 2: The Duchess Family of Shorthorns as Bred by Thomas Bates." *Journal of Heredity* 14 (1923): 405–22.

– "The Genetic Theory of Natural Selection." *Journal of Heredity* 21 (1930): 349–56.

– "The Relation of Livestock Breeding to Theories of Evolution." *Journal of Animal Science* 46 (1978): 1192–200.

– *Systems of Mating and Other Papers*. Ames: Iowa State College Press, 1958. [Reprint of "Systems of Mating." *Genetics* 6 (1921): 111–78; "Evolution in Mendelian Populations." *Genetics* 16 (1931): 97–159; "Correlation and Causation." *Journal of Agricultural Research* 20 (1921): 557–585; "The Method of Path Coefficients." *Annals of Mathematic Statistics* 5 (1934): 161–215].

Young, C.W. et al. "Estimates of Inbreeding and Relationship Among Registered Holstein Females in the United States." *Journal of Dairy Science* 79 (1996): 502–5.

Zarnecki, H.D. et al. "Heterosis for Growth and Yield Traits from Crosses of Friesian Strains." *Journal of Dairy Science* 76 (1993): 1661–70.

Farm Journals.

Agricultural Gazette of Canada.
American Poultry Journal
Breeder's Gazette
Canadian Breeder and Agricultural Review
Canadian Dairyman and Farming World
Canadian Live-Stock and Farm Journal
Canadian Poultry Chronicle
Canada Poultry Journal
Canadian Poultry Review
Farmer's Advocate
Farming
Farming World and Canadian Farm and Home
Farming World for Farmers and Stockmen
Hoard's Dairyman
Holstein World
Ontario Beef Farmer
Shorthorn Country

Secondary Sources

Books

Abott, H. "The Marketing of Livestock in Canada." MA thesis, University of Toronto, 1923.

Anderson, V.D. *Creatures of Empire: How Domestic Animals Transformed Early America*. Oxford: Oxford University Press, 2004.

Ash, H.B., revised by. *Marcus Porcius Cato on Agriculture; Marcus Terentius Varro on Agriculture*. Translated by W.D. Hooper. 1934. Reprint, Cambridge: Harvard University Press, 1960.

Atack, J., and F. Bateman. *To Their Own Soil: Agriculture in the Antebellum North*. Ames: Iowa State University Press, 1987.

Beauchamp, T.L. et al. *The Human Use of Animals: Case Studies in Ethical Practice*. Oxford: Oxford University Press, 2008.

Bidwell, P.W., and J.L. Falconer. *History of Agriculture in the Northern United States*. Washington: Carnegie Institute, 1925.

Bogue, A. *From Prairie to Corn Belt: Farming on the Illinois and Iowa Prairies in the Nineteenth Century*. Chicago: University of Chicago Press, 1963.

Bowler, P.J. *The Eclipse of Darwinism, Anti-Darwinism Evolutionary Theories in the Decades around 1900*. Baltimore: The Johns Hopkins University Press, 1983.
– *The Non-Darwinian Revolution, Reinterpreting a Historical Myth*. Baltimore: The Johns Hopkins University Press, 1988.
– *The History of an Idea*. Berkeley: University of California, 1984

Brackett, B.G. et al., eds. *New Technologies in Animal Breeding*. New York: Academic Press, 1981.

Bulmer, M. *Francis Galton: Pioneer of Heredity and Biometry*. Baltimore: The Johns Hopkins University Press, 2003.

Cassidy, R. *The Sport of Kings: Kinship, Class and Thoroughbred Breeding in Newmarket*. Cambridge: Cambridge University Press, 2002.

Chapman, A.B., ed. *General and Quantitative Genetics*. Amsterdam: Elsevier Science Publishers, 1985.

Coleman, W., and C. Limoges, eds. *Studies in History of Biology*. Baltimore: The Johns Hopkins University Press, 1979

Comstock, R.E. *Quantitative Genetics with Special Reference to Plant and Animal Breeding*. Ames: Iowa State University Press, 1996.

Crawford, R.D., ed. *Poultry Breeding and Genetics*. New York: Elsevier, 1990.

Crow, J.F. *Basic Concepts in Population, Quantitative, and Evolutionary Genetics*. New York: W.H. Freeman and Company, 1986.

Cunningham, E.P. *Quantitative Genetic Theory and Livestock Improvement*. University of New England, 1979.

Derry, M.E. *Art and Science in Breeding: Creating Better Chickens*. Toronto: University of Toronto Press, 2012.
– *Bred for Perfection: Shorthorn Cattle, Collies, and Arabian Horses since 1800*. Baltimore: The Johns Hopkins University Press, 2003.
– *Horses in Society: A Story of Breeding and Marketing Culture, 1800–1920*. Toronto: University of Toronto Press, 2006.
– *Ontario's Cattle Kingdom: Purebred Breeders and Their World, 1870–1920*. Toronto: University of Toronto Press, 2001.

Dreyer, P. *A Gardener Touched with Genius*. Los Angeles: University of California Press, 1985.

Dunn, L.C. *A Short History of Genetics: The Development of Some of the Main Lines of Thought, 1864–1939*. New York: McGraw-Hill, 1965.

Dunn, L.C., ed. *Genetics in the 20th Century*. New York: The MacMillan Company, 1951.

Ensminger, E.M. *Beef Cattle Science*. Danville: Interstate Printers and Publishers, 1987.

Falconer, D.S. *Introduction to Quantitative Genetics*. Edinburgh: Oliver and Boyd, 1960.

Falk, R. *Genetic Analysis: A History of Genetic Thinking*. Cambridge: Cambridge University Press, 2009

Fitzgerald, D. *The Business of Breeding: Hybrid Corn in Illinois, 1890–1940*. Ithaca: Cornell University Press, 1990.

– *Every Farm a Factory*. New Haven: Yale University Press, 2003.

Freeman, S., and J.C. Herron. *Evolutionary Analysis*. Upper Saddle River, NJ: Prentice Hall, 1998.

Frolov, I.T. *Philosophy and History of Genetics: the Inquiry and the Debates*. London: Macdonald, 1991.

Fudge, E. *Perceiving Animals: Humans and Beasts in Early Modern English Culture*. Basingstoke: MacMillan Press, 2000.

Futuyma, D.F. *Evolutionary Biology*, 3rd ed. Sunderland, MA: Sinauer Associates, 1998.

Gaudillière, Jean-Paul et al., eds. *From Molecular Genetics to Genomics: the Mapping Cultures of the Twentieth Century*. London: Routledge, 2004.

Gillham, N.R. *A Life of Sir Francis Galton: From African Exploration to the Birth of Eugenics*. Oxford: Oxford University Press, 2001

Glass, B., O. Temkin, and W.L. Straus Jr., eds. *Forerunners of Darwin, 1745–1859*. Baltimore: The Johns Hopkins University Press, 1951.

Goodall, D.M. *A History of Horse Breeding*. London: Robert Hall, 1977.

Green, E.L. *Genetics and Probability in Animal Breeding Experiments*. London: MacMillan, 1981.

Grothe, P.O. *Holstein Friesian: A Global Breed*. Netherlands: Misset Press, 1993 (English version 1994).

Hanke, O.A., ed. *American Poultry History, 1823–1973*. Madison, WI: American Poultry History Society, 1974.

Harman, O., and M.R. Dietrich, eds. *Rebels, Mavericks, and Heretics in Biology*. New Haven: Yale University Press, 2008.

Harrison, R. *Animal Machines: The New Factory Farming Industry*. London: V. Stuart, 1964.

Hartl, D.L. *A Primer of Population Genetics*. Sunderland, MA: Sinauer Associates, 1981.

Harwood, J. *Styles of Scientific Thought: The German Genetics Community, 1900–1933*. Chicago: University of Chicago Press, 1992.

– *Technology's Dilemma: Agricultural Colleges between Science and Practice in Germany, 1860–1934*. New York: Peter Lang Publishing Group, 2005.

Hemsworth, P.H., and G.J. Coleman. *Human-Livestock Interactions: The Stockperson and the Productivity and Welfare of Intensively Farmed Animals*. London: CAB International, 1998.

Herman, H.A. *Improving Cattle by the Millions: NAAB and the Development and Worldwide Application of Artificial Insemination*. Columbia: University of Missouri Press, 1981.

Innis, H., ed. *The Dairy Industry in Canada*. Toronto: Ryerson Press, 1937.

Introduction to Molecular Genetics and Genomics. www.bio-nica.info/biblioteca/AnonimoxxxIntroductionMolecularGEnetics.pdf

Johnson, P.C. *Farm Animals in the Making of America*. Des Moines: Wallace Homestead Book Company, 1975.

Jones, N. *The Plimsoll Sensation: the Great Campaign to Save Lives at Sea*. London: Little, Brown, 2006.

Jones, R. *History of Agriculture in Ohio to 1880*. Kent: Kent State University Press, 1983.

Jones, S.D. *Death in a Small Package: A Short History of Anthrax*. Baltimore: The Johns Hopkins University Press, 2010.

– *Valuing Animals: Veterinarians and Their Patients in Modern America*. Baltimore: The Johns Hopkins University Press, 2003.

Jordan, T. *North American Cattle Ranching Frontiers: Origins, Diffusion, and Differentiation: Histories of the American Frontier*. Albuquerque: University of New Mexico Press, 1993.

Kingsland, S.E. *Modelling Nature: Episodes in the History of Population Ecology*. Chicago: University of Chicago Press, 1995.

Kitcher, P. *In Mendel's Mirror*. Oxford: Oxford University Press, 2003.

Kloppenburg, J.R. *First the Seed: The Political Economy of Plant Biotechnology, 1492–2000*. Cambridge: Cambridge University Press, 1988.

Kuhn, T.S. *The Essential Tension*. Chicago: University of Chicago Press, 1977.

– *The Structure of Scientific Revolutions*, 4th ed. Chicago: University of Chicago Press, 2012.

Lampard, E.E. *The Rise of the Dairy Industry in Wisconsin: A Study of Agricultural Change, 1820–1920*. Madison: State Historical Society of Wisconsin, 1963.

MacEwan, G. *Highlights of Shorthorn History*. Winnipeg: Hignill Printing, 1982.

Mansfield, R.H. *Progress of the Breed: The History of US. Holsteins*. Holstein-Friesian World, 1985.

Marcus, A.I. *Agricultural Science and the Quest for Legitimacy: Farmers, Agricultural Colleges, and Experiment Stations, 1870–1890*. Ames: Iowa State University Press, 1985.

Marie, J. "The Importance of Place: A History of Genetics in 1930s Britain." PhD diss., University College London, 2004.

Matz, B. "Crafting Heredity: The Art and Science of Livestock Breeding in the United States and Germany, 1860-1914." PhD diss., Yale University, December 2011.

Mayr, E., and W.B. Provine. *The Evolutionary Synthesis: Perspectives on the Unification of Biology*. Cambridge: Harvard University Press, 1980

Mazumdar, P.M.H. *Eugenics, Human Genetics and Human Failings: The Eugenics Society, Its Sources and Critics in Britain*. London: Routledge, 1992.

– *Species and Specificity: An Interpretation of the History of Immunology*. Cambridge: Cambridge University Press, 1995

McCook, S. *States of Nature: Science, Agriculture, and the Environment in the Spanish Caribbean, 1760–1940*. Austin: University of Texas Press, 2002.

McCormick, V. *Farm Wife: A Self-Portrait, 1886–1896*. Ames: Iowa State University, 1990.

Mcelheny, V.K. *Watson and DNA: Making a Scientific Revolution*. Cambridge, MA: Persus Publishing, 2003.

McMurry, S. *Transforming Rural America: Dairying Families and Agricultural Change, 1820–1885*. Baltimore: The Johns Hopkins University Press, 1995.

Mingay, E., ed. Arthur Young and His Times. London: Macmillan Press, 1975.

Montcrieff, E., S. Joseph, and I. Joseph. *Farm Animal Portraits*. Woodbridge, UK: Antique Collectors' Club, 1996.

Olby, R. *Origins of Mendelism*. 2nd ed. Chicago: University of Chicago Press, 1985.

Olson, A., and J. Voss. *The Organization of Knowledge in America, 1860–1920*. Baltimore: The Johns Hopkins University Press, 1979.

Pawson, H.C. *Robert Bakewell Pioneer Livestock Breeder*. London: Crosby Lockwood, 1957

Persell, S.M. *Neo-Lamarckism and the Evolution Controversy in France, 1870–1920*. Lewiston: The Edwin Mellen Press, 1999

Peters, G.H. *The Plimsoll Line: the Story of Samuel Plimsoll, Member of Parliament for Derby, 1868 to 1880*. Chichester: Rose, 1975.

Philips, D. and S. Kingsland, eds. *New Perspectives on the History of the Life Sciences and Agriculture*. Springer, 2015.

Powell, F.W. *Bureau of Animal Industry: Its History, and Organization*. New York: AMS Press, 1974.

Provine, W.B. *The Origins of Theoretical Population Genetics*. Chicago: University of Chicago Press, 1971

– *Sewall Wright and Evolutionary Biology*. Chicago: University of Chicago, 1986.

Quirk, L. *Prof William Richard Graham: Poultryman of the Century*. Guelph: University of Guelph, 2005.

Rader, K. *Making Mice: Standardizing Animals for American Biomedical Research, 1900–1955*. Princeton: Princeton University Press, 2004.

Reaman, G.E. *History of the Holstein-Friesian Breed in Canada*. Toronto: Collins, 1946

Rifkin, J. *Beyond Beef: The Rise and Fall of the Cattle Culture*. New York: Penguin Books, 1992.

Ritvo, H. *The Animal Estate*. Cambridge: Harvard University Press, 1987.

– *The English and Other Creatures in the Victorian Age*. Cambridge: Harvard University Press, 1987.

– *The Platypus and the Mermaid and Other Figments of the Classifying Imagination*. Cambridge: Harvard University Press, 1997.

Rogers, E.M. *Diffusion of Innovations*. New York: The Free Press of Glencoe, 1962.

Rossiter, M. *The Emergence of Agricultural Science*, New Haven: Yale University Press, 1975.

Rothschild, M., and S. Newman, eds. *Intellectual Property Rights in Animal Breeding and Genetics*. New York: CABI Publishing, 2002.

Russell, N. *Like Engend'ring Like: Heredity and Animal Breeding in Early Modern England*. Cambridge: Cambridge University Press, 1986.

Ryder, M.L. *Sheep and Man*, London: Duckworth, 1983.

Sapp, J. *Beyond the Gene: Cytoplasmic Inheritance and the Struggle for Authority in Genetics*. Oxford: Oxford University Press, 1987.

Sawyer, G. *The Agribusiness Poultry Industry: A History of its Development*. New York: Exposition Press, 1971.

Schapsmeier, E.L., and F. H. Schapsmeier. *Henry A. Wallace of Iowa: The Agrarian Years, 1910–1940*. Ames: The Iowa State University Press, 1968.

Scott, R. *The Reluctant Farmer: The Rise of Agricultural Extension to 1914*. Chicago: University of Chicago Press, 1970.

Serafini, A. *The Epic History of Biology*. New York: Plenum Press, 1993.

Serpell, J. *In the Company of Animals: A Study of Human-Animal Relations*. Cambridge: Cambridge University Press, 1996.

Shepard, P. *The Others: How Animals Made Us Human*. Washington, DC: Island Press, 1995.

Shreeve, J. *The Genome War: How Craig Venter Tried to Capture the Code of Life and Save the World*. New York: Alfred A. Knopf, 2004.

Simm, G. *Genetic Improvement of Cattle and Sheep*. Ipswich, UK: Farming Press, 1998.

Smocovitis, V.B. *Unifying Biology: the Evolutionary Synthesis and Evolutionary Biology*. Princeton: Princeton University Press, 1996.

Swart, S. *Riding High: Horses, Humans and History in South Africa*, Johannesburg: Witswatersrand University Press, 2010.

Swart, S., and G. Bankoff, eds. *Breeds of Empire: The 'Invention' of the Horse in Southeast Asia and Southern Africa*. Copenhagen: Nordic Institute of Asian Studies Press, 2007.

Talbot, R.B. *The Chicken War: An International Trade Conflict between the United States and the European Economic Community, 1961–64*. Ames: Iowa State University Press, 1978.

Thomas, K. *Man and the Natural World: Changing Attitudes in England 1500–1800*. London: Allen Lane, 1983.

Thompson, J.A. *History of Livestock Raising in the United States, 1607–1860*. Washington: Bureau of Agricultural Economics, 1942.

Thurtle, P. *The Emergence of Genetic Rationality: Space, Time & Transformation in American Biological Science, 1870–1920*. Seattle: University of Washington Press, 2007.

Trow-Smith, R. *A History of British Livestock Husbandry, 1700–1900*. London: Routledge and Kegan Paul, 1959.

Tudge, C. *In Mendel's Footnotes: An Introduction to the Science and Technologies of Genes and Genetics from the Nineteenth Century to the Twenty-Second*. London: Jonathan Cape, 2000.

Venter, J.C. *A Life Decoded: My Genome, My Life*. New York: Viking, 2007.

Weir, B.S. et al., eds. *Proceedings of the Second International Conference on Quantitative Genetics*. Sunderland, MA: Sinauer Associates, 1988.

Wood, R.J. and V. Orel. *Genetic Prehistory in Selective Breeding: A Prelude to Mendel*. Oxford: Oxford University Press, 2001.

Woods, A. *Manufacturing Plague: The History of Foot-And-Mouth Disease in Britain*. London: Earthscan, 2004.

Articles

Allen, G.E. "Mendelian Genetics and Postgenomics: The Legacy for Today." *Ludus Vitalis* 21 (2004): 213–36.

Alter, S.G. "The Advantages of Obscurity: Charles Darwin's Negative Inference from the History of Domestic Breeds." *Annals of Science* 64 (2007): 235–250.

Barendse, W. "Development and Commercialization of a Genetic Marker for Marbling of Beef in Cattle: A Case Study." In *Intellectual Property Rights in Animal Breeding and Genetics*, edited by M. Rothschild and S. Newman, 197–212. New York: CABI Publishing, 2002.

Bateman, F. "Improvement in American Dairy Farming, 1850–1910: A Quantitative Analysis." *Journal of Economic History* 28 (1968): 255–73.

– "Labor Inputs and Productivity in American Dairying, 1850–1910." *Journal of Economic History* 29 (1969): 206–29.

Beckett, J.V. "Note on *Mathew and George Culley: Travel Journals and Letters, 1765–1798*." *English Historical Review* 118 (2003): 803–4.

Blackman, J. "The Cattle Trade and Agrarian Change on the Eve of the Railway Age." *Agriculture History Review* 23 (1975): 48–62.

Bonneuil, C. "Mendelism, Plant Breeding and Experimental Cultures: Agriculture and the Development of Genetics in France." *Journal of the History of Biology* 39 (2006): 281–308.

Bowie, G.G.S. "New Sheep for Old-Changes in Sheep Farming in Hampshire, 1792–1879." *Agricultural History Review* 35 (1987): 15–23.

Bowler, P.J. "The Changing Meaning of 'Evolution'." *Journal of the History of Ideas* 36 (1975): 95–104.

Boyd, P.J. "Making Meat: Science, Technology, and American Poultry Production." *Technology and Culture* 42 (2001): 631-64.

Brassley, P. "Cutting across Nature? The History of Artificial Insemination of Pigs in the United Kingdom." *Studies in History and Philosophy of Biological and Biomedical Sciences* 38 (2007): 442–61.

Brunger, E. "Dairying and Urban Development in New York State, 1850–1900." *Agricultural History* 29 (1955): 169–73.

Bugos, G.E. "Intellectual Property Protection in the American Chicken-Breeding Industry." *Business History Review* 66 (1992): 127–68.

Burke, J. "Dairywomen and Affectionate Wives: Women in the Irish Dairy Industry, 1890-1914." *Agricultural History Review* 38 (1990): 149–65.

Carlson, L. "Forging his own Path: William Jasper Spillman and Progressive Era Breeding and Genetics." *Agricultural History* 79 (2005): 50–73.

Cassidy, R. "Falling in Love with Horses: The International Thoroughbred Auction." *Society and Animals* 13 (2005): 51–67.

– "Turf Wars: Arab Dimensions to British Racehorse Breeding." *Anthropology Today* 19 (2003): 13–18.

Castonguay, S. "The Transformation of Agricultural Research in France: The Introduction of the American System." *Minerva* 43 (2005) 265–87.

Chapman, A.B. "Jay Lawrence Lush 1896–1982: A Brief Biography." *Journal of American Science* 69 (1991): 2671–76.

Charnley, B. "Experiments in Empire-building: Mendelian genetics as a national, Imperial, and global enterprise." *Studies in History and Philosophy of Science* Part A, 44 (2013): 292–300.

Churchill, R.B. "William Johannsen and the Genotype Concept." *Journal of the History of Biology* 7 (1974): 5–30.

Clutton-Brock, J. "Darwin and the Domestication of Animals." *Biologist* 29 (1982): 72–6.

Cooke, K.J. "From Science to Practice, or Practice to Science? Chickens and Eggs in Raymond Pearl's Agricultural Breeding Research, 1907-1916." *Isis* 88 (1997): 62–86.

Copus, A.K. "Changing Markets and the Development of Sheep Breeds in Southern England, 1750–1900." *Agricultural History Review* 37 (1999): 238–51

Cornell, J.F. "Analogy and Technology in Darwin's Vision of Nature." *Journal of the History of Biology* 17 (1984): 202–344.

Crew, F.A.E. "Reginald Crundall Punnett." *Genetics* 58 (1968): 1–7.

Crow, J.F. "Sewall Wright's Place in Twentieth-Century Biology." *Journal of the History of Biology* 23 (1990): 57–89.

Davis, B.D. "The Background: Classical to Molecular Genetics." In *The Genetic Revolution: Scientific Prospects and Public Perceptions*, edited by B.D. Davis. Baltimore: The Johns Hopkins University Press, 1991.

Denison, R.F., E.T. Kiers, and S.A. West. "Darwinian Agriculture: When Can Humans Find Solutions Beyond the Reach of Natural Selection?" *Quarterly Journal of Biology* 78 (2003): 145–68.

Derry, M.E. "Gender Conflicts in Dairying: Ontario's Butter Industry, 1880–1920." *Ontario History* 90 (1998): 31–47.

– "Chicken Breeding: The Complex Transition from Traditional to Genetic Methods in the United States." In *New Perspectives on the History of the Life Sciences and Agriculture*, edited by D. Philips and S. Kingsland, forthcoming. Springer, 2015.

Dunn, L.C. "Genetics at the Anikowo Station: A Russian Animal Breeding Center that has been Developed During the Reconstruction Period." *Journal of Heredity* 19 (1928): 281–4.

– "The Transformation of Biology: A Geneticist's Viewpoint." *Journal of Heredity* 57 (1966) 159–65.

Evans, L.T. "Darwin's Use of the Analogy between Artificial and Natural Selection." *Journal of the History of Biology* 17 (1984): 113–40.

Falconer, D.S. "Early Selection Experiments." *Annual Review of Genetics* 26 (1992), 1–14.

– "Quantitative Genetics in Edinburgh, 1947–1980." *Genetics* 133 (1993): 137–42.

Gasca, Ann Millán. "The Biology of Numbers: the Correspondence of Vito Volterra on Mathematical Biology." In *The Biology of Numbers: The Correspondence of Vito Volterra on Mathematical Biology*, edited by G. Israel and Ann Millán Gasca, 1-54. Vol. 26 of *Science Network: Historical Studies*. Berlin: Birkauser Verlag, 2002.

Gliboff, S. "Gregor Mendel and the Laws of Evolution." *History of Science* 37 (1999): 217–35.

Grasseni, C. "Designer Cows: The Practice of Cattle Breeding Between Skill and Standardization." *Society and Animals* 13 (2005): 33–49.

– "Managing Cows: An Ethnography of Breeding Practices and Uses of Reproductive Technology in Contemporary Dairy Farming in Lombardy

(Italy)." *Studies in History and Philosophy of Biological and Biomedical Sciences* 38 (2007): 488–510.

– "Skilled Vision. An Apprenticeship in Breeding Aesthetics." *Social Anthropology* 12 (2004) 41–55.

Grey, J. "History of Mathematics and History of Science Reunited?" *Isis* 102 (2011): 511–17.

Hacking, I. "Introductory Essay." In *The Structure of Scientific Revolutions.* 4th ed. Chicago: University of Chicago Press, 2012.

Hartmann, W. "From Mendel to Multi-national in Poultry Breeding." *World's Poultry Science* 45 (1989): 5–26.

Harwood, J. "Introduction to the Special Issue of Biology and Agriculture." *Journal of the History of Biology* 39 (2006) 237–9.

– "On the Genesis of Technoscience: A Case Study of German Agricultural Education." *Perspectives on Science* 13 (2005) 329–51.

Heleski, C.R., and J. Adroaldo. "Animal Science Student Attitudes to Farm Animal Welfare." *Anthrozoos* 19 (2006): 3–16.

Henlein, P. "Cattle Droving from the Ohio Country, 1800-1850." *Agricultural History* 28 (1954): 83–94.

Horowitz, R. "Making the Chicken of Tomorrow: Reworking Poultry as Commodities and as Creatures, 1945-1990." In *Industrializing Organisms: Introducing Evolutionary History,* edited by S.R. Schrepfer and P. Scranton. London: Routledge, 2004.

Kevles, D.J. "Patents, Protections, and Privileges." *Isis* 98 (2007): 323–31.

– "Protections, Privileges, and Patents: Intellectual Property in Animals and Plants since the Late Eighteenth Century." In *Contexts of Invention,* edited by M. Biagioli, P. Jaszi, and M. Woodmansee. Chicago: University of Chicago Press, 2010.

– "The Advent of Animal Patents: Innovation and Controversy in the Engineering and Ownership of Life" In Rothschild, M., and S. Newman, eds. *Intellectual Property Rights in Animal Breeding and Genetics.* New York: CABI Publishing, 2002.

Johnson, K. "Iowa Dairying at the Turn of the Century: the New Agriculture and Progressivism." *Agricultural History* 45 (1971): 95–110.

Johnson, L.P.V. "Dr. W.J. Spillman's Discoveries in Genetics: An Evaluation of His Pre-Mendelian Experiments with Wheat." *Journal of Heredity* 39 (1948): 247–54.

Kimmelman, B. "The American Breeders' Association: Genetics and Eugenics in an Agricultural Context, 1903-1913." *Social Studies of Science* 13 (1983): 163–204.

– "Mr. Blakeslee Builds his Dream House: Agricultural Institutions, Genetics, and Careers 1900-1945." *Journal of the History of Biology* 39 (2006): 241–80.

Kingsland, S.E. "The Battling Botanist: Daniel Trembly MacDougal, Mutation theory, and the Rise of Experimental Evolutionary Biology in America, 1900-1912." *Isis* 82 (1991): 479–509.

– "Mathematical Figments, Biological Facts: Population Ecology in the Thirties." *Journal of the History of Biology* 19 (1986): 235–56.

Loew, F.M. "Animal Agriculture." In *The Genetic Revolution: Scientific Prospects and Public Perceptions*, edited by B.D. Davis. Baltimore: The Johns Hopkins University Press, 1991.

Marie, J. "For Science, Love and Money: The Social Worlds of Poultry and Rabbit Breeding in Britain, 1900-1940." *Social Studies of Science* 38 (2008): 919–36.

Matz, B. "Crossing, grading, and keeping pure: animal breeding and exchange around 1860." *Endeavour* 35 (2011): 7–15.

Mayr, E. "The Nature of the Darwinian Revolution." *Science* 176 (June 1972): 981–9.

Mayr, O. "The Science-Technology Relationship as a Historiographic Problem." *Technology and Culture* 17 (1976): 663–73.

McCormick, V. "Butter and Egg Business: Implications from the Records of a Nineteenth Century Farm Wife." *Ohio History* 100 (1991): 329–35.

Mendelsohn, J.A. "'Like All That Lives'": Biology, Medicine and Bacteria in the Age of Pasteur and Koch." *History and Philosophy of the Life Sciences* 24 (2002): 3–36.

Müller-Wille, S. and V. Orel. "From Linnaean Species to Mendelian Factors: Elements of Hybridism, 1751-1870." *Annals of Science* 64 (2007): 171–215.

Nebel, H. et al, "Patenting and Sequencing the Genome." In *Intellectual Property Rights in Animal Breeding and Genetics*, ed. Rothschild and Newman, 119-146.

Olby, R. "Mendel no Mendelist?" *History of Science* 12 (1979): 53–72.

Ollivier, L. "Jay Lush: Reflections on the Past." *Lohmann Information* 43 (2008): 3–12.

Orel, V. "Cloning, Inbreeding, and History." *Quarterly Review of Biology* 72 (1997): 437–40.

– "Commemoration of the N.I. Vavilov Centennial at Brno." *Folia Mendelianna* 23 (1988): 37–50.

– "Selection Practice and Theory of Heredity in Moravia Before Mendel." *Folia Mendelianna* 12 (1977): 179–99.

– "The Spectre of Inbreeding in the Early Investigation of Heredity." *History and Philosophy of the Life Sciences* 19 (1997): 315–30.

Orel, V., and R.J. Wood. "Early Development in Artificial Selection as a Background to Mendel's Research." *History and Philosophy of the Life Sciences* 3 (1981): 145–70.

Orel, V., and R.J. Wood. "Scientific Animal Breeding in Moravia Before and After the Discovery of Mendel's Theory." *Quarterly Review of Biology* 75 (2000): 149–57.

Orland, B. "Turbo-Cows: Producing a Competitive Animal in the Nineteenth and Early Twentieth Centuries." In *Industrializing Organisms: Introducing Evolutionary History*, edited by S.R. Schrepfer and P. Scranton. London: Routledge, 2004.

Palladino, P. "Between Craft and Science: Plant Breeding, Mendelian Genetics, and British Universities, 1900-1920." *Technology and Culture* 34 (1993): 300–23.

Paul, D.B., and B. Kimmelman. "Mendel in America: Theory and Practice, 1900-1919." In *The American Development of Biology*, edited by R. Rainger, K. Benson, and J. Maienschein. Philadelphia: University of Pennsylvania Press, 1988.

Pauly, P.J. "The Appearance of Academic Biology in Late Nineteenth-Century America." *Journal of the History of Biology* 17 (1984): 369–97.

Pawson, H.C. "Some Agricultural History Salvaged." *Agricultural History Review* 7 (1959): 6–13.

Perry, P.J. "The Shorthorn Comes of Age, 1822-1843." *Agricultural History* 56 (1982): 560–6.

Preston, R. "The Genome Warrior." *The New Yorker*, 12 June 2000.

Punnett, R.C. "Early Days of Genetics." *Heredity* 4 (1950): 1–10.

Quinn, M.S. "Corpulent Cows and Milk Machines: Nature, Art and the Ideal Type." *Society and Animals* 1 (1993): 145–57.

Rheinberger, H-J. and P. McLaughlin. "Darwin's Experimental Natural History." *Journal of the History of Biology* 17 (1984): 345–68.

Richards, R.A. "Darwin and the Inefficacy of Artificial Selection." *Studies in History and Philosophy of Science* 28 (1997): 75–97.

Ritvo, H. "Possessing Mother Nature: Genetic Capital in Eighteenth-Century Britain." In *Early Modern Conceptions of Capital*, edited by J. Brewer and S. Staves, 413–26. London: Routledge, 1995.

Roll-Hansen, N. "Theory and Practice: the Impact of Mendelism on Agriculture." *C.R. Acad. Sci. Paris, Sciences de la Vie/Life Sciences* 323 (2000): 1107–16.

Rowe, D.J. "The Culleys, Northumberland Farmers 1767–1813." *Agricultural History Review* 19 (1971): 156–74.

Ruse, M. "Charles Darwin and Artificial Selection." *Journal of the History of Ideas* 36 (1975): 339–50.

Ryder, M.L. "The History of Sheep Breeds in Britain." *Agricultural History Review* 12 (1964): Part 1: 1–12, Part 2: 79–97.

Sapp, Jan. "The Nine Lives of Gregor Mendel." In *Experimental Inquiries: Historical, Philosophical and Social Studies of Experimentation in Science,* edited by H.E. Le Grand, Dordrecht, 137–166. Kluwer Academic Publishers, 1990.

– "The Struggle for Authority in the Field of Heredity, 1900-1932: New Perspectives on the Rise of Genetics." *Journal of the History of Biology* 16 (1983): 311–342.

Secord, J.A. "Nature's Fancy: Charles Darwin and the Breeding of Pigeons." *Isis* 72 (1981): 163–186.

Sloan, P.R. "Essay Review: Ernst Mayr on the History of Biology." *Journal of the History of Biology* 18 (1985): 145–153.

Stamhuis, I.H. et al. "Hugo de Vries on Heredity, 1889–1903: Statistics, Mendelian Laws, Pangenes, Mutations." *Isis* 90 (1999): 238–67.

Sterrett, S.G. "Darwin's Analogy between Artificial and Natural Selection: How does it go?" *Studies in History and Philosophy of Biological and Biomedical Sciences* 33 (2002): 151–68.

Taylor, D. "The English Dairy Industry, 1860-1930." *Economic History Review* 29, no. 2 (1976): 585–601.

– "London's Milk Supply, 1850-1900: A Reinterpretation." *Agricultural History* 45 (1971): 33–8.

Theunissen, T. "Breeding for Nobility or Production? Cultures of Dairy Cattle Breeding in the Netherlands, 1945-1995." *Isis* 103 (2012): 278–309.

– "Breeding Without Mendelism: Theory and Practice of Dairy Cattle Breeding in the Netherlands, 1900-1950." *Journal of the History of Biology* 41 (2008): 637–76.

– "Closing the Door on Hugo de Vries' Mendelism." *Annals of Science* 51 (1994): 225–48.

– "Connecting Genetics, Evolutionary Theory and Practical Animal Breeding: Arend L. Hagadoorn (1885–1953)." Unpublished manuscript, 2012. Microsoft Word file.

– "Darwin and his Pigeons: The Analogy Between Artificial and Natural Selection Revisited." *Journal of the History of Biology*, Online 2011.

– "Knowledge is Power: Hugo de Vries on Science, Heredity and Social Progress." *British Journal for the History of Science* 27 (1994): 291–331.

Thurtle, P. "Harnessing Heredity in Gilded Age America: Middle Class Mores and Industrial Breeding in a Cultural Context." *Journal of the History of Biology* 35 (2002): 43–78.

Van Riper, W. "Aesthetic Notions in Animal Breeding." *Quarterly Review of Biology* 7 (1932): 84-7.

Vyas, M.A., ed. *Issues in Ethics and Animal Rights.* Kindle, 2013.

Walton, J. "The Diffusion of Improved Shorthorn Cattle in Britain During the Eighteenth and Nineteenth Centuries." *Transactions of the Institute of British Geographers* 9 (1984): 22–36.

– "Pedigree and the National Cattle Herd circa 1750-1950." *Agricultural History Review* 34 (1986): 149–70.

Wieland, T. "Scientific Theory and Agricultural Practice: Plant Breeding in Germany from the late 19th to the Early 20th Century." *Journal of the History of Biology* 39 (2006): 309–43.

Wilmot, S. "From Public Service to Artificial Insemination: Animal Breeding Science, 1890-1951." Full Research Report, ESRC End of Award Report, Reference No. Res-000–23–0390 (2007): 14–22.

– "From 'Public Service' to Artificial Insemination: Animal Breeding Science and Reproductive Research in Early 20th Century Britain." *Studies in History and Philosophy of Biological and Biomedical Sciences* 38 (2007): 411–41.

– "Between the Farm and the Clinic: Agricultural and Reproductive Technology in the Twentieth Century." *Studies in History and Philosophy of Biological and Biomedical Sciences* 38 (2007): 303–15.

Whetham, E. "The Trade in Pedigree Livestock, 1850-1910." *Agricultural History Review* 27 (1979): 47–50.

Wood, R.J. "Robert Bakewell (1725–1795) Pioneer Animal Breeder and His Influence on Charles Darwin." *Folia Mendelianna* 8 (1973): 231–42.

Wood, R.J., and V. Orel, "Scientific Breeding in Central Europe during the Early Nineteenth Century: Background to Mendel's Later Work." *Journal of the History of Biology* 38 (2005): 239–72.

Wykes, D. "Robert Bakewell (1725–1795) of Dishley: farmer and livestock improver." *Agricultural History Review* 52 (2004): 38–55.

Yi, Doogab. "Who Owns What? Private Ownership and the Public Interest in Recombinant DNA Technology in the 1970s." *Isis* 102 (2011): 446–74.

Other

Cluny Exports. "Crossbreeding Holstein-Friesian x Jersey Cattle." Accessed 28 December 2010. www.clunyexports.com.au/media/crossbreedjhfjv11 .pdf,.

Hy-Line International website. Accessed 8 March 2007. http://www.hyline .com.

"Poultry Marketplace – Overview of the Primary Poultry Breeder Industry." Accessed 21 April 2006. http://www.agr.ca/poultry/brov-elap_e.htm.

"Primary Poultry Breeders: Company Profiles." Accessed 8 March 2007. http://www.agr.gc.ca/poultry/brpr-elpr_m.htm.

Select Sires, "Crossbreeding 101: What Every Producer Should Know." Accessed 27 December 2010. www.selectsires.com/dairy/ crossbreeding_101.pdf.

– "Crossbreeding 103: A Further Update on Crossbreeding Results." Accessed 27 December 2010. www.selectsires.com/dairy/crossbreeding_103_ aug2006.pdf.

Semex Alliance, Canada. Access/modified date. http://www.semex.com.

"Sequencing of the Chicken Genome: An Overview." December 2004 article in *Nature*, reproduced on *The Poultry Site*. Accessed 20 April 2006. http://www.thepoultrysite.com/FeaturedArticle/FAType.asp?AREA.

The Chicken: Its Biological, Social, Cultural, and Industrial History from Neolithic Middens to McNuggets. International Chicken Conference, Agrarian Studies, Yale University, May, 2002. Accessed 25 April 2006. http://www.yale.edu/agrarianstudies/chicken/description.html.

"The Impact of Genetics on Breeder Management." May 2005 article in *Hybro B.V.*, reproduced on *The Poultry Site*. Accessed 20 April 2006. http://www.thepoultrysite.com/FeaturedArticle/FAType.asp?AREA.

United Poultry Concerns, Inc. Accessed 19 April 2006. http://www.upc-online.org.

Index

Accelerated Genetics, 118
agricultural practices/science
  connection. *See* science/practice
  connection
AI. *See* artificial insemination
alleles, defined, 235
American Cornish, 76, 85
American Leghorn, 76, 83, 89, 161, 166
American Poultry Association:
  beauty/utility issue, 75–8, 87,
  183; breeder/producer divide,
  80; cooperation with other
  associations, 103; and scientists,
  127; standards, 75–8, 80, 87, 183,
  187. *See also* chickens
Analysis of Variance (ANOVA), 62
ancestry breeding, defined, 235
Angus, 88–90, 101, 123, 154, 168–9
animal breeding. *See* artificial
  selection
Animal Breeding (Hagedoorn), 66–7
Animal Breeding Plans (Lush), 67
Animal Machines (Harrison), 175
Animal Suffering (Dawkin), 176
animal welfare, 160–78; about, 7–8,
  173–4, 177–8, 187–8; activism,
  8, 168–78, 187–8; animal

sentience, 176–8; cloning, 131,
  176, 223n6; confinement, 163,
  177–8; dehorning, 8, 167–73,
  178, 187–8; five freedoms, 175–6,
  187; legislation, 168–72, 174, 178;
  natural behaviour, 164, 175–6;
  pain control, 168, 169–73, 177–8;
  reproductive technology, 176;
  travel conditions, 163, 174–5; veal
  industry, 163. *See also* ethics and
  industry-needs
ANOVA (Analysis of Variance), 62
Arabian horses, 14–16, 27, 29, 104, 189
Argentina, cattle, 97
Art and Science in Breeding (Derry),
  10–11
artificial insemination, 94–128;
  about, 6–7, 94, 130, 181–2, 187;
  beef cattle, 94, 105, 122–4, 141,
  155; BLUP, 111–12, 114, 115, 123;
  chickens, 105, 125–6, 146–7; costs,
  105; data collection, 98–100,
  127; dependence on existing
  structures, 6–7, 104, 122, 146,
  182, 184–5, 186; ET and MOET,
  130–1; frozen semen, 105; future
  trends, 190; historical background,

104–7; horses, 105, 189; intellectual property, 186; male *vs.* female focus, 130–1, 151–2; natural service *vs.* AI, 104, 112, 130, 164, 189; pigs, 105, 124–5, 146–7; progeny testing, 6–7, 107, 109, 111, 181–3; purebred breeding, 95–6, 106; regulation of breeding males, 6, 94–6; research review on, 10; science/practice connection, 121–3, 127–8, 182, 186–7; sexed semen, 164–6, 178; sheep, 105; statistics, 105. *See also* dairy cattle and artificial insemination; globalization of AI in dairy cattle; reproductive technology

artificial insemination organizations: about, 6–7, 94, 118, 186; breeding associations, 104–5, 106–7, 109, 117; corporations, 109, 118, 121, 124–5, 130, 186; data collection and registries, 98, 127, 141, 181–2; genomic research, 145–9; intellectual property, 145, 186; international cooperation, 119–20, 149–51; progeny testing, 141–2, 147. *See also* Genex; Interbull; Select Sires

artificial selection (generally): about, 3–5, 187, 192; animal breeding as hybrid, 4–5, 13; future trends, 8, 189–92; historical overview, 13–14, 179–88; plant *vs.* animal research, 68–9, 82–3; research needed, 8, 189–92; research review on, 9–11; science/practice connection, 4–5, 8–10; and technology, 4–5, 8, 160–1, 178, 187. *See also* art *vs.* science; ethics and industry-needs; history of artificial selection; science/ practice connection

art *vs.* science: about, 192; in dairy and chicken industries, 125; and experimental research, 54; and Mendelism, 180; practical breeding as purebred "art," 44. *See also* science/practice connection

Australia, 66–7, 137, 142, 188

Ayrshires, 98, 99

Babcock, E.B., 49

Babcock test, 99

Bakewell, Robert, 17, 199n51

Bakewellian system: about, 17–20, 179–80; "breed" creation, 18–19, 60; defined, 235; in-and-in breeding, 17, 19–20; inbreeding, 17, 19–20, 22, 23, 60; influence of, 23, 35; progeny testing, 17–20, 23; purebred breeding, 28, 32, 46; registries, 33; Shorthorns, 24–5, 179; true lines *vs.* hybrid breeding, 23. *See also* inbreeding

Barred Plymouth Rock breed, 73, 85, 166

Bates, Thomas, 25–7, 59

Bateson, William, 37–8, 41–2, 52–4, 204n25

BeadChip technology, 145, 147–59. *See also* genomics

beef cattle: about, 5, 88–91, 92–3; AI use, 94, 122–4, 141, 155; associations, 122–4, 154–5; biodiversity, 156; biological lock, 89–90; breeder/feeder structure, 88–90, 92, 123; breeds, 88–90, 154; crossbreeding, 88–90, 92, 122, 155; dairy/beef divide, 154, 162; dehorning, 8, 167–73, 187–8; excess male calves, 161–3; feedlots, 123, 168; historical background,

88–93; hornless breeds, 172, 178; purebreds, 89–90, 93, 122–3; quantitative genetics, 122–4; registries, 90–1; science/practice connection, 93, 122–3

beef cattle, breeds: Angus, 88–90, 101, 123, 154, 168–9; Charolais, 90; Galloway, 25, 88, 168; Herefords, 33, 88, 89, 90; Limousin, 90; Shaver Blend, 89–91, 122. *See also* Shorthorn cattle

beef cattle and molecular genetics: BLUP systems, 123; bovine genome, 140, 184; cloning, 131, 223n6; dependence on existing structures, 154–5; DNA markers, 137, 143, 184; EPDs, 123; genomic testing, 145, 154–5; HapMaps, 184; intellectual property, 145. *See also* molecular genetics

Beef Improvement Federation, 123, 155

Belgium, 57, 150

Berlin, Germany, research, 56–7

biodiversity, 156

bioinformatics, 8, 191, 235

biological lock: beef cattle, 89–90; and breeder/producer divide, 73; chickens, 73, 75, 84–5, 181; corn, 73; defined, 235; double mating system, 73; hybrid breeding, 88, 186–7; as natural patent, 71, 73, 75, 88, 181, 186–7; pigs, 91–2, 124, 126. *See also* intellectual property

biomathematics, 36, 64

biometry: about, 5, 35–6, 180, 184; continuous inheritance, 41–2, 180, 184; Darwinism, 5, 35–6, 38, 42, 180; defined, 235; Galton's theories, 35–8, 42;

math/biology dilemma, 37–8, 63–4, 114; Mendelism/biometry divide, 40–3, 53, 180; Mendelism/biometry synthesis, 40, 58, 60–2, 69–70; natural selection, 36; quantitative traits, 41, 53. *See also* Darwinism; quantitative genetics

Black and White Friesian cattle, 67, 98, 106, 116–19. *See also* Holsteins

black box thinking: about, 38–9, 184, 190–1; and genomics, 143–4, 153, 158–9, 184, 190–1; and molecular genetics, 136–7; and quantitative genetics, 38–9, 133, 158–9, 184

Blueprint for USDA Efforts in Agricultural Animal Genomics (USDA), 158

BLUP (best linear unbiased prediction), 111–12, 114, 115, 123, 235. *See also* MACE; sire indexing

boar, defined, 235

bovines. *See* cattle

Bowlker, T.J., 101

breed true. *See* true breeding

Britain: animal welfare, 163, 173–6; beauty/utility issue, 75, 77; beef breeder/feeder structure, 88–9; biology/math dilemma, 64; chickens, 66, 75–7, 157–8; double mating system, 73; genetics research, 55–6, 66, 107–8; genomic research, 140, 157–8; historical background, 13, 17, 21, 34–5, 55–6; horses, 14–15, 34–5, 95–7; Mendelism/biometry divide, 41–2; naturalists, 21–2; pigs, 124–5, 140; purebred breeding, 24–8; registries, 15–16, 27–8; regulation of breeding males, 95–7; science/practice connection,

55–6, 66; sheep, 188; transgenics,
130. *See also* Bakewellian system;
Darwinism; Shorthorn cattle;
Thoroughbreds
Britain, dairy cattle: AI and
inbreeding, 120; AI research,
104–5, 107, 181–2; associations,
lack of, 97, 107; dehorning, 168;
Holsteins, 118; international
comparisons, 97, 118, 119; progeny
testing, 107; regulation of breeding
males, 97; veal exports, 163. *See
also* dairy cattle
Brown Swiss dairy cattle, 98, 145
bulls. *See* beef cattle; cattle;
dairy cattle

Canada, beef cattle: dehorning,
168–72, 187–8; DNA markers, 138;
genomic testing, 154; hornless
breeds, 172, 178. *See also* beef cattle
Canada, chickens: associations,
75; beauty/utility issue,
75–7, 82; biological lock, 85;
breeder/scientist interaction,
80–2; corporations, 84, 88;
crossbreeding, 74–5, 85;
government standards, 77–8;
import chicks, 84; meat breeding,
84–5. *See also* American Poultry
Association; chickens
Canada, dairy cattle: animal welfare,
161–2, 168–74, 178; associations,
97–8, 102–4, 127; compared to
American, 114, 116, 117, 149–50,
183; data collection, 98–100, 141,
149; dehorning, 168–73, 178, 187–8;
excess male calves, 161–3; feed,
70; genomic testing, 145, 148–52;
government support, 97–8, 102–3;

Hanover Hill bulls, 114–16, 120,
148, 183; historical background,
70, 97; international comparisons,
118, 119; progeny testing, 115,
148–51; purebreds *vs.* genetics,
114–16, 183; sire indexing, 114–15,
127, 183. *See also* dairy cattle;
Holsteins
Canada, dairy cattle and AI:
associations (studs), 97, 115, 141–2,
149; costs, 141–2; frozen semen,
105; inbreeding, 119–20; practices,
114–15, 148, 183; progeny
testing, 141–2. *See also* artificial
insemination
Canada, history (before 1950):
animal welfare, 161–2, 168–73,
174; biology/math dilemma,
64, 114; crossbreeding, 74–5;
eugenics, 36; experiment stations,
43; government funding for
research, 42–3; plant hybridizers,
34; productivity increase, 70;
purebred breeding (19th c.),
28–9
Castle, William E., 6, 52–4, 58–60, 69
cattle: biodiversity, 156; bovine
genome, 140, 184; cloning, 131,
176, 223n6; dehorning, 8, 168–73,
178, 187–8; frozen semen, 105;
purebred breeding, 126; research
review on, 10. *See also* beef cattle;
dairy cattle
CDDR (North American
Collaborative Dairy DNA
Repository), 149
Centre d'Insemination Artificeille
du Quebec (CIAQ), 115
Charolais, 90
Chenery, Winthrop, 116

chickens: about, 5–6, 86–8, 92–3;
AI use, 105, 125–6, 146–7;
animal welfare, 7–8, 166–7, 176;
associations, 75–7, 103, 181;
beauty/utility issue, 75–8, 82,
86–7, 181, 183; biodiversity,
155–6; biological lock, 73, 84–5,
126, 181; breeder/producer
divide, 79–80, 86–8, 125–6, 181;
comparison with dairy cattle
industry, 125–7; corporations,
82–6, 88, 146–7, 181, 186;
crossbreeding, 74–5, 85; ethics
and excess male young, 7–8;
excess male young, 161, 166–7;
experimental research qualities,
68; flock size, 83; frozen semen,
105; future trends, 189–90;
government standards, 77–8;
hatcheries, 79, 84, 125, 181;
historical background, 29;
inbreeding/outcrossing system,
72–3; intellectual property, 10,
126, 134, 138, 181, 185; markets,
6; meat industry, 75, 84–6,
161, 166–7; progeny testing,
46–7, 72, 81; quantitative
genetics, 134; science/practice
connection, 6, 45–6, 66, 80–2,
182; sexing methods, 66, 166–7;
standardbreds, 29, 76–8, 80,
82, 84–5, 87, 103, 183. *See also*
American Poultry Association;
Canada, chickens; United States,
chickens
chickens, breeds: American Cornish,
76, 85; American Leghorn, 76, 83,
89, 161, 166; Barred Plymouth
Rock, 73, 85, 166; New Hampshire,
85; Red Danish, 101; Rhode Island

Red, 76, 85; White Plymouth Rock,
76, 85, 166
chickens, history (before 1950):
about, 6, 71, 86–8, 92–3, 181;
beauty/utility issue, 75–8, 86–7,
181; "best to the best," 46–8, 73;
biological lock, 84, 181; breeder/
producer divide, 73, 79–80, 84,
86–8; breeding methods (19th c.),
72–5; breeds, 76, 85; continuous/
discontinuous inheritance,
45–9; corporate role, 82–6, 88,
92; crossbreeding, 85; double
mating system, 73; feed, 70; flock
size, 83; government standards,
77–8; Hagedoorn's research,
66–7; hatcheries, 79, 84, 181;
human gender roles, 79–80, 181;
inbreeding systems, 68, 72–5,
83–4; intellectual property, 80;
markets, 6, 83–4; mass selection *vs.*
ancestry, 47, 72, 86; meat breeding,
84–6; Mendelism, 54–5, 81;
productivity increase, 70; progeny
testing, 46–7, 72, 81, 86; science/
practice connection, 45–6, 66, 80–2,
85–6; sexing methods, 66, 166;
standardbreds, 29, 76–8, 80, 82,
84–5, 87, 103, 183
chickens and molecular genetics:
about, 155–8, 184–5; biodiversity,
155–6; chicken genome, 140,
155–7, 184, 185; dependence on
existing structures, 146, 185;
disease resistance, 155–6, 158;
DNA historical comparison
research, 157–8; DNA markers,
138–9; genomic testing, 145, 155–8;
intellectual property, 156. *See also*
molecular genetics

chickens and standards: about, 75–8;
American Poultry Association's
influence, 75–8, 80, 87, 183,
187; beauty/utility issue, 75–8,
82, 86–7, 181, 183; Chicken-of-
Tomorrow contest, 84–5; Record
of Performance, 77–80, 84, 87, 187;
Standard of Perfection, 75–8, 87, 183
Chicken-of-Tomorrow contest, 84–5
China, 61, 140
chromosome, defined, 235
classical genetics. See genetics,
classical
Clausen, R.E., 49
cloning, 131, 176, 223n6, 236
Coates, George, 25
cockerel, defined, 236
Cole, J.B., 154, 159
Cole, L.J., 101
Colling, Robert and Charles, 24–6,
28, 59, 179, 199n51. See also
Shorthorn cattle
Commission on Horse Breeding
reports, 95–6
corn: biological lock, 73;
corporations, 82–3, 88; hybrid
corn, 9, 69–70, 73, 82–4;
inbreeding, 51, 61; science/
practice connection, 50–2, 68–70.
See also plants
Cornish, 76, 85
corporations: artificial insemination,
109, 118, 121, 124–5, 130, 186;
chickens, 82–6, 88, 146–7, 181, 186;
ET and MOET, 130; government
role, 85–6, 186; hybrid corn, 82–3,
88; intellectual property, 88, 181,
186–7; pigs, 91–2, 124–5, 186
corporations and molecular genetics:
costs, 138–9; dairy cattle, 130;

DNA markers, 138; funding,
144–5; genomics, 186; intellectual
property, 138; research contraints,
138; SNP chips, 144–5. See also
Genex; Illumina, Inc.; molecular
genetics
cows. See beef cattle; cattle;
dairy cattle
Crew, A.E., 107
Crick, Francis, 131
crossbreeding: about, 21–2, 192;
and AI inbreeding, 120–2; beef
cattle, 88–9, 122, 155; chickens,
74–5, 83–5; dairy cattle, 100–2,
120–2; defined, 236; double-cross
system, 51–2; heterosis, 50–1, 122;
historical background, 21, 50–1,
53–4; hybrid vigour, 61, 74, 85; and
inbreeding, 22, 61, 120, 192; and
inbreeding/outcrossing, 72–5, 84;
naturalists, 21–2; pigs, 91; progeny
traits, 54
culture/science connection. See
art vs. science; science/practice
connection

dairy cattle: about, 5, 7, 125–8,
182–3, 187; associations, 99–100,
102–3, 127; beauty/utility issue,
67, 183; beef/dairy divide, 154,
162; biodiversity, 156; comparison
with chicken industry, 125–7,
187; corporations, 127; crossbreds
vs. purebreds, 100–2, 190; data
collection on, 98–100, 187;
dehorning, 8, 168–73, 178, 187–8;
excess calves, 7–8, 161–7; feed,
70; future trends, 190–2; heifer
growers, 163–4; inbreeding/
line-crossing method, 101–2;

MACE, 119, 150; natural behaviour, 164; population *vs.* quantitative genetics, 112–14; progeny testing, 105, 106–7, 111, 125; purebreds, 67, 99–104, 117, 126–7; purebreds *vs.* genetics, 114–16, 126–8, 182–3; registries, 98–9, 109; science/practice connection, 102, 127–8, 182–3, 192; sire indexing, 110, 114–15, 125, 127, 183; standards, 99, 187; trends in inbreeding, 119–22. *See also* Britain, dairy cattle; Canada, dairy cattle; United States, dairy cattle

dairy cattle, breeds: Ayrshires, 98, 99; Brown Swiss, 98, 145; Friesian Black and Whites, 67, 98, 106, 116–19; Guernseys, 98, 101; Jerseys, 98–9, 101, 119, 120, 145–6, 149, 153. *See also* Holsteins

dairy cattle, history (before 1950): crossbreds *vs.* purebreds, 100–1; data collection, 98–100, 103, 127; early history, 13; excess male calves, 161–3; markets and costs, 103; poor quality of breeding males, 103; purebred breeders, 99–104; science/practice connection, 66, 103; standards, 98–9; testing associations, 99–100, 127

dairy cattle and artificial insemination: about, 6–7, 94, 182–3; associations (studs), 97, 104–5, 114, 125–6, 127, 141–2, 147; BLUP methods, 114; bull selection, 106, 108–9; comparison with chicken industry, 125–7; corporations, 118, 121, 124–5, 186; costs, 141–2, 147; crossbreeding,

120–2; historical background, 97, 107–8, 126–8; Holstein dominance, 116–19, 141–2; inbreeding, 119–22, 190; Interbull, 119–20, 149–51; international comparisons, 118; progeny testing, 107–9, 111, 141–2; purebred breeders, 105, 125–6; purebreds *vs.* genetics, 114–16, 183; quantitative genetics, 108–9, 122–8, 159, 186; regulation of breeding males, 94–8; science/ practice connection, 121, 127–8, 182; sexed semen, 164–6, 178; use of AI, 141, 164. *See also* artificial insemination; Canada, dairy cattle and AI; globalization of AI in dairy cattle; United States, dairy cattle and AI

dairy cattle and molecular genetics: about, 159, 184–5; BeadChips, 147–52; bovine genome, 140; breed-specific testing, 152–3; cloning, 131, 223n6; corporations, 130; dependence on existing structures, 7, 146, 154, 184–5; ET and MOET, 130; female testing, 151–2, 185; genomic testing, 148–52, 154; historical background, 127; inbreeding, 130; international cooperation, 149–51; MACE and GMACE, 150; progeny testing, 141–2; progeny *vs.* DNA testing, 147–9, 184–5; research questions, 147; science/practice connection, 152, 184–5; single-step evaluation, 185. *See also* molecular genetics

Dairy Herd Improvement Associations (DHIA), 99–100, 101, 103, 109

Darwin, Charles, 3, 32–3, 200n70

Darwinism: about, 3, 14, 33–5;
and biometry, 5, 35–6, 38, 42;
and breeders, 41; continuous
inheritance, 42, 48; historical
background, 5, 14, 34, 35;
Mendelism/Darwinism divide,
42; Mendelism/Darwinism
synthesis, 58, 61–2, 180; natural
selection, 3, 11, 13–14, 33–5, 48
Dawkin, Marian, 176
Dean, H.H., 102
definitions, 235–40
Denmark: AI practices, 104, 124, 142;
associations, 99, 104; dairy cattle,
99, 101, 104, 119, 150; genomic
testing, 150; pigs, 124; veal
industry, 163
de Vries, Hugo, 34, 66
DHIA. See Dairy Herd Improvement
Associations
Dickerson, G.E., 107
diseases: genomic research, 155–6,
158, 191; prevention by AI, 106
DNA double helix, 131–2, 236. See
also genomics; molecular genetics
dogs, 13, 29, 75, 145
Donald, H., 93
double-cross hybrid breeding
system, 51–2
double helix of DNA, 131–2, 236. See
also molecular genetics
draft horses, 29–32
Dunn, L.C., 55–8, 66

Elton, C.S., 64
England. See Britain
EPD (estimated progeny difference),
123, 188, 236
ET (embryo transfer), 130, 236. See
also reproductive technology

ethics and industry-needs, 160–78;
about, 7–8, 160–1, 173, 187–8;
breeding issues, 173–8; cloning,
176; dehorning, 8, 167–73, 178,
187–8; excess male young, 7–8,
161–7; five freedoms, 175–6, 187;
market solutions, 167; natural
behaviour, 175–6; pain control,
168, 169–73, 177–8; sexed semen,
164–6; transgenics, 130–1, 176. See
also animal welfare
eugenics: about, 35–7; and animal
breeding, 16–17, 36, 50; biometry,
35–6, 38; Fisher's interest in, 58–9,
61–2, 69; Galton's theories, 35–7;
population genetics, 61–2; social
class and genealogy, 16–17. See also
human molecular genetics
EuroGenomics, 150
Europe: AI practices, 98, 106, 124,
142; animal welfare, 175–7;
early history, 13, 22, 28–30; early
regulation of breeding males,
95–8; ET and MOET, 130–1; ethics
and male chicks, 167; genetics
research, 135; genomic testing,
142–3, 149–50; Holsteins (Friesian
Black and White), 67, 98, 106,
116–19; international comparisons,
118, 119; Moravian sheep, 10, 22–4;
research review on, 10; sheep, 188;
Standardbred horses, 28–30. See
also specific countries
Europe, history (1900–1950): AI
practices, 97, 98; animal genetic
research, 55–8; comparison
with U.S. research, 67; dairy
cattle crossbreds vs. purebreds,
100–2; Dunn's reports, 55–8, 66;
experimental animal breeding,

54–62, 65–6; experiment stations, 43; farm conditions as constraint, 67, 70; government funding for research, 42–3; Hagedoorn's research, 57–8, 65–8; progeny testing, 67; science/practice connection, 57–8, 65–8, 70; Wriedt's research, 66, 67

evolution theory: about, 3, 5, 33–6, 40–2, 180; biometry, 5, 36, 38, 41, 180; continuous/discontinuous inheritance, 41–9, 180; and genomics, 191; Mendelism, 40–1, 180; mutation theory, 3, 40–2, 52–3, 180; naturalists, 5, 180; natural selection, 3, 11, 13–14, 33–5, 48; science/practice connection, 180. *See also* biometry; Darwinism; Mendelism

experiment stations: about, 34, 42–3; Bakewellian system, 45; chickens, 45–6, 80–2, 85; crossbreeding, 101; dairy cattle, 102; experiments *vs.* theories, 54–5; government role, 42–3, 85; Mendelism/biometry divide, 45–6; population genetics, 65; progeny testing, 45; science/practice connection, 45–6, 80–2

farm/science connection. *See* science/practice connection
Felch, I.K., 74, 80, 83, 86, 209n8
Finland, 150, 163
Fisher, R.A.: about, 6, 58–9, 61–2, 180, 182; continuous variation, 113; and eugenics, 58, 61–2, 69; influence of, 63–4, 69, 108, 180; livestock genetics, 180; Mendelism/Darwinism synthesis, 58, 61–2; quantitative genetics,

58–9, 61–2, 112; science/practice connection, 62, 69; true lines *vs.* inbreeding/line crossing, 108. *See also* infinitesimal model
five freedoms of animal welfare, 175–6, 187
fowl. *See* chickens
France: AI progeny testing, 142; draft horse breeding, 30–2; genetic research (1920s), 57; genomic testing, 150; historical background, 21; pigs and AI, 124; regulation of stallions, 95; veal industry, 163. *See also* Europe
Frateur, Leopold, 57
Friesian Black and White cattle, 67, 106, 116–19. *See also* Holsteins
functional genomics, 8, 191, 235

Galloway, 25, 88, 168
Galton, Francis, 35–8, 42
gender, human: and AI gender bias, 18; chicken breeders/producers, 79–80, 181
gene, defined, 236
General Stud Book (GSB), 15–16, 27–8. *See also* registries
genetics, classical, 5–7, 42, 236. *See also* biometry; Mendelism; population genetics; quantitative genetics
genetics, molecular, 7–8, 238. *See also* genomics; molecular genetics
Genex, 118, 121, 147, 148
genomics: about, 7, 139–41, 159; animal genomes, 139–41, 184–6; BeadChips, 142, 147–59; beef breeding, 154–5; black box thinking, 143–4, 159; breed-specific testing, 145, 152–3, 155, 159;

breed variation and testing, 152–4; chickens, 155–8; data storage and processing, 153–4, 159; defined, 236; dependence on existing structures, 7, 146, 159, 184–5; female testing, 151–2, 185; functional genomics, 8, 191; funding for, 139–40, 144; future trends, 8, 190–2; GMACE, 150; haplotypes, 140, 143, 153; HapMap, 140, 143, 155–6, 184; human genome, 139, 184, 191–2; international cooperation, 149–51; next-generation sequencing, 144; progeny vs. DNA testing, 147–51, 184–5; and quantitative genetics, 140–1, 143, 146, 158–9; research questions, 141; science/practice connection, 152, 154; shotgun sequencing, 140, 144; single-step evaluation, 185, 186. See also corporations and molecular genetics; molecular genetics; SNPs

genotype, defined, 236

Germany: AI practices, 124, 142; dairy cattle associations, 99; Darwin's influence, 34, 35; ethics and male chicks, 167; genetic research (1920s), 56–7; genomic testing, 150; historical background, 10, 21, 34, 56–7; Holsteins, 116–17, 119, 150; pigs, 34, 124; research review on, 10. See also Europe

gilt, defined, 236

globalization: cooperation of organizations, 119–20, 139–41, 149–51; genome consortia, 140–1, 150–1, 184; GMACE, 150; knowledge distribution, 61. See also intellectual property

globalization of AI in dairy cattle: corporations, 118; crossbreeding, 120–2; Holsteins, 116–19; inbreeding, 119–20; Interbull, 119–20, 149–51; progeny testing, 109, 111; research, 118, 119; Starbuck as sire, 115–16

GMACE (genomic multiple across-country evaluations), 150, 236. See also MACE

goat semen, 105

Goodale, H.D., 81

government role: about, 42–3; chicken standards, 77–8; and corporations, 85–6; dairy cattle, 97–8, 102–3; Lush's support for, 65; Mendelism/biometry divide, 42–3; science/practice connection, 85–6; shift to corporations, 85–6, 186. See also experiment stations

Gowell, G.M., 81

Graham, William, 77

Great Britain. See Britain

Guernseys, 98, 101

Hagedoorn, Arend L., 57–8, 65–8

Haldane, J.B.S., 61, 107, 180

Hanover Hill Holsteins, 114–16, 120, 183

Hanson, J.A., 81

Hansson, Nils, 100

haplotypes, 140, 143, 153, 236

HapMap projects, 140, 143, 155–6, 184

Harris, D.L., 133–4

Harrison, Ruth, 175

Hazel, Lanoy N., 107, 109–12, 182

Heape, Walter H., 43–5

Heffering, Peter, 114–16, 120, 148, 183

heifer, defined, 236

Henderson, Charles R., 109–12, 114, 182
Heredity in Live Stock (Wriedt), 66, 67
Herefords, 33, 88, 89, 90
heterosis: about, 50–1; AI and dairy cattle, 122; beef cattle, 91; chicken breeding, 86; defined, 122, 236; DNA markers to predict, 138. *See also* hybrid vigour
heterozygosis, 50, 237
Hill, W.C., 107, 112, 113
HIR (Herd Improvement Registry), 100, 109
history of artificial selection (before 1900): about, 5, 14, 179–81; ancient and early history, 13, 17; Bakewellian system, 17–20, 23, 28, 179–80; biometry, 35–9; black box thinking, 38–9; "breed" creation, 18–19; crossbreeding, 21; Darwin's influence, 32–5; draft horse breeding, 30–2; eugenics, 35–7; experiments *vs.* theories, 20–4, 34, 36; genealogy *vs.* progeny test, 19–20, 24, 26; hybridizing, 23–4; inbreeding, 17–20, 23, 25–7; markets, 30–2; Moravian sheep, 10, 22–4; naturalists, 5, 13, 20–4; plant experiments, 21; population *vs.* individual, 18–19; progeny testing, 17–20, 23; purebreds, 6, 19, 24–31, 34; registries, 25–7; science/practice connection, 6, 22–3; Shorthorns, 24–8; Standardbred horses, 5, 28–30; Thoroughbreds, 14–17, 24, 26, 30, 189; true lines *vs.* hybrid breeding, 18, 22, 23. *See also* Canada, history (before 1950); United States, history (before 1900)

history of artificial selection (1900–1950): about, 69–70, 180–2; agricultural colleges, 43; "best to the best," 46–7; biometry/Mendelism divide, 40–3; continuous/discontinuous inheritance, 41–9; eugenics, 50, 58; experiment stations, 42–3; experiments *vs.* theories, 54–5, 61–2; farm conditions as constraint, 70; government role, 42–3; mass selection *vs.* ancestry, 47; math/biology dilemma, 37–8, 63–4, 69; Mendelism, 40–3, 54–5; Mendelism/biometry synthesis, 58; Mendelism/ Darwinism synthesis, 58, 61–2, 69, 180; mutation theory, 52–3; plant *vs.* animal genetics, 68–9; productivity increase, 70; progeny testing, 46–7; purebred culture, 44–5; quantitative genetics, 58–60; science/practice connection, 50–2, 57–8, 59–60, 64–7; science/practice divide, 41–50. *See also* Canada, history (before 1950); chickens, history (before 1950); dairy cattle, history (before 1950); Europe, history (1900–1950); science/ practice connection, history (1900–1950); United States, history (1900–1950)
history of artificial selection (1950–2000): about, 6–7, 129, 181–4; artificial insemination, 94, 181–2, 187; quantitative genetics, 134, 182–4; reproductive technology, 130–1. *See also* artificial insemination; quantitative genetics; reproductive technology

history of artificial selection (2000s): about, 7, 159, 183–6; genomics, 139–41, 159; molecular genetics, 131–2, 159, 183–6. *See also* genomics; molecular genetics

Holland. *See* Netherlands

Holsteins, 114–19; AI and inbreeding, 119–22; AI practices, 106, 114–19; associations, 99, 117, 118, 150–1; beauty/utility issue, 67; comparison of U.S. and Canadian, 114, 116, 117; data collection, 98–9, 141, 150; female genomic testing, 151–2, 185; genomic testing, 145–6, 149–52; Hanover Hill bulls, 114–16, 120, 148, 183; history of Friesian Black and White cattle, 67, 98, 106, 116–19; inbreeding/line-crossing, 101–2; international comparisons, 118, 150–1; linear systems, 117; progeny testing, 67, 106–7, 141–2; purebreds, 98–9, 101–2, 114–17, 182–3; purebreds *vs.* crossbreds, 101–2; purebreds *vs.* genetics, 114–16, 182–3; registries, 98–9; science/practice connection, 152; sire indexing, 114; standards, 99, 117. *See also* dairy cattle

homozygosis, 60, 237

horses: about, 13–16, 29–30, 189; AI practices, 105, 189; associations, 96; beauty/utility issue, 75, 189; cloning, 223n6; crossbreeding, 29–30, 189; frozen semen, 105; inbreeding, 15–16; molecular genetics, 145, 189; natural service, 189; purebreds, 95, 189; registries, 15–16, 27–8; regulation of stallions, 95–6, 189; research review on, 11; sexed semen, 164

horses, breeds: Arabians, 14–16, 27, 29, 104, 189; draft horses, 29–32; Percherons, 30–2; Standardbred trotting/pacing horse, 29–30; terminology, 30; warmbloods, 29. *See also* Thoroughbreds

human inheritance. *See* eugenics

human molecular genetics: about, 131; human genome, 139, 184, 191. *See also* eugenics

humane societies. *See* animal welfare

hybrid breeding: about, 21–2, 91, 181; biological lock, 88, 186–7; defined, 237; historical background, 21–4; markets, 28; plants, 21–2, 69. *See also* biological lock; crossbreeding

hybrid vigour: crossbreeding for, 61, 74; defined, 237; double-cross system, 51–2; Wright's work on, 61. *See also* heterosis

Illumina, Inc., 144–5, 147–57

in and in breeding: in Bakewellian system, 17, 19–20; defined, 237

inbreeding: about, 21–2, 181, 192; AI's impact on dairy cattle, 119–22, 190; Bakewellian system, 17, 19–20, 60; and crossbreeding, 26, 120, 192; Darwin's study of, 33–4; defined, 17, 237; future trends, 190–2; historical background, 18–22, 24, 53; hybridizing *vs.* inbreeding, 23; inbreeding depression, 51; inbreeding/line-crossing, 54, 101–2; inbreeding/outcrossing, 72–3; plants, 21–2, 51; and population genetics, 113;

progeny testing, 24; as purity, 25–6; and quantitative genetics, 60–2, 190; registries to prevent, 15–16; science/practice connection, 53–4. *See also* Bakewellian system; path coefficient

individual worth, defined, 237

infinitesimal model of inheritance: about, 58–9, 129, 191; defined, 237; and genomics, 143, 191; and molecular genetics, 136–7, 153; pigs, 91. *See also* quantitative genetics

intellectual property: about, 71; artificial insemination, 145, 186; beef cattle, 89–90; biological lock as patent, 71, 73, 88, 181, 186–7; chickens, 10, 73, 80, 126, 138, 156, 181; corporations, 181, 186–7; dairy cattle, 126; genomic research, 145; historical background, 71; molecular genetics, 138, 156; open access *vs.* restraints, 134, 138, 145, 156, 181, 186; plants, 126; registries, 15; sheep, 71. *See also* biological lock

Interbull (International Bull Evaluation Service), 119–20, 142, 149–51

Ireland, 168

Italy, 64, 95, 142

Ivanov, E.I., 104

Jerseys, 98–9, 101, 119, 120, 145–6, 149, 153

Johannsen, William, 38, 49–50

Johnson, M., 81

Jones, D.F., 51, 61

Kempthorne, O., 113–14, 133

Kingsland, Sharon, 9

Knight, T.A., 21–2

Köller, Martin, 23

Kölreuter, J.G., 21

Laurvigen, Ahlefeldt, 101

LD (linkage disequilibrium), 136, 156–7, 237. *See also* SNPs

Leghorn, American, 76, 83, 89, 161, 166

Lerner, I.M., 93, 108, 113

Lewontin, R.C., 39, 113, 133

Limousin, 90

Linnaeus, Carl, 21

livestock genetics: about, 180–2, 184; and biometry, 42, 184; Castle's influence, 52–4; historical background, 42, 50–2; Mendelism/Darwinism synthesis, 58, 61–2, 180, 184; Wright/Fisher influences on, 61–2, 180–2. *See also* biometry; Fisher, R.A.; Lush, Jay L.; population genetics; Wright, Sewall

locus, defined, 237

Lush, Jay L.: about, 6, 62–9, 180; *Animal Breeding Plans,* 67; breeding methodology, 65; chicken industry, 82; dairy cattle crossbreds *vs.* purebreds, 101; Fisher/Wright influences, 62–5, 112, 180–2; inbreeding/crossing research, 101; inbreeding theory, 62–3, 65, 113; influence of, 69, 108, 113, 182; livestock genetics, 180–1; population genetics, 62–3, 65, 112–13; science/practice connection, 62–3, 64–5, 69; sexed semen, 164; on sire indexing, 111

MACE (multiple across-country evaluations), 119, 150, 237. *See also* BLUP; GMACE; sire indexing

Marshall, William, 16
mass selection, 47, 72, 237
mathematics and science: biology/
    math dilemma, 37–8, 63–4, 114;
    biometry, 35–9; research needed,
    8; rise of statistics (1930s), 69,
    182; science/practice connection,
    133–4; statistics and quantitative
    genetics, 109–12, 133–4; statistics
    and sire indexing, 110–11;
    statistics/genetics synthesis, 63–4.
    See also biometry; Fisher, R.A.;
    quantitative genetics
McCook, S., 11
Mendelism: about, 5–6, 40–3,
    180, 184; biometry/Mendelism
    divide, 5–6, 40–3, 180; biometry/
    Mendelism synthesis, 40, 58,
    60–2, 180; chickens, 54–5, 81;
    corn, 50–2, 69; Darwinism, 40,
    42; defined, 237; discontinuous
    inheritance, 41–4, 48, 180;
    historical background, 24, 40–2,
    54–5, 180; hybrid vigour, 50;
    inbreeding/crossing method,
    40, 50–1; mutation factors, 40–2;
    qualitative traits, 40, 180; research
    review on, 9, 11; science/practice
    connection, 40, 43, 48–50, 180, 184;
    unit character theory, 41, 55, 58–9,
    136. See also Bateson, William
Meuwissen, T.H.E., 142–4
microsatellites, 137, 156, 238
Miller, Gerritt S., 116
MOET (multiple ovulation embryo
    transfer), 130–1, 236. See also
    reproductive technology
molecular genetics: about, 7,
    131–2, 159, 183–6; biodiversity,
    156; bioinformatics, 8, 191;
    cloning, 131, 176, 223n6, 236;
    defined, 238; dependence on
    existing structures, 7, 146,
    184–5; DNA markers, 137–9, 140,
    142–4, 184–5; DNA structure,
    131, 135–9; future trends, 8,
    190–1; historical background,
    129, 131–2; microsatellites, 137,
    156; and quantitative genetics,
    132–4, 135–7, 158–9, 183–4;
    shotgun sequencing, 139–40,
    144; transgenics, 130–1, 176. See
    also corporations and molecular
    genetics; genomics; human
    molecular genetics; LD; QTL;
    recombinant DNA; reproductive
    technology; RFLPs; SNPs
molecular genetics and specific
    livestock. See beef cattle and
    molecular genetics; chickens and
    molecular genetics; dairy cattle
    and molecular genetics
morality. See animal welfare; ethics
    and industry-needs
Moravian sheep breeding, 10, 22–4
Morman, J.G., 49
Moscow, Russia, 56–7
Muir, B., 149

NAAB (National Association of
    Animal Artificial Breeders), 105, 117
Nathusius, Herman von, 34, 35
National Association of Animal
    (Artificial) Breeders (NAAB),
    105, 117
National Inheritance (Galton), 37
naturalists: about, 5, 14, 20–1,
    179–81; Bakewellian system,
    23; double-cross hybrid system,
    51–2; experimentation, 20–1;

natural selection, 21, 180; science/practice connection, 14, 22. *See also* Bakewellian system; Darwinism
Nestler, J.K., 23–4
Netherlands: AI practices, 98, 106, 124, 142; beauty/utility issue, 67; dairy cattle, 10, 67, 98, 150; genetic research (1920s), 57–8; genomic testing, 150; Hagedoorn's research, 65–8; Holsteins (Friesian Black and White), 67, 106, 116–19, 150; pigs, 124; plant hybridizers, 34; population genetics, 182; progeny testing, 67, 106–7; science/practice connection, 10, 67; veal industry, 163. *See also* Europe
New Perspectives on the History of the Life Sciences and Agriculture (Kingsland and Phillips), 9
New Zealand, 118, 120, 142, 188
next-generation sequencing, 144, 238
North American Collaborative Dairy DNA Repository (CDDR), 149
Norway, 56, 66, 67, 100, 124, 163

organizations, AI. *See* artificial insemination organizations
Origin of Species (Darwin), 33
outcrossing, defined, 238. *See also* crossbreeding

patents. *See* intellectual property
path coefficient theory, 60–1, 74, 238
Pearl, Raymond, 45–9, 53, 55, 101
Pearson, Karl, 37–8, 40–2, 49, 53, 60
pedigree-keeping systems. *See* registries
Percheron horses, 30–2
Perry, E.J., 104
phenotype, defined, 238

Phillips, Denise, 9
pigeons, 33, 56
pigs: about, 5, 92–3, 124–5, 185–6; AI practices, 105, 124–5, 146–7; animal welfare, 176, 178; biological lock, 91–2, 124, 126; cloning, 176, 223n6; confinement, 178; corporations, 91–2, 124–5, 147, 186; excess young, 161; frozen semen, 105; hybrid breeding, 91–3, 124–5, 186; infinitesimal model, 91; intellectual property, 126; pig genome, 140, 185–6; population genetics, 64–5; purebreds, 92, 93, 186; research review on, 10; sexed semen, 164; single-step evaluation, 186; SNP testing, 145, 146–7; transgenics, 130, 176
Pirchner, F., 113
plants: about, 9, 11; animal *vs.* plant research, 68–9, 82–3; and genomics, 159; historical background, 9, 11, 21–2, 50–2, 68–9; hybrids, 21–2, 34, 69–70; inbreeding, 21–2; inbreeding/crossing, 50–1; intellectual property, 126; Mendelism, 40, 69; QTL research, 135; science/practice connection, 10–11, 50–2. *See also* corn
Plymouth Rock, 73, 76, 85, 166
Poland, 118, 189
population genetics: about, 61, 180–2; defined, 238; and genomics, 158–9; influence of Haldane, Fisher, and Wright, 58, 60–1, 112–13, 180; influence of Lush, 62–3, 65, 112–13; influence of Stoddard, 73; quantitative *vs.* population, 112–14, 182. *See also* quantitative genetics

Population Genetics in Animal Breeding (Pirchner), 113
poultry: intellectual property, 126; quantitative genetics, 134; as research animal, 68. *See also* American Poultry Association; chickens
Poultry Breeding (Hagedoorn and Sykes), 66
Poultry Science Association, 103
practice/science connection. *See* science/practice connection
Prentice, E. Parmalee, 81
progeny testing: about, 6; AI's influence on, 6–7; in Bakewellian system, 17–20, 23; crossbreeding, 54; defined, 238; DNA testing *vs.* progeny testing, 147–51, 184–5; historical background, 6, 23–4, 46, 67; records and registries, 23. *See also* EPD
pullet, defined, 238
Punnett, R.C., 66
purebred breeding: about, 14, 16, 25–8, 93, 179–83; associations, 96–7, 102–4; Bakewellian system, 28, 46; beauty/utility issue, 67, 183; Darwin on, 32–3; defined, 26–7, 238; future trends, 8; genealogy, 27; historical background, 24–8, 93, 95, 97; inbreeding, 25–7; markets, 30–2; purebreds *vs.* genetics, 114–16, 126–8, 182–3; purity, 16, 25–7; race constancy, 28; registries, 25–7, 33; regulation of stallions, 95–6; science/practice connection, 14, 70, 93; and social class, 16–17, 26; terminology, 30; Thoroughbred influences, 16, 25–7, 30. *See also* registries; standardbred breeding

QTL (quantitative trait loci): defined, 238; and DNA markers, 137, 139, 142–3; and genomic testing, 153, 158; and quantitative genetics, 135–7, 153; and RFLPs, 135
qualitative traits, defined, 239
quantitative genetics: about, 7, 134, 182–4; Bakewellian system, 18; beef cattle, 122–3; black box thinking, 38–9, 133, 137; BLUP research, 111–12, 114, 115, 123; dairy cattle, 108–12; decline of, 132–4, 158–9; defined, 238; EPD research, 123; as experimental, 114; Fisher's influence, 58–9, 112, 113; and genomics, 7, 140–1, 143, 146, 158–9; Hazel and Henderson's work, 109–12, 182; historical background, 7, 18, 35–9, 42, 58–9; and inbreeding, 60–2, 113, 190; Lush's influence, 63, 111, 112, 113; math/biology dilemma, 63–4, 114; and molecular genetics, 132–4, 135–7, 158–9, 183–4; multiple trait selection, 110; path coefficient theory, 60–1, 74; population *vs.* quantitative, 112–14, 182; purebreds, 114–16; QTL research, 135–7; sire indexing, 110–11, 183; statistics use, 109–12, 182; Wright's influence, 112, 113. *See also* black box thinking; infinitesimal model; MACE; Robertson, Alan
quantitative traits, defined, 239–40

race constancy, 28
recombinant DNA: about, 132; animal welfare issues, 176; defined, 238; DNA markers, 137; next-generation sequencing,

144. *See also* cloning; molecular genetics; RFLPs; SNPs

Record of Merit (ROM), 98–9

Record of Performance (ROP): chickens, 77–80, 84, 87, 187; dairy cattle, 98–9, 187

Red Danish, 101

registries: about, 15–16, 25–8; beef cattle, 90–1; as breeding regulation, 27; dairy cattle, 98–9, 109; horses, 15–16, 27, 29–32; inbreeding, 15–16, 27, 33; intellectual property, 15; markets, 15, 33; pedigrees, 27–8, 33; progeny testing, 23; purebreds, 15–16, 25–7, 33; Shorthorns, 25–7, 33; Standardbred horses, 29–30; terminology, 30

Rendel, J.M., 108–9

reproductive technology: about, 130–1; animal welfare, 176, 178; chicken sexing, 166–7; cloning, 131, 176, 223n6; corporations, 130, 186; ET and MOET, 130–1; female focus, 130–1, 151–2, 185; frozen semen, 105; sexed semen, 8, 164–6, 178. *See also* artificial insemination; genomics; molecular genetics

RFLPs (restriction fragment length polymorphisms), 132, 135, 137, 155, 239

Rhode Island Red, 76, 85

rights, animal. *See* animal welfare

Robertson, Alan, 108–9, 113, 135–6, 182

ROP. *See* Record of Performance (ROP)

Roslin Institute, Edinburgh, 131, 223n6

Russia, 56–7, 95, 104

Sanger, F., 132

science/practice connection: about, 4–10, 187, 192; AI practices, 121–3, 127–8, 182, 186–7; continuous/discontinuous inheritance, 41–9; corporations, 85–6, 124–5, 186–7; dairy cattle breeding, 102, 127–8; Darwin on, 3; early history, 14, 22–3, 37; eugenics, 37; future trends, 8, 189–92; genomic testing, 152, 154; government role, 85–6; major factors, 4–6, 187, 192; Mendelism/biometry divide, 180; naturalists, 22–3; overview of history, 179–88; plants, 8, 50–2; purebreds, 93; research review on, 8–11; technology and, 187. *See also* art *vs.* science

science/practice connection, history (1900–1950): about, 43–5; animal population size as constraint, 70; breeder/scientist divide, 41–50; chickens, 45–6, 66, 85–6; continuous/discontinuous inheritance, 41–9; corporate role, 85–6; dairy cattle, 66; government role, 85–6; Lush's influence, 64–5; Mendelism, 49–50; plants, 50–2; population genetics, 62, 64–5; purebreds, 70; sheep, 66–7

Scotland, 56, 168. *See also* University of Edinburgh

Sebright, Sir John, 19–20, 22, 46, 60

Select Sires, 118, 121, 164

Semex Canada, 115, 126

sexed semen, 8, 164–6, 178. *See also* reproductive technology

Shadybrook Farm, 137–8

Shaver Blend cattle, 89–91, 122

sheep: about, 188; AI practices, 105; cloning, 131; historical background, 10, 22–4, 56, 63, 188; intellectual property, 71; Moravian sheep, 10, 22–4; quantitative traits, 63; science/practice connection, 66–7; sheep genome, 140; SNP testing, 145

Shorthorn cattle: about, 24–6, 28, 179; AI use, 155; Bakewellian influence, 24–5; biometric research, 38; colour coats, 38, 113; crossbreeding, 25, 88, 155; dehorning, 169; DNA markers, 137–8; genomic testing, 155; inbreeding, 25, 59–60; markets, 27, 33; population genetics, 113; purebred breeding, 24–6, 28, 179; registries, 25–7, 33; Wright's research, 59, 61, 63, 113. *See also* beef cattle

shotgun sequencing, 139–40, 144, 239

Shull, G.H., 50, 61

single-step evaluation in genomics, 185, 186

sire indexing, 110–11, 114, 125, 127, 183, 239. *See also* BLUP; MACE

Slocum, Rob R., 45–6

Snedecor, G.W., 63–4, 67

SNPs (single nucleotide polymorphisms): about testing, 132, 142–3, 159, 184–5; black box thinking, 143, 153, 158–9; breed-specific testing, 145, 152–3, 155, 159; chickens, 145–6, 156–7; dairy cattle, 145–9; defined, 239; DNA chip technology, 142–59; future trends, 191; haplotypes, 140; LD, 136, 156–7; next-generation sequencing, 144;

pigs, 145, 146–7; research, 144–5. *See also* genomics

Sorensen, Edward, 104

sows. *See* pigs

standardbred breeding: about, 28–30, 70, 179–80, 183; beauty/utility issue, 76, 82, 183; dairy cattle, 98–9; defined, 239; science/practice connection, 70, 82; Standardbred horses, 5, 28–30; terminology, 30. *See also* chickens and standards

Standard of Perfection for chickens, 75–8, 87, 183

Starbuck, Hanoverhill, 115–16, 120, 183

States of Nature (McCook), 11

Statistical Methods Applied to Experiments in Agriculture (Snedecor), 63

statistics. *See* mathematics and science; quantitative genetics

Stoddard, H.H., 46–9, 72–3, 80, 81, 86

Stolzman, Maria, 118

stud books. *See* registries

studs. *See* artificial insemination organizations

Sweden, 56, 142, 150, 163

swine. *See* pigs

Switzerland, 98

Tancred, D., 81

terminology, 235–40

theory/practice. *See* science/practice connection

Thompson, E.B., 82

Thompson, W.R., 64

Thoroughbreds: about, 14–17, 30, 179, 189; AI use, 189;

and Arabians, 15, 27, 189;
crossbreeding, 29; defined, 15,
239; inbreeding, 15–16; purebred
breeding, 16, 25–7, 126, 189; race
constancy, 28; registries, 14, 15–16,
27; and Shorthorns, 25–6; and
social class, 16–17; terminology,
30. *See also* horses
trade secrets. *See* intellectual property
training population, defined, 239
transgenics, 130–1, 176, 240
Trevena, Kenneth, 114–15
trotting/pacing horses, 29–30
true breeding: about, 18, 22; defined,
240; historical background, 20–2;
hybrids *vs.* true lines, 18, 22, 23;
inbreeding, 23; inbreeding/line
crossing, 108; purebred breeding,
26; science/practice connection,
22–3

unit character theory, 41, 55, 58–9,
136, 240. *See also* Mendelism
United States, chickens: breeder/
scientist interaction, 80–2;
breeding methods, 72–7; breeds,
76; corporations, 82–6, 88, 181,
186; ethics and excess male chicks,
167; government standards, 77–8;
meat breeding, 84–6. *See also*
chickens
United States, dairy cattle:
associations, 97, 99–100, 102–4,
127; compared to Canadian,
114, 116, 117, 149–51, 183;
crossbreeding, 100–2; data
collection, 98–100, 127, 141;
genomic testing, 149–52;
government support, 97, 102–3;
imports from Argentina, 97;

purebreds, 98–102; science/
practice connection, 102, 127–8.
See also dairy cattle
United States, dairy cattle and AI:
associations, 97, 104–5, 114, 125–6;
inbreeding, 119–20; progeny
testing, 111, 142. *See also* artificial
insemination
United States, history (before 1900):
animal welfare, 173–4; draft
horse breeding, 30–2; purebred
breeding, 28–9; registries, 27–8,
30–1; Thoroughbreds, 27–8. *See
also* history of artificial selection
(before 1900)
United States, history (1900–1950):
beauty/utility issue, 75–8, 183;
biometry, 42; chicken breeding,
72–80, 86–8; continuous/
discontinuous inheritance,
45–6; double-cross hybrid
system, 51–2; double mating
system, 73; experimental animal
breeding, 54–5; government
role, 42–3; Lush and livestock
genetics, 62–5; math/biology
dilemma, 63–4; Mendelism, 54–5;
Mendelism/biometry divide,
42, 53; purebred culture, 70;
regulation of breeding males,
95–7; research constraints, 70;
science/practice connection,
68–70. *See also entries beginning
with* history of artificial selection
University College, London, 107
University of Cambridge, 55–7, 66,
107, 124. *See also* Fisher, R.A.
University of Edinburgh, 55–6, 66,
107–9, 131, 171. *See also* Robertson,
Alan

Variation of Animals and Plants
  (Darwin), 33
veal industry: about, 8, 162–3, 188;
  confined spaces, 163, 178; sexed
  semen, 8, 165–6
Volterra, Vito, 64
von Sachs, Julius, 34

Wales, 97
Wallace, Henry A., 82–3
Walsh, B., 112, 140
warmbloods (horses), 29
Watson, James, 131, 139
Weldon, Walter F.R., 37, 41–2
White Plymouth Rock, 76, 85, 166

Wriedt, Christian, 56, 66, 67
Wright, Sewall: about, 6, 59–64, 69;
  biometry/Mendelism synthesis,
  58, 60, 69–70; Castle's influence
  on, 54, 60, 69; hybrid vigour, 61;
  inbreeding quantification, 6, 59–62,
  69, 74, 113; influence of, 63, 64, 108,
  180, 182; Mendelism/Darwinism
  synthesis, 61–2; path coefficient
  theory, 60–2, 74; population
  genetics, 61, 113, 180; science/
  practice connection, 59–60, 62, 69;
  Shorthorns, 59, 61, 63, 113

Yule, G. Udny, 58